NATURE GUIDE
STARS AND PLANETS

SMITHSONIAN
NATURE GUIDE
STARS AND PLANETS

Robert Dinwiddie • Will Gater
Giles Sparrow • Carole Stott

LONDON, NEW YORK, MELBOURNE, MUNICH, AND DELHI

DORLING KINDERSLEY

Senior Editor Peter Frances	**Project Art Editor** Anna Hall
Editors Lizzie Munsey, Martyn Page, Andrew Szudek	**Designer** Fiona Macdonald
US Editors Jill Hamilton, Rebecca Warren	**Design assistant** Jonny Burrows
Jacket Editor Manisha Majithia	**Jacket Designer** Laura Brim
Production Editor Rebekah Parsons-King	**Picture Researchers** Jo Walton, Julia Harris-Voss
Managing Editor Camilla Hallinan	**Production Controller** Erika Pepe
Publisher Sarah Larter	**Managing Art Editor** Michelle Baxter
Associate Publishing Director Liz Wheeler	**Art Director** Philip Ormerod
Publishing Director Jonathan Metcalf	

DK INDIA

Managing Editor Rohan Sinha	**Deputy Managing Art Editor** Mitun Banerjee
Deputy Managing Editor Alka Thakur Hazarika	**Consultant Art Director** Shefali Upadhyay
Senior Editor Alka Ranjan	**Art editors** Pushpak Tyagi, Manish Upreti
Editors Mona Joshi, Rupa Rao	**Senior DTP Designers** Harish Aggarwal
DTP Designers Vishal Bhatia, Manish Upreti	**CTS Manager** Sunil Sharma

CONSULTANT

Andrew K. Johnston, Geographer, Center for Earth and Planetary Studies, National Air and Space Museum, Smithsonian Institution, USA.

First American Edition, 2012
Published in the United States by DK Publishing
375 Hudson Street, New York, New York, 10014

12 13 14   10 9 8 7 6 5 4 3 2 1

001 – 181826 – Jun/2012

Published in Great Britain by Dorling Kindersley Limited.

A catalog record for this book is available from the Library of Congress.

ISBN 978-0-7566-9040-3

DK books are available at special discounts when purchased in bulk for sales promotions, premiums, fund-raising, or educational use. For details, contact: DK Publishing Special Markets, 375 Hudson Street, New York, New York, 10014 or SpecialSales@dk.com

Printed and bound in China by Leo Paper Products

Discover more at **www.dk.com**

Half-title page a total solar eclipse **Title page** the Orion Nebula

CONTENTS

KEY

- Average distance from the Sun
- Orbital period (length of year)
- Rotational period (length of day)
- Diameter at equator
- Number of moons
- Maximum magnitude
- Size ranking from 1 (largest) to 88 (smallest)
- Width, measured in parts of an outstretched hand
- Depth, measured in parts of an outstretched hand
- Brightest star
- Months during which a constellation is highest in the sky at 10 pm

THE NIGHT SKY

Stars, gas, and dust near the center of the Milky Way

WHAT IS THE UNIVERSE?

The Universe is everything that exists—space, time, energy, and matter, from the largest galaxy clusters to the smallest particles. It has been expanding since it formed, and it is likely that parts of it will never be seen from Earth.

SIZE AND STRUCTURE

The size of the Universe is unknown and may be infinite, although the part we can observe is finite in size (see opposite). The visible matter of the Universe is clumped into strings of galaxy clusters separated by enormous empty spaces called voids, at a scale of hundreds of millions of light-years (a light-year is 5,878 billion miles, or 9,460 billion km). Galaxy clusters are composed of many separate galaxies, each containing billions of stars, smaller colder bodies such as planets and comets, vast amounts of gas and dust, and some extreme concentrations of matter called black holes.

Galaxy supercluster
Galaxies group together into clusters, and clusters group into superclusters

The Oort Cloud
This is a vast, spherical region in the outer reaches of the Solar System that contains more than 1 trillion frozen comets

The Milky Way
The Sun is only one of between 200 and 400 billion stars in the Milky Way galaxy

The Solar System
Our solar system consists of a number of planets that all orbit the Sun

Planet Earth
Earth is one of four rocky planets in our local planetary system

The scale of the Universe
Although the Universe is unknowably vast, one way to get an idea of its scale is to use a series of stepping stones—from a planet, to the Solar System, the Milky Way galaxy, and far beyond.

Filaments and voids
Galaxy superclusters form a network of filaments and voids. These structures are thought to have their origins in the very first moments of the Universe.

ORIGINS OF THE UNIVERSE

Light and other radiation from distant galaxies indicate that these galaxies are moving away from us. In the 1920s, it was discovered that the farther away a galaxy is, the faster it is receding. The only logical explanation for this is that the Universe must be expanding.

Astronomers now believe that the Universe came into existence in an exceedingly hot, compact state around 13.7 billion years ago, in an event called the Big Bang. The expansion that has been occurring ever since is a gradual inflation of the whole of space, not an expansion into space—there is no recognizable "edge" to the Universe.

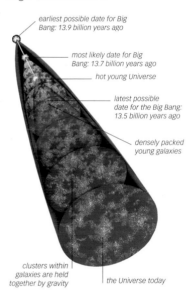

earliest possible date for Big Bang: 13.9 billion years ago

most likely date for Big Bang: 13.7 billion years ago

hot young Universe

latest possible date for the Big Bang: 13.5 billion years ago

densely packed young galaxies

clusters within galaxies are held together by gravity

the Universe today

The expanding Universe
Since the Big Bang, between 13.5 and 13.9 billion years ago, the Universe has been expanding. Over the same period, it has cooled from an exceedingly hot initial state.

THE OBSERVABLE UNIVERSE

Although the Universe may be infinite, the part of it that astronomers can actually observe and know anything about is finite. This is the "observable" Universe, the region from which light has had time to reach us since the Big Bang. It is a vast, spherical region around Earth, with a diameter of over 90 billion light-years. Light from galaxies near the edge of the observable Universe has taken about 13.2 billion years to reach us. Consequently, we see these galaxies as they were when they were very young, no more than a few hundred million years old. It is likely that there are more distant galaxies that we will never be able to observe, because the expansion of the Universe is carrying them away from us faster than the speed of light.

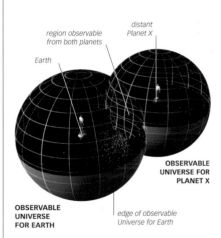

distant Planet X

region observable from both planets

Earth

OBSERVABLE UNIVERSE FOR PLANET X

OBSERVABLE UNIVERSE FOR EARTH

edge of observable Universe for Earth

Overlapping observable Universes
The part of the Universe visible from Earth is different from the part that would be seen from another, extremely distant planet. The two observable regions might overlap, as shown here, or they might not.

MATTER AND ENERGY

The Universe is made of matter and energy, which are interconvertible. Together they make up its total mass–energy. in addition to ordinary matter (made of atoms, which form the planets, stars, and galaxies), the Universe contains "dark" matter—so called because it does not emit light—and dark energy, whose nature is unknown.

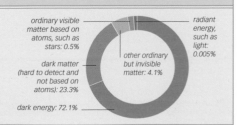

ordinary visible matter based on atoms, such as stars: 0.5%

radiant energy, such as light: 0.005%

other ordinary but invisible matter: 4.1%

dark matter (hard to detect and not based on atoms): 23.3%

dark energy: 72.1%

OBJECTS IN THE SKY

Many types of celestial object can be seen from Earth. The most obvious are stars, the Moon, and planets; but other bodies are also present. Some appear as smudges of light, others are only visible through binoculars or a telescope.

STARS

By far the most numerous objects in the night sky, stars are hot, distant balls of plasma (ionized gas), which generate energy through nuclear fusion reactions in their cores. Stars form from the condensation of clumps of gas and dust and sometimes occur in pairs or clusters. They exist in a range of sizes and vary in their color, brightness, and likely life span (pp.124–125). Apart from the Sun and a few particularly close, large stars that have been imaged by the Hubble Space Telescope, stars appear as pinpricks of light, even when highly magnified.

⊢ Earth–Sun distance

Betelgeuse
This red supergiant star is large enough and close enough to Earth to have been imaged by the Hubble Space Telescope as a disk.

Star variety
This Hubble image of part of the constellation of Sagittarius shows stars in a variety of colors and a wide range of sizes and brightness.

THE SUN

By day, we see a single star in the sky. This is our local star, the Sun (pp.66–71). The Sun is a 4.6-billion-year-old, medium-sized yellow star. It is by far the largest object in the Solar System, making up 99.86 percent of the total mass. When observing the Sun (which should only be attempted using adequate safety precautions), what we see is a specific light-emitting layer within it, the photosphere, which is tens to hundreds of miles thick. This bright layer has a temperature of about 10,000°F (5,500°C). Outside the photosphere are other layers of the Sun's atmosphere, which only become visible during total solar eclipses.

Our local star
The Sun shines many tens of thousands of times brighter than any other celestial object.

THE MILKY WAY

All the stars we can observe individually in the night sky are part of our local galaxy, the Milky Way (pp.138–141). However, the vast bulk of stars within our galaxy are not easily picked out as distinct pinpoints of light. Instead, they appear on a clear night as a broad, irregular "milky" band across the sky. We see this band of light because of our galaxy's shape: a disk with a bulge at its center. When looking along the plane of this disk, a high concentration of stars falls along the line of sight, whereas far fewer stars are present above or below the plane.

Milky Way from Earth
On a clear night, the Milky Way appears as a band across the sky. This image shows a view of the Milky Way above the Dolomites in Italy.

THE MOON

When it is present and visible, Earth's natural satellite, the Moon (pp.72–81), is the most prominent night-time celestial object. A cold, dry, lifeless ball of rock with an insubstantial atmosphere, the Moon has a diameter about one quarter of Earth's. Nevertheless, it looms large in our sky, shining by reflecting the light of the Sun. As our nearest neighbor, the Moon was an early target for space exploration, and it is the only world apart from Earth that people have walked on. It is locked to Earth in synchronous spin mode, which means it completes one spin on its axis in the same time it makes one orbit of Earth. Consequently, the same side of the Moon always faces Earth. Both the visible and far sides of the Moon are heavily cratered from the impact of asteroids and meteoroids (space rocks) over the past 4.5 billion years.

The Moon from Earth
The Moon can be viewed for a few hours or more on about 70 percent of clear nights. When the sky is clear and the Moon full, our satellite is observable for most or all of the night.

›› PLANETS

A planet is an object that orbits a star and is large enough for its own gravity to have pulled it into a spherical shape, but not so massive that nuclear fusion reactions occur in its core (in contrast to stars). It must also have swept the region of its own orbit free of small rocky objects other than its own satellites.

Eight planets orbit the Sun, shining by reflecting sunlight. Of these, Mercury (pp.86–87), Venus (pp.88–89), and Mars (pp.90–93) are small rocky planets, like Earth, that orbit close to the Sun. Jupiter (pp.94–97), Saturn (pp.98–101), Uranus (pp.102–103), and Neptune (pp.104–105) are larger, gas-shrouded planets, which orbit farther away from the Sun. Depending on the time of year, several

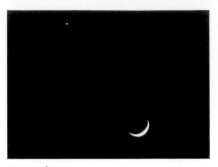

Moon and Venus
When present, the planet Venus is usually the brightest object in the night sky apart from the Moon. Here, Venus and the crescent Moon are seen in the western sky, an hour or so after sunset.

planets may be observable in a single night. All the planets except Neptune are visible to the naked eye, although Uranus is only just visible. Venus is typically brighter than any star, while Jupiter and Mars look like bright stars.

Surface of Mars
This image of the surface of Mars taken by the Spirit Rover in 2005 shows the typical rust color of the Mars terrain, along with scattered volcanic rocks.

PLANETARY MOONS AND RINGS

All the planets except for Venus and Mercury have natural satellites, which are called moons. The four outer, gas-shrouded planets have many moons, as well as ring systems. Ring systems are made up of debris, and each ring system is divided into several bands. The composition of the rings varies—some are made of dust, others of rock or ice. Some moons and rings are observable using amateur equipment. For example, the four largest moons of Jupiter and Saturn's rings are bright enough to be visible through a small telescope.

Uranus with its moons
This image shows the planet Uranus (with ring structure) and its five largest moons. Diagonally from top left these are Titania, Umbriel, Miranda, Ariel, and Oberon.

DWARF PLANETS AND ASTEROIDS

Planets that are too small to have cleared their orbits of other bodies are called dwarf planets. So far, only five have been identified. One, Ceres, can be seen with binoculars. The others can only be seen with a telescope. Most dwarf planets are in the Kuiper belt (pp.106–107), in the outer Solar System. Asteroids (pp.114–117) are small, rocky bodies, usually with irregular shapes. Many orbit the Sun between Mars and Jupiter in a group called the Main Belt. Several asteroids are visible through binoculars.

Pluto and its moons
This image shows the dwarf planet Pluto (center) with its largest moon Charon (below right). To Charon's right are two of Pluto's smaller moons, Nix (top) and Hydra.

Asteroid Ida
Ida is a typically irregular asteroid, around 36 miles (58km) long and 14 miles (23km) wide. Its heavily cratered surface suggests that it is not a young object.

COMETS

Chunks of ice, rock, and dust that orbit the Sun are called comets (p.108–111). As a comet nears the Sun, it starts to vaporize in the heat, releasing gas and trapped dust to form a bright head or "coma.". Radiation and streams of particles emanating from the Sun (the solar wind) then push the gas and dust away from this coma to produce long comet tails. Bright comets are rare—only a handful occur each century—but many fainter comets appear every year. Comets appear as fuzzy patches of light, sometimes with visible tails. If a comet is observed over several nights, it will be seen to be move against the background stars. Many comets are believed to originate in the Oort cloud (p.109), a vast spherical region thought to contain trillions of comets.

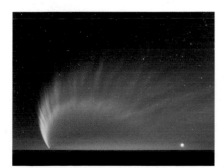

Comet McNaught
In 2006/2007, this great comet, with its stunning dust tail, was the brightest seen in the Southern Hemisphere for 40 years. It is now heading off to the outer parts of the Solar System and is expected never to return.

METEORS

Popularly known as shooting stars, meteors are linear trails of light-radiating material, produced in Earth's upper atmosphere by the impact of small, dusty fragments of comets or asteroids. About 1 million visible meteors are produced in Earth's skies each day.

Meteor showers, with frequent meteors all emanating from the same point in the sky, occur on particular dates in the year. They result from Earth crossing a stream of debris in space that has accumulated in the orbit of a comet. Meteor showers are best viewed with the naked eye, in as dark a site as possible, from a reclining chair.

The Leonids
Every year, the Leonid meteors are seen around November 17. They appear to shoot in all directions out of the constellation of Leo.

⟫ STAR CLUSTERS

Many stars in the Milky Way are grouped into clusters (pp.134–137), of which there are two types: globular and open. A globular cluster is a densely-packed ball of anything from 10,000 to several million extremely old stars. Open clusters are typically smaller and more loosely grouped. A number of examples of both types are visible to the naked eye. These include open clusters such as the Pleiades, Hyades, and Jewel Box clusters, as well as globular clusters such as Omega Centauri, 47 Tucanae, M13, and M15.

Jewel Box Cluster
Lying in the southern constellation of Crux, the Jewel Box is an open cluster dominated by hot blue stars, with one cooler red supergiant.

Globular Cluster M15
At the limit of naked-eye visibility, M15 lies in the constellation of Pegasus, and is one of the densest globular clusters in the Milky Way.

Lagoon Nebula
This highly luminous, oval-shaped emission nebula is one of the largest in the sky.

NEBULAE

Most of the Milky Way galaxy is filled with an extremely thin gas, consisting mainly of hydrogen and helium, mixed with a little dust. It permeates the space between the stars, so it is known as the interstellar medium. Denser clumps of gas and dust in space are known as nebulae (pp.128–131). Many nebulae are visible to us as light or dark smudges in the night sky. They fall into a number of different types. Emission nebulae are clouds of gas that produce a brilliant glow as they absorb radiant energy from a nearby star or cluster of stars, and then reradiate the energy as light. In contrast, dark nebulae are visible only as smudges that block out starlight. Two other types are reflection nebulae—clouds of dust that reflect light from nearby stars—and planetary nebulae, which are heated haloes of material shed by dying low-mass stars.

SUPERNOVAE

Stars that undergo cataclysmic explosions produce temporary bright objects in the sky called supernovae. These are rare—so rare that no supernova has been observed with certainty in the Milky Way since 1604, although a bright supernova occurred in the Large Magellanic Cloud, a companion galaxy to the Milky Way, in 1987. Known as supernova 1987A, it was caused by the explosion of a blue supergiant star and was visible to the naked eye for two months. The expanding cloud of material after a supernova is called a supernova remnant. This can remain as a faint, gradually expanding feature in the sky for centuries.

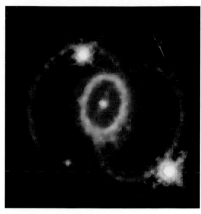

Supernova 1987A
This Hubble Space Telescope image shows the remnant of supernova 1987A, in the large Magellanic Cloud, ten years after the original explosion.

GALAXIES

Until about a century ago the Milky Way was thought to comprise the whole Universe. Today we know that just the observable part of our Universe contains more than 100 billion galaxies (pp.142–145). These vary from dwarf galaxies, a few hundred light-years across and containing a few million stars, to giants spanning several hundred thousand light-years, with several trillion stars. A number of these galaxies are large and close enough to be viewed with amateur equipment. One—the Andromeda Galaxy—is visible to the naked eye. Galaxies are bound by gravity to form clusters, which can contain from 20 to several thousand galaxies. Chains of a dozen or so galaxy clusters are loosely linked to make up superclusters, which can be up to 200 million light-years across. Clusters and superclusters can only be observed through powerful telescopes.

Southern Pinwheel Galaxy
A magnificent face-on spiral galaxy, M83 lies on the borders of the constellations Hydra and Centaurus.

Galaxy cluster Abell 1689
This massive galaxy cluster is 2.2 billion light-years away. It contains hundreds of individual galaxies.

THE CELESTIAL SPHERE

When recording and tracking the positions of objects in the sky, it is useful to think of the sky as an imaginary hollow sphere around Earth, which all objects are either attached to or move across. This imaginary globe is known as the celestial sphere.

SURFACE FEATURES

The celestial sphere has surface features related to the real sphere of the Earth. These include the north and south celestial poles, which lie directly above Earth's North and South Poles. The celestial sphere also has a celestial equator, which sits above Earth's equator. Very distant objects, such as stars and galaxies, have more or less "fixed" positions on the celestial sphere, although these positions do change extremely slowly over time as a result of precession (opposite). Solar System objects, such as the Sun and planets, continually shift in their positions, in most cases staying on or close to a circular path on the sphere's surface called the ecliptic (pp.20–21).

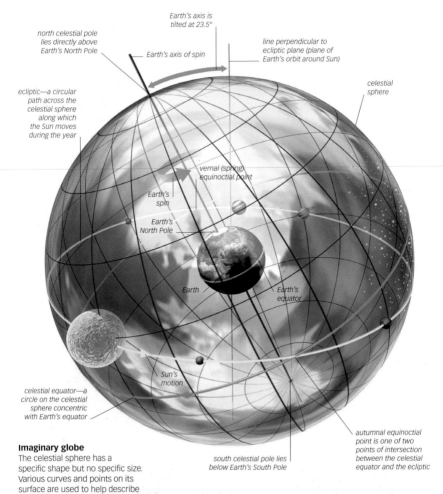

north celestial pole lies directly above Earth's North Pole

Earth's axis is tilted at 23.5°

Earth's axis of spin

line perpendicular to ecliptic plane (plane of Earth's orbit around Sun)

celestial sphere

ecliptic—a circular path across the celestial sphere along which the Sun moves during the year

vernal (spring) equinoctial point

Earth's spin

Earth's North Pole

Earth

Earth's equator

Sun's motion

celestial equator—a circle on the celestial sphere concentric with Earth's equator

Imaginary globe
The celestial sphere has a specific shape but no specific size. Various curves and points on its surface are used to help describe the positions of objects in the sky.

south celestial pole lies below Earth's South Pole

autumnal equinoctial point is one of two points of intersection between the celestial equator and the ecliptic

CELESTIAL COORDINATES

Coordinates called declination (dec) and right ascension (RA) are used to define positions on the celestial sphere. Dec is measured in degrees and arcminutes north (+) or south (-) of the celestial equator (1° is equal to 60 arcminutes). RA can be measured in hours (1 hour = 15°) or in degrees and arcminutes east of the celestial meridian—a curved line that passes through both celestial poles and the vernal equinoctial point.

Recording a star's position
The star shown here has a declination of +45° and a right ascension of 15°. Declination is equivalent to latitude on Earth, and right ascension is equivalent to longitude.

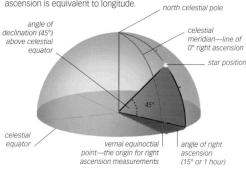

north celestial pole

angle of declination (45°) above celestial equator

celestial meridian—line of 0° right ascension

star position

45°

celestial equator

vernal equinoctial point—the origin for right ascension measurements

angle of right ascension (15° or 1 hour)

VISIBLE REGION

As Earth orbits the Sun, the Sun seems to move against the stars, washing out light from successive parts of the sky. At the same time, the part of the celestial sphere on the opposite side to Earth from the Sun—the part visible at night—continually changes. The visible area of the sky changes significantly over the year, except for observers at or close to Earth's poles.

June and December skies
A night-time observer on the Earth's equator sees one half of the celestial sphere in June, and the opposite half in December.

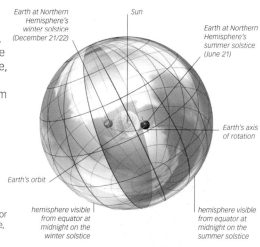

Earth at Northern Hemisphere's winter solstice (December 21/22)

Sun

Earth at Northern Hemisphere's summer solstice (June 21)

Earth's axis of rotation

Earth's orbit

hemisphere visible from equator at midnight on the winter solstice

hemisphere visible from equator at midnight on the summer solstice

PRECESSION

Earth executes a slow "wobble" called precession, which alters the direction of its spin axis over a 25,800-year cycle. As a result, the positions of the celestial poles, celestial equator, and equinoctial points gradually shift, and even the coordinates of "fixed" objects such as stars change very slowly.

Earth's wobble
Due to precession, the north celestial pole traces out a circular path in a 25,800-year cycle, shown here against a starry background.

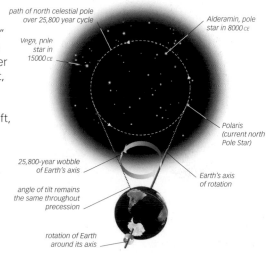

path of north celestial pole over 25,800 year cycle

Alderamin, pole star in 8000 CE

Vega, pole star in 15000 CE

Polaris (current north Pole Star)

25,800-year wobble of Earth's axis

angle of tilt remains the same throughout precession

Earth's axis of rotation

rotation of Earth around its axis

CONSTELLATIONS AND ASTERISMS

Since ancient times, people have imagined shapes among the stars in the night sky, and have linked stars to form constellations and smaller groups called asterisms. These figures are named after the shapes they are thought to resemble—usually animals, mythological figures, or objects.

CONSTELLATIONS

Over thousands of years, many different groupings of stars have been suggested as constellations. A definitive list has been established gradually, starting with the ancient astronomer Ptolemy in the 2nd century. Since 1922, an internationally agreed system has divided the celestial sphere into 88 areas, called standard constellations. Therefore, to an astronomer the term constellation now denotes a part of the sky containing a figure, rather than the figure itself.

Atlas Coelestis
This illustration from *Atlas Coelestis*, published in London in 1729, shows representations of several constellations visible from the Northern Hemisphere.

Ursa Major
This chart shows, at center, the constellation of Ursa Major, including the constellation figure (pattern of lines joining bright stars).

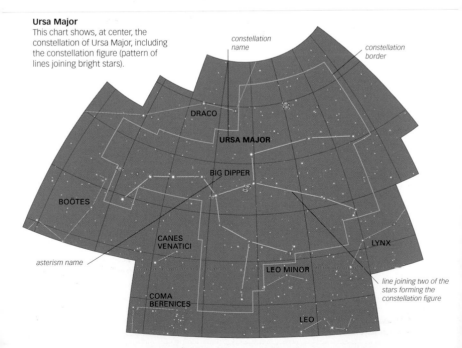

constellation name

constellation border

DRACO

URSA MAJOR

BIG DIPPER

BOÖTES

CANES VENATICI

LYNX

asterism name

LEO MINOR

line joining two of the stars forming the constellation figure

COMA BERENICES

LEO

ASTERISMS

Small distinctive groups of stars within constellations are called asterisms. One of the best known is the Big Dipper (below) in the constellation of Ursa Major. Other examples are the Sickle asterism in the constellation Leo, and the three bright stars that make up Orion's belt. The stars in an asterism are usually not particularly close to each other in three-dimensional space, despite their seeming proximity when viewed from Earth. However, there are exceptions. Five of the stars in the Big Dipper really are quite close to one another, and travel through space at a similar speed and in the same direction.

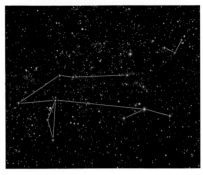

Leo and the Sickle
Six stars in the constellation of Leo make up the Sickle asterism, shown here in red. This asterism looks like a back-to-front or mirror-image question mark.

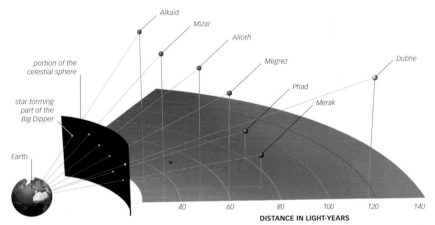

Line of sight
An asterism such as the Big Dipper is a two-dimensional view of what may be a widely scattered sample of stars. Although the stars seem to lie in the same plane, they are at different distances from Earth.

SHIFTING STARS

Stars that are close to each other in the sky often have different "proper" motions (long-term movements in particular directions relative to Earth). As a result, the shapes of constellations and asterisms can change gradually over time periods measured in thousands of years.

2. BIG DIPPER IN 2000 CE

1. BIG DIPPER IN 100,000 BCE

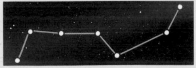

3. BIG DIPPER IN 100,000 CE

SEASONS AND THE ZODIAC

The tilt in Earth's spin axis, combined with its orbit around the Sun, causes seasons. As the seasons pass, the Sun itself traces a path across the celestial sphere, known as the ecliptic. As it does so, it moves through a band of sky called the zodiac.

THE SEASONS

As Earth orbits the Sun, it maintains a 23.5° tilt. Consequently, at different times in the year Earth's two hemispheres are either tilted toward or away from the Sun. This produces the seasons, because the hemisphere tilted toward the Sun receives a higher intensity of solar radiation and becomes warmer.

Each year, the Northern Hemisphere reaches its maximum tilt in the Sun's direction on June 20 or 21—summer solstice in the Northern Hemisphere and winter solstice in the Southern Hemisphere. For a few weeks to months around this date, the northern polar region is sunlit all day, while the south pole is in darkness. On December 21 or 22, the situation is reversed. Midway between the solstices are the equinoxes, when Earth's axis is broadside to the Sun, and the periods of daylight and darkness are roughly equal for all points on Earth.

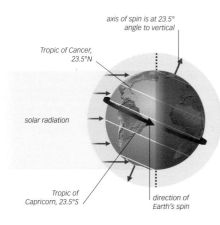

axis of spin is at 23.5° angle to vertical

Tropic of Cancer, 23.5°N

solar radiation

Tropic of Capricorn, 23.5°S

direction of Earth's spin

Sunlight intensity
The intensity of solar radiation is greatest within the tropics. Toward the poles, the Sun's rays must pass through a greater thickness of atmosphere and are spread out over a wider area, giving a less intense heating effect.

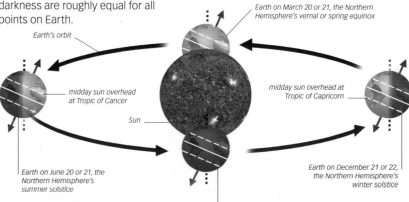

Earth's orbit

Earth on March 20 or 21, the Northern Hemisphere's vernal or spring equinox

midday sun overhead at Tropic of Cancer

midday sun overhead at Tropic of Capricorn

Sun

Earth on June 20 or 21, the Northern Hemisphere's summer solstice

Earth on December 21 or 22, the Northern Hemisphere's winter solstice

Earth on September 22 or 23, the Northern Hemisphere's autumnal equinox

Solstices and equinoxes
At the solstices (extreme left and right), one hemisphere is maximally tilted toward the Sun. Around the equinoxes (top and bottom), neither hemisphere tilts toward it.

THE ECLIPTIC PLANE

The Earth can be thought to move around the Sun on an imaginary flat surface known as the ecliptic plane. The ecliptic plane is so-named because a solar eclipse can occur only when the Moon passes through this plane. The planets orbit the Sun on or close to the ecliptic plane. The intersection between the ecliptic plane and the celestial sphere (p.16) creates a great circle on the celestial sphere, called the ecliptic. The ecliptic is also the path that the Sun appears to follow across the sky when viewed from Earth.

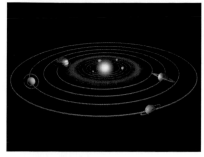

Plane of planetary orbits
All the planets orbit the Sun on, or close to, the same (ecliptic) plane. This is because the whole Solar System originally formed from a large rotating disk of gas and dust.

THE ZODIAC

Extending about 8° to either side of the ecliptic is a band of the celestial sphere called the zodiac. This band of sky contains all or part of 24 constellations (pp.18–19). Due to the Sun's glare, its movement along the ecliptic and through the zodiac cannot be seen easily, but if it could it would be seen that each year the Sun passes through 13 constellations within the zodiac, of which 12 have the same names as the astrological "signs of the zodiac." The Sun spends a variable number of days in each of these 13 constellations. It currently passes through each one on dates very different from traditional astrological dates, partly due to the phenomenon of precession (p.17). The Sun is not the only celestial body whose orbit is confined to the zodiac. Because planets orbit on or close to the ecliptic plane, their positions and movements are always restricted to this region of the celestial sphere.

Constellations of the zodiac
The band of sky known as the zodiac has 13 constellations arranged around its central part, of which 12 are star-sign constellations. The 13th constellation is Ophiuchus, part of which lies between Scorpius and Sagittarius.

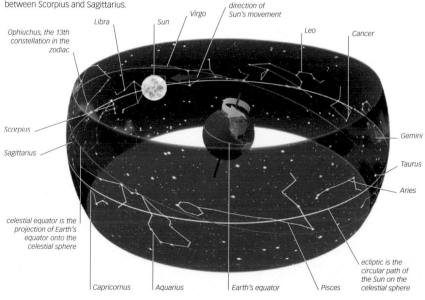

CELESTIAL MOTIONS

Although objects like stars have near-constant positions on the celestial sphere, they still undergo daily movements due to Earth's spin. Solar System objects undergo the same daily motion, but with additional longer-term movement patterns.

EARTH'S ROTATION

Earth's spin produces a continuous pattern of movement across the sky for all celestial objects—including the stars, Moon, Sun, and planets. This movement is predominantly from east to west and continues day and night, although for most objects it can only be observed at night. Earth's rotation creates the dominant and most obvious of all celestial motions. The way this movement looks to an observer varies significantly depending on his or her position on Earth's surface (see below). The rate of movement is a 1° shift of the whole sky every 4 minutes. The effects can be seen clearly in long-exposure photographs, in the form of long star "trails" that describe perfect circles or part-circles around a celestial pole.

THE EFFECTS OF LATITUDE

The daily pattern of movement of sky objects caused by Earth's spin appears different for observers on the equator, at the poles, or between the equator and the poles (at mid-latitudes). At the North Pole, an observer can see all the stars in the northern sky, and these stars and other objects seem to wheel around a single point, directly overhead. No star ever rises or sinks below the horizon. At the equator, the view is very different—Earth's spin brings some stars into view as they rise, and hides others as they set. At mid-latitudes, an intermediate pattern of movement is seen.

Observer's position	Apparent star movement	Observer's view	Star trails seen
NORTH POLE equator	For an observer looking up from the north pole, the stars and other celestial objects seem to circle counter-clockwise around a point directly overhead. At the south pole the movement is clockwise.	north celestial pole	
MID-LATITUDE	For an observer at mid-latitudes—that is between about 25° and 65°, either north or south of the equator—celestial objects rise in the east, cross the sky obliquely, and set in the west.	area around the pole	
EQUATOR	For an observer on or near the equator, celestial objects appear to rise vertically or near-vertically in the east, swing overhead, and then drop vertically down again in the west.	apparent star movement	

LUNAR MOTIONS

The Moon crosses the sky from east to west each day as a result of Earth's spin. However, because it is so close to Earth and orbits our planet, the Moon exhibits some additional movements, which occur relative to the celestial sphere and are not just due to Earth's spin. The most important of these is a continuous west-to-east motion, involving a shift of about 12° across the sky each day. This movement accompanies regular changes in the Moon's phases, and means that it both rises and sets about an hour later on each successive day. Other, more complex cycles in the Moon's celestial motion result from the fact that it orbits the Earth at an angle to the ecliptic plane (p.21), and from various subtle perturbations caused by its interaction with Earth and the Sun.

Movement over a two-week period
This image shows the Moon gradually moving west-to-east over nearly a fortnight. Each photograph was taken about an hour before dawn.

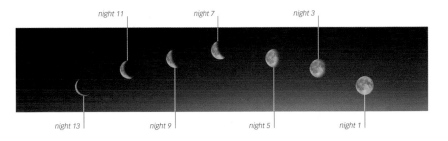

night 11   night 7   night 3

night 13   night 9   night 5   night 1

SHIFTING SUN AND PLANETS

The Sun traces a path across the sky from east to west each day, due to Earth's spin. But a combination of Earth's tilt and its solar orbit means that, for an observer at a fixed location, the daily path the Sun traces varies over the year (see right). A further complication affecting the Sun's motion is a small variation in Earth's speed in its elliptical orbit around the Sun. Together, these factors mean that if the Sun is observed at the same time of day many times over

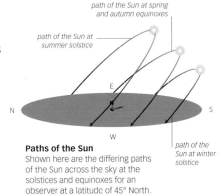

path of the Sun at spring and autumn equinoxes

path of the Sun at summer solstice

E

N   S

W

path of the Sun at winter solstice

Paths of the Sun
Shown here are the differing paths of the Sun across the sky at the solstices and equinoxes for an observer at a latitude of 45° North.

The Sun's analemma
To produce this image of the Sun's analemma, the Sun's position in the sky was recorded at the same time of day on 38 occasions over a year.

a year, it traces a figure-eight path in the sky, called an analemma (see left). Earth's companion planets have complex wandering motions on the celestial sphere, partly influenced by Earth's own solar orbit because this continuously changes the point of view of Earth-bound observers. In general, the closer the planet, the more complex its movements appear. For example, Mars regularly goes through zig-zagging or looping movements, called retrograde motion, over the course of several months.

NAMING SKY OBJECTS

In ancient times, there were only a few hundred easily distinguishable sky objects that required names. Today, more than a billion have been identified—newly discovered objects are systematically named using universally agreed conventions.

STARS

About 350 or so of the brightest stars have ancient names, mostly of Arabic or Latin origin. The first systematic method for naming stars was introduced in 1603 by German astronomer Johann Bayer. Bayer distinguished up to 24 stars in each constellation, labeling them with Greek letters in roughly decreasing order of brightness. In 1712, English astronomer John Flamsteed introduced an alternative system, numbering stars in order of their right ascension—that is, from west to east across each constellation. Both systems are still in use, but faint stars—the majority—are not listed in either system so they are designated simply by a number. Specialized systems have been devised for cataloguing variable, binary, and multiple stars.

Alkaid
85 η

Mizar
79 ζ

Alioth
77 ε

Megrez
69 δ

Dubhe
50 α

Phad
64 γ

Merak
45 β

Bayer and Flamsteed systems
This diagram of the Big Dipper asterism shows each star's ancient name, Bayer letter (a Greek letter), and Flamsteed number. Thus, Merak is also called Beta (β) Ursae Majoris (Bayer) or 45 Ursae Majoris (Flamsteed).

DEEP-SKY OBJECTS

Many night-sky objects appear to the naked eye as fuzzy patches of light. These include star clusters, nebulae, galaxies, and supernova remnants. In the 18th century Charles Messier made a catalog of 110 of these objects. Each is denoted by the letter "M" and a number. Later, two larger catalogs were compiled, the NGC (New General Catalog) and IC (Index Catalog). Each object in these lists is denoted by "NGC" or "IC" followed by a number. Most nebulous objects just have a catalog number, but some also have traditional names, such as the Pleiades star cluster, or more recent names linked to their appearance, like the Sombrero galaxy.

Crab Nebula
This spectacular nebulous object—the remnant of a supernova observed in 1084—is named the Crab Nebula because of its shape, and M1 because it was the first object listed in Messier's catalog.

SOLAR SYSTEM BODIES

Some solar system objects, such as the Moon and most of the planets, have ancient names. Various conventions are used for naming newly discovered comets and asteroids. Comets are named after their discoverers and also given a formal designation that indicates (among other information) the type of comet and the half-month of its discovery. Asteroids are designated by a three- to five-digit number indicating roughly where the asteroid lies in the historical order of asteroid discoveries, usually followed by a name. In the early days of asteroid identification, mythological names were used. This system broke down because of a lack of names, and today asteroids are often named after famous scientists, writers, and musicians.

Comet Hale–Bopp C/1995 O1
This comet was named after its co-discoverers, the Americans Alan Hale and Tom Bopp. The formal designation C/1995 O1 indicates that it is a non-periodic comet discovered in August 1995.

253 Mathilde
The 253 in the name of this asteroid, discovered in 1885, indicates that it was (approximately) the 253rd asteroid identified. "Mathilde" was in honor of a French astronomer's wife.

DISTANCES AND MAGNITUDES

Three main units are used for expressing distances in astronomy: the astronomical unit or AU, the light-year, and the parsec. The magnitudes (brightnesses) of celestial objects are expressed as either apparent or absolute magnitude.

DISTANCE MEASURES

The astronomical unit—the average distance between Earth and Sun—is used for expressing distances within the Solar System. Otherwise, it is more common to use the light-year—the distance light travels through a vacuum in a year—or the parsec, the distance at which a star would have an angular parallax shift of one sixtieth of a degree (see right). One tricky area is the expression of distances to very remote galaxies, complicated by the fact that the expansion of the Universe is carrying them away very fast. Different measures are used to express such distances; one, called the "proper" distance, is the true distance between Earth and the remote galaxy at the present moment in time.

TABLE OF DISTANCES

Unit	Distance
Astronomical	93 million miles unit (AU) 150 million km
Light-year	5,879 billion miles 9,461 billion km
Parsec	19,173 billion miles 30,857 billion km 3.26 light-years

Parallax shift
When Star A is observed from opposite sides of Earth's orbit, its shift is greater than that of more distant Star B. From this shift, an observer can calculate the parallax angle between the star and the two positions of Earth. The star's distance can be determined from this angle.

Cosmic distances
The chart below uses a logarithmic scale. The first division represents 6,200 miles (10,000 km). Each further division marks a tenfold increase in distance on the previous one.

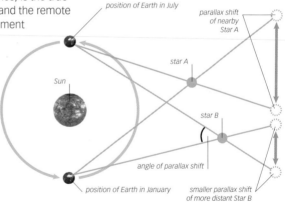

position of Earth in July

parallax shift of nearby Star A

star A

Sun

star B

angle of parallax shift

position of Earth in January

smaller parallax shift of more distant Star B

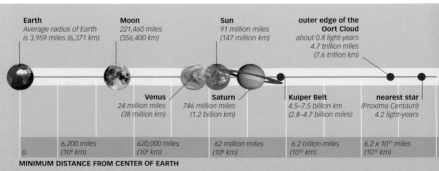

Earth
Average radius of Earth is 3,959 miles (6,371 km)

Moon
221,460 miles (356,400 km)

Sun
91 million miles (147 million km)

outer edge of the Oort Cloud
about 0.8 light-years 4.7 trillion miles (7.6 trillion km)

Venus
24 million miles (38 million km)

Saturn
746 million miles (1.2 billion km)

Kuiper Belt
4.5–7.5 billion km (2.8–4.7 billion miles)

nearest star
(Proxima Centauri) 4.2 light-years

| 0 | 6,200 miles (10^4 km) | 620,000 miles (10^6 km) | 62 million miles (10^8 km) | 6.2 billion miles (10^{10} km) | 6.2 x 10^{11} miles (10^{12} km) |

MINIMUM DISTANCE FROM CENTER OF EARTH

BRIGHTNESS OF SKY OBJECTS

Knowing the measured brightness of an object is an aid to identifying it in the night sky. It is important to distinguish between an object's absolute magnitude—how bright it really is—and its apparent magnitude, which is how bright it looks from Earth. Apparent magnitude differs from absolute magnitude because the farther away a star is, the fainter it looks. Apparent magnitude is the measure most commonly used by stargazers. Both the apparent magnitude and absolute magnitude scales are arranged in such a way that faint objects have high numerical values, while bright objects have low or negative values. A decrease in the numerical magnitude value of 1 roughly corresponds to a 2.5 increase in brightness.

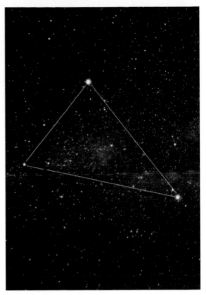

Summer Triangle
Of the triangle of stars Deneb (left), Vega (top), and Altair, Deneb is plainly the faintest—its apparent magnitude is the least. However, Deneb is over 50 times more distant than the others, so it is actually thousands of times more luminous, with a much higher absolute magnitude.

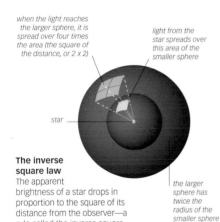

when the light reaches the larger sphere, it is spread over four times the area (the square of the distance, or 2 x 2)

light from the star spreads over this area of the smaller sphere

star

The inverse square law
The apparent brightness of a star drops in proportion to the square of its distance from the observer—a rule called the inverse square law. This happens because light energy spreads out over a progressively larger area as it travels away from a star.

the larger sphere has twice the radius of the smaller sphere

COMPARING BRIGHTNESS

Object	Maximum apparent magnitude
The Sun	-26.7
The Moon	-12.9
Venus	-4.9
Sirius (brightest night-time star)	-1.4
Saturn	-0.5
Vega (star)	0.0
Ganymede (moon of Jupiter)	4.4
Vesta (brightest asteroid)	5.1
Uranus	5.3
Faintest naked-eye objects	around 6.0
Pluto (dwarf planet)	13.6

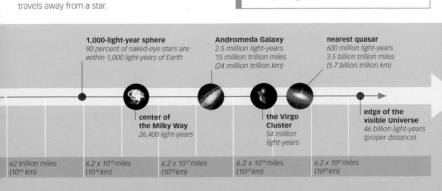

1,000-light-year sphere
90 percent of naked-eye stars are within 1,000 light-years of Earth

Andromeda Galaxy
*2.5 million light-years
15 million trillion miles
(24 million trillion km)*

nearest quasar
*600 million light-years
3.5 billion trillion miles
(5.7 billion trillion km)*

center of the Milky Way
26,400 light-years

the Virgo Cluster
54 million light-years

edge of the visible Universe
*46 billion light-years
(proper distance)*

62 trillion miles (10^{14} km)

6.2×10^{15} miles (10^{16} km)

6.2×10^{17} miles (10^{18} km)

6.2×10^{19} miles (10^{20} km)

6.2×10^{21} miles (10^{22} km)

TOOLS AND
TECHNIQUES

A star party in Iran, organized to look for deep-sky objects in the Messier catalog

THE BASICS

It is thrilling to look up and see stars in the night sky—even more so if you can also identify them. Simple preparations, such as knowing what you might see, can help you get the best from a night's observing.

GETTING STARTED

For many, there is no better way of observing than with the naked eye, for this allows the stars to be seen against the backdrop of the entire sky, a context that is lost when viewing through binoculars or a telescope. Indeed, familiarizing yourself with the sky as a whole is just as important as viewing particular objects in detail, so optical equipment should be considered optional at first. What you must take, however, is a compass for getting your bearings; a planisphere (see below); a flashlight; a list of the objects you hope to see; and food, if you are observing for a long time. Warm clothing is also essential; you will soon get cold in the night air, even on a summer night.

THINKING AHEAD

Before going out, use a planisphere, a star chart, or computer software to plan your viewing. Make a note of what you can see for your time and location. Initially, aim to identify a handful of the more prominent constellations and the brightest stars. Familiarizing yourself with the constellation shapes before going outside will help with recognition. Remember, if the Moon is full it will flood the sky with light and make some objects impossible to see. Form a plan of action and make a list of what to see first, second, and so on. If using a telescope, check it is working, cool it to the night temperature, and allow time to set it up.

Using a planisphere
A planisphere is a device that shows the positions of the stars throughout the year at a given latitude. Changing the date and time on the planisphere changes the portion of the night sky that it reveals.

Electronic star charts
Computer software that enables you to see the night sky at the time and location you are interested in is readily available. Alternatively, astronomy applications can be bought for smartphones and tablets (left); these allow you to hold the screen up to the sky and to see the stars mapped out before you.

OBSERVING TOGETHER

A great way to learn about the night sky is to spend time with a practiced observer. If you don't know one personally, seek out your local astronomical society. Its members will be happy to share their astronomical knowledge, and the society may have a telescope you can use.

Most societies hold regular lecture evenings and practical demonstrations of equipment, as well as star parties at which astronomers gather to observe. Also, look out for specially organized observing events across the country or farther afield, such as solar eclipse holidays and tours with expert guides.

Star parties
All around the world, astronomers gather together at dark-sky sites, well away from light pollution, to share their passion for the night sky. It is a great opportunity to learn observing techniques and get tips from others.

WHAT TO EXPECT

Once outside, give your eyes time to adjust to the dark As time passes, more stars will appear. Before 30 minutes have passed, your eyes will be fully dark-adapted and you will see stars of varied brightness. With the naked eye, it is possible to see a range of objects: the

Moon, star clusters, planets, comets, and galaxies. The faintest of these cannot be seen when viewed straight on due to a "blind spot" in the human eye; look slightly away from these objects and they will appear in your peripheral vision. Binoculars or a telescope will reveal more objects and greater detail.

WHAT CAN I SEE?

An observer using a 12 in (300 mm) reflecting telescope in fine observing conditions can expect to see the objects listed here.

THE MOON Craters, mare (seas), mountains, and other surface features can all be seen in fine detail.

THE PLANETS Mars appears the size of a baseball seen at a distance of 10 ft (3 m); surface details are only visible when Mars is closest to Earth. Jupiter's upper atmosphere cloudbands are visible, as are the planet's four largest moons. Uranus is a tiny dot.

STARS The stars look much brighter, but they still only appear as points of light.

DEEP-SKY OBJECTS Galaxies and nebulae appear as smudges of light. Stars and star clusters are bright and jewel-like.

Naked eye view
The planets and the Moon all move within the band of constellations known as the zodiac (p.21) and so can often be viewed together. Here, the Moon and Venus appear close together in the dawn sky.

WHEN AND WHERE TO OBSERVE

A cloud-free night is an opportunity to observe, but what you see in the night sky depends on the time and place of observation. Atmospheric conditions, such as light pollution and air turbulence, can also affect your viewing.

SHIFTING SKIES

Because Earth spins on its axis once every 23.9 hours, it takes 23.9 hours for the stars to complete a 360° circuit of the sky. To observers standing at the north and south poles, the stars neither rise nor set but trace circles around the north and south celestial poles respectively (p.16). If the observers move toward the equator, the effect of Earth's rotation becomes very different, and the stars begin to rise above and fall beneath the horizon (p.22). Because Earth also orbits around the Sun, the portion of the sky that is visible to the observer changes during the year. An observer looking south in the Northern Hemisphere (or north in the

1 MARCH, 10PM

15 MARCH, 10PM

Fixed location, changing sky
An observer at 50°N looking south at the same time on successive nights sees Orion lower and lower in the sky as March progresses. By the end of the month, it has almost disappeared below the western horizon.

Southern Hemisphere) will see that the constellations are not in the same place at the same time each night but are gradually sinking and drifting westward.

CHOOSING A LOCATION

The quality of a night's observing depends on the quality of the sky, and this varies from place to place. Stars, planets, nebulae, and galaxies are all best seen in a dark sky, but most people live in well-lit towns or cities where it is never truly dark. Nevertheless, even in a city the brightest stars are visible, and it is possible to make out constellations; if the sky is clear and moonless, about 300 stars are visible. A darker, rural sky reveals about 1,000 stars, while the darkest country location reveals some 3,000 stars. Although dim objects are easier to see out in the country, the constellation patterns are harder to distinguish among so many stars.

Light and dark skies
Sirius is the brightest star in the night sky. It is usually visible even in a bright city center (top). However, it is best viewed from rural areas (bottom), where it can be seen surrounded by other stars.

ATMOSPHERIC CONDITIONS

For the best view of an object, it pays to wait until it reaches its highest point in the sky. In this position, the object's light passes through a thinner section of Earth's atmosphere than when it is near the horizon and so suffers from fewer of the distorting effects caused by

Overhead advantage
Observing an object when it is directly overhead reduces the thickness of the atmosphere through which its light must travel, and so improves the view.

atmospheric conditions. One distortion is due to different parts of the atmosphere being at different temperatures, which causes the image seen through the telescope to ripple. A steady atmosphere is one that astronomers say has good "seeing," a quality that is often rated on a scale of one to ten, ten being perfect seeing. A second distortion is caused by the amount of water vapor and dust in the atmosphere. An atmosphere that has little of either of these is said to have good "transparency." This quality improves with altitude.

Red light
Red cellophane placed over the end of a flashlight softens its light. Red light allows you to see what you are doing but keeps your eyes adapted to the dark.

LATITUDE AND LONGITUDE

Changing your position on the globe— even by thousands of miles—does not necessarily change your view of the night sky. Observers at the same latitude but on opposite sides of the world see the same sky. If Orion is visible from Albuquerque, USA, at 11pm local time, it will also be visible in Tokyo, Japan, at 11pm local time, because both cities are on the same latitude. Changing your latitude, however, by moving north or south, can make a huge difference to your view. If you travel from the North Pole to the South Pole, new constellations appear and the entire night sky becomes inverted.

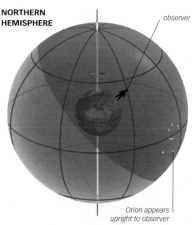

NORTHERN HEMISPHERE

observer

Orion appears upright to observer

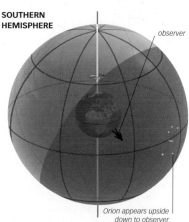

SOUTHERN HEMISPHERE

observer

Orion appears upside down to observer

Inverted skies
These simplified celestial spheres show how Orion appears when viewed from the Northern Hemisphere and the Southern Hemisphere. The darkened area represents the sky above the observer's horizon.

ORIENTATION

Finding your way around the night sky can be easy and fun. All you need are a few simple techniques to get you started. With a little practice, judging the angles, sizes, and locations of celestial objects soon becomes second nature.

ANGLES IN THE SKY

The position of an object in the sky is often expressed in terms of its altitude and azimuth. An object's altitude is its angle above the horizon from any given latitude; a star directly overhead therefore has an altitude of 90°. An object's azimuth is its angle from the north point; a star directly behind you as you look north therefore has an azimuth of 180°. These angles are easily measured using your arms, and soon you will be able to judge them by sight alone. If in unfamiliar surroundings, use a compass to find north, or use the "star-hopping" method shown opposite.

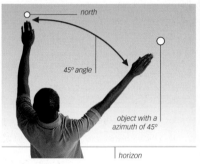

Measuring altitude
To determine an object's altitude, stretch one arm above your head to make a 90° angle with the horizon. Then point your other arm at the object and judge its angle above the horizon.

Measuring azimuth
To determine an object's azimuth, extend one arm in the direction of due north and swing the other arm around to point at the object. East is 90° to the right; northeast is 45°.

SIZING THINGS UP

The size of an object in the sky, as well as the distance between objects, is often expressed in degrees or parts of a degree. The constellation Leo, for example, is about 40° across, measured from the lion's nose to its tail. These and smaller sizes can be measured using your hands, a method that helps you relate a map of a constellation to the real stars in the sky.

Finger width
A finger held at arm's length covers about 1° of sky and is useful for measuring small distances between stars. The full Moon measures half a degree across.

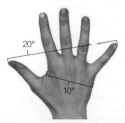

Hand spans
The width of an outstretched hand measures about 20° of sky. A fist (the distance across the base of the fingers) measures some 10°.

Finger joints
The three joints of a finger can measure small distances across the sky. The first joint is about 3° wide, the second roughly 4°, and the third about 6°.

STAR-HOPPING

Many constellations contain stars that act as pointers to other parts of the night sky, and these can help you orient yourself. Three such "star-hopping" routes are shown here. The first begins in the Big Dipper, which is an asterism (p.19) in Ursa Major, a constellation in the Northern Hemisphere. The second starts in Orion, an equatorial constellation that is fully visible from most latitudes on Earth. The third features the constellations Crux and Centaurus, which can only be seen in the Southern Hemisphere.

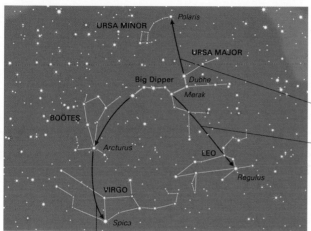

The Big Dipper
The pan-shaped pattern called the Big Dipper is one of the most familiar patterns in the northern sky. Two of its seven stars point the way to Polaris, which marks the north celestial pole.

five times the distance from Merak to Dubhe takes you to Polaris

extending a line from the two stars forming the edge of the pan bowl nearest to the handle leads to the bright star Regulus in Leo

the Big Dipper's handle makes a curve that can be extended to Arcturus in Boötes and Spica in Virgo

the belt stars point through Orion's hand to Aldebaran in Taurus

a line leading from Orion's belt and past his lower body points to the brilliant star Sirius in Canis Major

Orion
Extending imaginary lines from the three stars in Orion's belt takes you to the constellations Canis Major and Taurus.

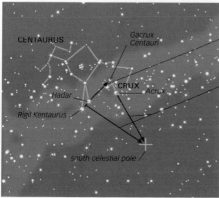

three times the distance from Rigil Kentaurus to Hadar takes you to Gacrux

extending one line from Crux's longer axis and a second line from between Hadar and Rigil Kentaurus creates a pair of pointers that converge on the south celestial pole

Crux and Centaurus
Tiny Crux and much larger Centaurus are embedded within the southern-sky path of the Milky Way. Stars within these constellations point to the south celestial pole, the point around which the southern stars revolve.

TYPES OF BINOCULARS

Binoculars are a great instrument for both the novice and the seasoned observer alike. They are easy to use and provide outstanding views, increasing both the number and clarity of the objects seen in the night sky.

WHY BINOCULARS?

Binoculars are small, affordable, portable, and uncomplicated. They allow you to use both eyes for viewing, and the image they produce is automatically in the correct orientation, as opposed to the upside-down and left-to-right view produced by many telescopes. They provide a wide field of view, so they are ideal for making visual tours across large swathes of sky and for looking at large objects, such as the Orion Nebula and comets with long tails. Their relatively low magnification also makes them ideal for studying the surface of the Moon.

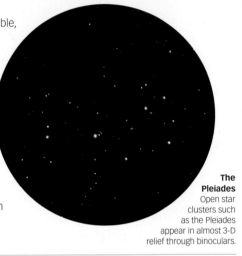

The Pleiades
Open star clusters such as the Pleiades appear in almost 3-D relief through binoculars.

MAGNIFICATION AND APERTURE

The two main qualities of binoculars are their ability to collect more light than the naked eye and their ability to magnify images. These two qualities are given numerical values that are used to describe a pair of binoculars. For example, an 8x50 ("eight by fifty") pair of binoculars is one that magnifies a naked-eye image eight times and has a objective lens that is 50mm in diameter—a measurement expressed in millimeters (mm) and known as the aperture (p.44). This standard pair of binoculars (shown right) will reveal about 200 times more stars than are visible to the naked eye. When choosing binoculars, select a pair with an aperture that is at least five times greater than the magnification figure; if it is smaller than this, the view through the binoculars will be too dim to be useful for astronomy.

↖ Aperture

COMPACT BINOCULARS

STANDARD BINOCULARS

objective lens

Objective lenses
The aperture of a pair of binoculars (shown here on the two main types of binoculars) determines its light-gathering ability. Larger apertures collect more light and reveal more stars.

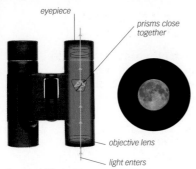

Compact binoculars
The prisms in compact binoculars are set close together, which allows the light to take a straight path through the optical tubes. This makes for a smaller, more compact design of binoculars.

BINOCULAR DESIGN

Binoculars are essentially two identical, low-powered telescopes joined together. Their bodies contain glass lenses and prisms that turn the formed image the right way around. They have two basic designs: compact (or roof-prism) and standard (or Porro-prism). The latter have the classic dog-leg shape and are bulkier than compact binoculars. In general, standard binoculars produce better, brighter images than compact binoculars and are better value for money. Large binoculars are a larger version of the standard binocular design.

Standard binoculars
The prisms in standard binoculars are set apart from each other, allowing for a longer light path and so greater magnification. The lenses are typically 50 mm wide and produce bright images.

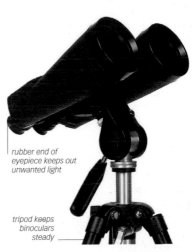

Large binoculars
The larger the aperture, the bigger, heavier, and more cumbersome the binoculars are. Binoculars with an aperture of 70 mm or more need supporting on a tripod to achieve effective, shake-free viewing.

CHOOSING A PAIR OF BINOCULARS

Type	Advantages	Disadvantages
Compact	Small size and weight make these highly portable and easy to handle. They are good for terrestrial use.	The small objective lens allows only a small amount of light to be collected, so these are the least effective for astronomy.
Standard	These show clear, bright images. Their large objective lenses are good for observing faint objects.	Standard binoculars are blind to objects fainter than magnitude 11 and will have image shake with magnifications greater than 10.
Large	The large lens reveals more celestial objects. Magnifications of 15x allow close study of the Moon and planets.	Large binoculars are too big and too heavy to be held by hand and so must be supported by a tripod for effective viewing.

USING BINOCULARS

Binoculars enhance your view because they collect more light than the human eye. They are very rewarding, being easy to use and requiring little in the way of preparation. Choose a dark, sheltered observing site, make yourself comfortable, focus the binoculars, and then begin to observe.

SEEING WITH BINOCULARS

To get the most from your binoculars, allow time for your eyes to adjust to the dark and then get the binoculars in focus. Make sure you support the binoculars (see opposite); they may seem light at the start of your observing, but you will soon tire of holding them up. A common problem is finding the target object in the field of view, even when it is visible to the naked eye. One method is to work up from a recognizable feature on the horizon; another is to locate an easier object and use this to navigate to your target.

Sitting comfortably
A deckchair supports the back and neck, allowing you to look up in comfort. Rest your elbows on your upper body as you hold the binoculars.

FOCUSING BINOCULARS

The Moon, the planets, and the stars are a long way off, so binoculars need to be focused on the far distance. Because your left and right eyes may differ in their strength of vision, each eyepiece is focused separately. Start by pivoting the two optical tubes about the central bar to align the eyepieces with your eyes. Next, seek out a bright star or the Moon and use it for focusing (see panel, right). If using a star, it will appear as a bright, sharp point of light through the eyepieces. Once focused, scan the sky to become familiar with the greater detail now visible. You can then start to navigate the sky, hopping from one celestial object to another via different stars.

Focusing the eyepieces
Turn the central focusing ring to focus the nonadjustable eyepiece (left) and focus the adjustable eyepiece using its own ring (right).

STEP-BY-STEP FOCUSING

Once the eyepieces are aligned with your eyes, use the two focus rings—one on the central bar and a second on one of the eyepieces—to focus the two eyepieces. It is usually the right-hand eyepiece that has its own independent focus ring.

1 Find which eyepiece can focus independently. Close that eye and look through the other.

2 Rotate the binoculars' main central focusing ring until the image is in sharp focus.

3 Open only the other eye. Use the eyepiece focusing ring to bring the image into focus.

4 Both eyepieces should now be in focus. Open both eyes and start observing.

SUPPORTING BINOCULARS

It is important to support binoculars, not only because you will be more comfortable and able to observe for longer but also to reduce wobble. The field of view typically seen through a pair of 10x magnification is about 6°. As you hold them, you will wobble by about 1.4°.

handle for adjusting direction

Binocular mount
A binocular or camera tripod mount provides the best platform for supporting large binoculars. Above are 25 x 100 binoculars used for deep-sky viewing.

Support your whole body by reclining in a deckchair or lounger or on a waterproof rug on the ground. All of these allow you to view regions of the sky at high angles without straining your neck. Alternatively, rest your elbows on the top of a gate or fence, or lean back against a wall. Even a piece of string can help. Tie one end to the binoculars and loop the other around your foot. Pull up gently with the binoculars to keep them steady.

Steadying binoculars
Sitting with your elbows resting on your knees helps support the weight of the binoculars and keeps them steady.

WHAT TO LOOK AT

Binoculars are particularly suitable for scattered star clusters that are too expansive for a telescopic field of view. Nearby galaxies, such as Andromeda and Triangulum, appear as small, gray, misty patches. The Orion Nebula and other nebulae look similar. But if the sky is clear and free of light pollution, the Orion Nebula is bright enough and large enough for some structure to be visible. Binoculars add clarity to the Moon's surface, bringing prominent mountain ranges into view as well as the smooth maria and the largest craters. A standard pair of binoculars will reveal Jupiter's four largest moons.

OBJECTS TO SEEK OUT WITH BINOCULARS

Object Name	Object Type	Constellation
M13	Globular cluster	Hercules
M22	Globular cluster	Sagittarius
M31	Galaxy	Andromeda
M33	Galaxy	Triangulum
M42	Emission nebula	Orion
M45	Open cluster	Taurus
NGC 869	Open cluster	Perseus
NGC 884	Open cluster	Perseus
NGC 5139	Globular cluster	Centaurus

Milky Way
When turned on the Milky Way, even low-power binoculars show that this cloudy band is in fact made of individual stars.

Orion Nebula
The Orion Nebula is striking when seen through medium-power binoculars; its gaseous halo is clearly visible.

TYPES OF TELESCOPE

Telescopes are astronomers' primary tools. Traditionally they were of two varieties—the refractor and the reflector—but today a hybrid design is also available. All have a range far greater than the most powerful binoculars.

THE TELESCOPE

All telescopes are designed to collect light from distant objects, to bring the light to a focus, and to produce an image of those objects. Once magnified by an eyepiece, this image is viewed or recorded. The refractor uses a lens to collect the light, the reflector uses a mirror, and the hybrid, or catadioptric, uses a combination of the two. Each type consists of three basic parts: the tube assembly containing the optics (the lenses, mirrors, or both); the mount, which supports the tube and allows the telescope to be pointed precisely; and a solid base onto which the mount is secured. The base can be a tripod, which is portable, or a fixed pier.

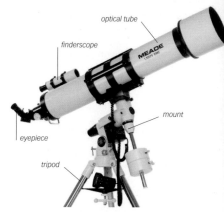

optical tube

finderscope

mount

eyepiece

tripod

Telescope features
The telescope tube contains the optics. This model is a refractor, so it contains a lens at its top end. The tube and its mount, which turns the telescope to a target in the sky, are fixed on a tripod.

REFRACTOR TELESCOPE CAPABILITIES

STRENGTHS

The refractor is robust and needs little maintenance. This makes it suitable for children to use and for moving from place to place.

The images produced are sharp and have good definition, making the refractor ideal for looking at the Moon, the planets, and double stars.

There are no mirrors or obstructions within the telescope tube, allowing for high-contrast images and greater detail. The refractor is also good for daytime use.

WEAKNESSES

Because lenses are more expensive to make than mirrors, refractor telescopes are more expensive than reflectors of the same aperture.

The refractor is prone to an imperfection called chromatic aberration. This occurs when the objective lens fails to focus all the colors of an object on the same point—giving stars, for example, a faint halo of color.

The low position of the eyepiece can make it awkward to observe objects that are located high in the sky, unless a diagonal is used (p.53).

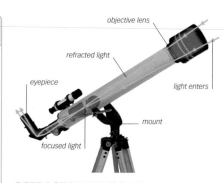

objective lens

refracted light

eyepiece

light enters

mount

focused light

REFRACTOR TELESCOPE

The refractor uses a lens, or a series of lenses, to collect and focus the light from distant objects. It is the classic long-tube astronomical telescope and the direct descendant of the instruments built and used by Galileo. In its simplest form, it consists of a tube with a large lens (the objective) at the top end. The objective collects light and focuses it down the tube, where a smaller lens, the eyepiece, magnifies the image, which the observer then sees.

REFLECTOR TELESCOPE CAPABILITIES

STRENGTHS

Telescope mirrors are easier and cheaper to manufacture than lenses. This means the reflector is inexpensive compared to a refractor of the same aperture.

The eyepiece is positioned near the top end of the telescope tube, making the reflector easy and comfortable to use.

This type of telescope is good for viewing deep-sky objects, such as dim nebulae and galaxies. The reflector is also free from chromatic aberration.

WEAKNESSES

The reflector's mirrors need to be checked periodically to see that they are correctly aligned. Also, the mirror coatings degrade and need to be replaced every few years.

The reflector is prone to coma, an effect whereby stars at the edge of the field of view appear to be wedge-shaped.

The secondary mirror obstructs the light traveling from the object being observed to the main mirror. This can result in "spikes" appearing to emanate from bright objects.

eyepiece

light enters

secondary mirror

reflected light

objective mirror

mount

REFLECTOR TELESCOPE

The reflector uses mirrors to collect and focus light from distant objects. The light is collected by a concave main mirror (the objective) at the bottom end of the telescope tube. The light is then reflected to a smaller mirror (the secondary), which is suspended much higher up the tube. This sends the light to the side of the tube, where an eyepiece magnifies the image.

CATADIOPTRIC TELESCOPE

This telescope uses both lenses and mirrors. At the top end of the tube is a lens known as the corrector plate, which compensates for optical imperfections. The light passes through this to the main mirror at the bottom end. From here it is reflected to a secondary mirror fixed to the rear of the plate. It is then sent back down the tube and through a hole in the center of the main mirror to the eyepiece.

corrector plate

light enters

convex secondary mirror

concave primary mirror

eyepiece

CATADIOPTRIC TELESCOPE CAPABILITIES

STRENGTHS

The catadioptric is compact and relatively light, making it highly portable. It also offers a long focal length (p.44) in a short tube.

It is far less prone to the refractor's chromatic aberration, or to the reflector's coma problems.

When used with a computerized motor, the catadioptric automatically finds and tracks objects. It is ideal for taking photographs.

WEAKNESSES

The catadioptric's more complex design tends to make it more costly than refractors or reflectors of equivalent aperture size.

Because light is sent back and forth through the telescope tube, some of the light is lost before reaching the eyepiece. The result is an image that is dimmer than it would be from an equivalent refractor or reflector. However, special coatings applied to the mirrors help reduce light loss.

This telescope is designed as a sealed unit and cannot be taken apart should major adjustments be needed.

TELESCOPE MOUNTS

A telescope mount is an integral part of a telescope system and care should be taken when selecting one. It provides a stable platform for the optical tube and allows for accurate aiming of the instrument. There are two basic types to choose from.

ALTITUDE-AZIMUTH MOUNT

The simpler of the two mounts is the altitude-azimuth (or alt-az) mount. Lighter and more compact than the alternative equatorial mount, it has the advantage of requiring no special setting up. Like the mounting of a camera tripod, it allows the telescope to move around two axes: in altitude (up and down through 90°) and in azimuth (left to right through 360°). It is a handy mount for casual observing, but its drawback is that stars do not move in rigid altitude-azimuth trajectories. This makes it difficult to handle when tracking celestial objects.

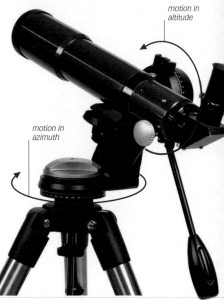

motion in altitude

motion in azimuth

Alt-az-mounted telescope
To find an object using an alt-az mount, the observer must know the object's angle of elevation above the horizon (its altitude) and its compass bearing (azimuth).

VARIANT ALT-AZ MOUNT DESIGNS

The Dobsonian mount and the fork mount are two of the most popular alt-az designs. The Dobsonian is a simple model in which the telescope tube, typically a refractor, pivots up and down on bearings set in a box that swings left and right. Fork mounts have a more open design, featuring one or two supports for the telescope tube. These have become the standard mount for automated go-to telescopes (pp.50–51).

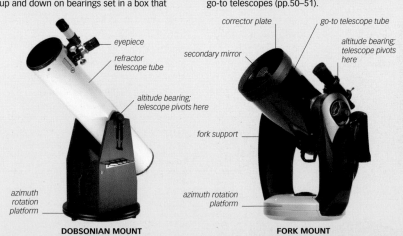

eyepiece

refractor telescope tube

altitude bearing; telescope pivots here

azimuth rotation platform

DOBSONIAN MOUNT

corrector plate

go-to telescope tube

secondary mirror

altitude bearing; telescope pivots here

fork support

azimuth rotation platform

FORK MOUNT

EQUATORIAL MOUNT

The equatorial mount also moves up and down and left to right, but it does so relative to the celestial equator rather than to Earth's surface. In other words, the entire mount is tilted so that it is aligned with the celestial pole (p.49) and moves on axes known as right ascension (RA) and declination (DEC). These axes, which correspond to the sky's equivalent of longitude and latitude, allow the observer to find celestial objects by using the celestial coordinates system (p.17). If an object's celestial coordinates are known and the mount is correctly aligned, the observer simply adjusts the RA and DEC settings on the equatorial mount to bring the object into the telescope's field of view.

motion in declination

motion in right ascension

Equatorial-mounted telescope
A telescope on an equatorial mount can move in tandem with celestial objects. The latitude of an object on the celestial sphere (pp.16–17) is its declination; its longitude is its right ascension.

TRACKING AN OBJECT

Earth's spin on its axis makes the sky appear to rotate around the celestial poles. This movement means that an object in the sky moves quickly out of a telescope's field of view. To keep it in view, the telescope needs to move with the sky, in the opposite direction to Earth's rotation.

To keep an object in view with an alt-az mount, the observer must move the telescope by hand, both in azimuth and in altitude, in an awkward series of up-and-across, down-and-across steps. The equatorial mount, by contrast, can

Motorized tracking
An electric motor drive attached to the RA axis of an equatorial mount allows the telescope to track an object in the night sky. Adjustments in position can be made using the motor's control unit.

track an object smoothly. This is because it is aligned with the celestial pole and is free to move on its RA axis only. The object can then be tracked by hand, or automatically if a motor is attached to the mount. Because of this smooth tracking ability, the equatorial mount is popular among photographers; deep-sky objects take several minutes to photograph, requiring the object and the camera to be perfectly synchronized.

Tracking stars
Stars appear to follow circular paths around the celestial poles. An alt-az-mounted telescope follows a stepped path when tracking a star; the equatorial mount allows for a smooth, near-circular motion.

——————— **Apparent star path**
——————— **Alt-azimuth mount path**
——————— **Equatorial mount path**

TELESCOPE PROPERTIES

The function of a telescope is to collect light from distant objects, to form it into an image, and then to magnify the image for viewing. How well an individual telescope does this depends on its optical properties. An understanding of these is therefore essential for knowing which telescope is best for you.

APERTURE

The job of collecting light is done by the telescope's main lens or mirror, called the objective. The bigger the objective, the more light it collects, and the greater detail you will see. By doubling the objective's diameter, you quadruple the light collected and double the distance you can see. The diameter of the objective is known as the aperture, and it is usually measured in inches or millimeters. A telescope with a 2.4 in (60mm) main lens gathers 70 times as much light as the naked eye.

Objective sizes
The aperture is the most important specification of a telescope. The larger the aperture, the more magnification the telescope can use. Using more magnification than an aperture allows produces a blurred, indistinct image.

4.7 IN (120 MM) APERTURE   **2.6 IN (66 MM) APERTURE**

position of
objective lens

focal length

finderscope

aperture

mount

tripod
head

focal point

eyepiece

FOCAL LENGTH

The distance from the main lens or mirror to the focal point (where the light comes to a focus) is the telescope's focal length. A telescope with a long focal length will produce a large but faint image at its focal point. One with a short focal length will produce a smaller but brighter image. Focal length divided by aperture gives a telescope's focal ratio, or f-number. Telescopes with a large f number, greater than f/9, are best suited to viewing close, bright objects, such as the Moon and planets. Those with a smaller f-number, such as about f/5, are better for viewing faint, distant objects.

Focal ratio
This small refracting telescope has an aperture of 4.7 in (120mm) and a focal length of 39.4 in (1,000mm). The focal length divided by the aperture gives it a focal ratio of f/8.3.

MAGNIFICATION

The degree of magnification offered by a telescope depends on the strength of its eyepiece. An eyepiece is described in terms of its focal length, and this is marked on its side. The focal length is an indication of its magnifying power; the shorter the focal length, the greater the magnification. A 1.0 in (25 mm) eyepiece is good for general use, and a 1.6 in (40 mm) is good for star clusters and nebulae. A wide-angle eyepiece gives a wider field of view, and is also used for star clusters and nebulae.

0.4 IN (9 MM) EYEPIECE

1.0 IN (25 MM) EYEPIECE

Greater detail
Switching eyepieces changes the magnifying power of a telescope. A higher-magnification eyepiece (left) gives a closer view of an object at the expense of contrast and brightness.

Eyepieces
This collection of eyepieces offers a good range of magnifications. The Barlow lens is inserted in front of an eyepiece and increases its magnification by a given number of times.

0.4 IN (9 MM) **1.0 IN (25 MM)** **1.6 IN (40 MM)** **2X BARLOW** **WIDE-ANGLE**

TELESCOPE CAPABILITIES

The optical heart of a telescope consists of the tube and the eyepiece. The tube, blackened on the inside to prevent reflection, houses the objective, which focuses light to form an image at the focal point. The eyepiece acts as a magnifying glass and magnifies this image. The level of magnification is calculated by dividing the focal length of the telescope by thc focal length of the eyepiece. The resultant figure (25x, 50x, 100x, and so on) describes how big an object appears in the eyepiece compared to the naked-eye view. Although changing the eyepiece changes magnification, each telescope has limits set by its own aperture.

The aperture also defines the resolving power of a telescope. This is the amount of detail an instrument can reveal; for example, whether it can distinguish between the two stars of a binary system. Resolving power is measured in tiny angles separating objects known as arcseconds (identified by the symbol ").

APERTURE AND MAGNIFICATION

The maximum magnification of a telescope depends on the size of its aperture. As a general rule, the highest usable magnification of a particular telescope is about twice the size of the telescope's aperture in millimeters.

Aperture inches (mm)	Highest usable magnification	Faintest objects visible	Resolving power
2.4 (60)	120x	mag. 11.6	1.9"
3.1 (80)	160x	mag. 12.2	1.5"
3.5 (90)	180x	mag. 12.5	1.3"
3.9 (100)	200x	mag. 12.7	1.2"
4.7 (120)	240x	mag. 13.1	1.0"
5.9 (150)	(300x)	mag. 13.6	0.8"
7.9 (200)	(400x)	mag. 14.2	0.6"
9.8 (250)	(500x)	mag. 14.7	0.5"

SETTING UP A TELESCOPE

The details of setting up a telescope vary from model to model, but the basic process is the same for all. This process is outlined here, using a reflector on a motorized equatorial mount as an example. If the telescope is left untouched between observing sessions, few steps will need to be repeated.

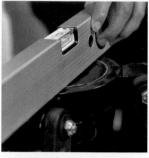

1 Level the tripod
Set up your tripod on solid, level ground. Use a spirit level to check that the tripod's top plate is horizontal.

2 Adjust the legs
Avoid extending the tripod legs fully, which reduces stability and gives no scope for fine adjustment of height. Secure the leg locks.

3 Place the mount
Gently place the mount onto the tripod (left), ensuring that the protrusion on the mount slots into the hole in the tripod (below).

4 Secure the mount
Tighten the mounting screw beneath the tripod head. Make sure it is secure and that the mount is firmly fixed to the tripod.

5 Attach the motor drive
Attach the motor drive to the mount and ensure that the gears of the motor are correctly engaged with those of the mount.

6 Polar alignment
If using an equatorial mount, check that the right ascension (RA) axis (the long part of the central "T" of the mount) is pointing roughly toward the north or south celestial pole (p.49).

7 Add the counterweights
Slot the weights onto the counterweight shaft and secure them in position with the nut. Attach the extra safety screw if provided.

8 Mount the telescope
Place the telescope tube inside the mounting rings that are part of the mount. Use the screws to clamp the rings tight around the tube.

10 Add the drive controller
Plug the drive controller into the mount's motor drive, but do not connect it to the power supply.

9 Add the fine adjustment cables
Screw in the fine adjustment cables. These allow you to make small changes to the RA and declination (DEC) axes when observing.

11 Fit the finder and eyepiece
Fit the desired eyepiece and finderscope (or red dot finder) onto the telescope. Secure with the nuts. Align the finder with the telescope (p.48).

12 Balance right ascension
Loosen the RA axis locking handle (p.49). Undo the nut securing the weights and slide the weights until they balance the telescope. Secure the nut at this point.

FULLY ASSEMBLED TELESCOPE

13 Balance declination
Support the telescope and loosen the DEC axis locking handle (p.49). Loosen the rings holding the telescope. Slide the tube until balanced. Secure the rings and locking handle.

14 Power up
Connect the motor to the power supply and fine tune the polar alignment of the tube (p.49). The telescope is now ready to use.

ALIGNING A TELESCOPE

Locating an object through the magnified view of a telescope takes practice. But help is at hand in the form of a finder, which sits on the side of the main telescope. Once it is in position and the mount is aligned, the telescope is ready for the night sky.

FINDING OBJECTS

There are two main types of finder. The first, a finderscope, is a small refracting telescope that gives a wide field of view. Once your object is visible in the center of its eyepiece, it will be detectable through the main telescope. By contrast, a red dot finder gives an unmagnified view of the sky, which is seen through a screen of transparent glass or plastic. A red dot projected onto the screen indicates where the telescope is pointing.

RED DOT FINDER

sight

FINDERSCOPE

eyepiece

Types of finder
Locating a faint star in the narrow field of view afforded by a telescope can be difficult and time-consuming. Finders solve this by showing the star in its surrounding context.

ALIGNING A FINDER

A finder must first be aligned with the main telescope. This can be done during daylight hours using a distant terrestrial object. Waiting until night and aligning the finder with a star is less reliable because the night sky moves relative to the observer. The steps for aligning a finderscope are outlined below, but they are the same for aligning a red dot finder. Follow this procedure periodically to keep the finderscope and telescope in alignment, and remember to keep the adjustment wheels tight.

1 Aim the telescope
During daylight hours, choose a distant object, such as a lamppost, and move your main telescope carefully so that your target object sits in the middle of your field of view.

2 Check the finderscope
Make sure the barrel of your finderscope is roughly parallel with your telescope. Look through the finderscope's eyepiece. The target object will probably be off center.

3 Adjust the finderscope
Use the adjustment screws or wheels on the finderscope to make small changes in the alignment of the finderscope tube. Keep checking the view through the two instruments.

4 Finalize alignment
Once the target object is in the center of the field of view of both the finderscope and the main telescope, tighten the adjustment screws on the finderscope.

POLAR ALIGNMENT

Since the stars revolve around Earth's rotational axis, a telescope must be aligned with the same axis for an observer to track the stars. To achieve this in the Northern Hemisphere, lift the telescope (which must be equatorially mounted) so that the polar axis of the mount is pointing north. Next, set your latitude on the scale at the base of the mount; this will adjust the telescope until it is aligned roughly with the Pole Star. To align a telescope in the Southern Hemisphere, set your latitude when pointing the telescope south; it will now be aligned roughly with the south celestial pole.

A more accurate alignment can be made with a polarscope, a small telescope set within the polar axis of many mounts. Once the view through the polarscope matches the constellations engraved onto its eyepiece, the telescope is aligned.

Setting latitude to find the celestial pole
Once the mount's polar axis (top right) is pointing north (in the Northern Hemisphere) or south (in the Southern Hemisphere), the observing latitude can be set on the scale (center) at the base of the mount.

USING THE SETTING CIRCLES

The telescope is ready for use once set-up and alignment are complete. Double-check that all is correct by pointing the telescope to a star you can readily identify and whose celestial coordinates you know (p.17). Center the star in your eyepiece and set the right ascension (RA) setting circle to the star's RA coordinate. If the telescope is set up correctly, the declination (DEC) setting circle will show the star's correct declination. To see another object, look up its coordinates, and adjust the settings accordingly.

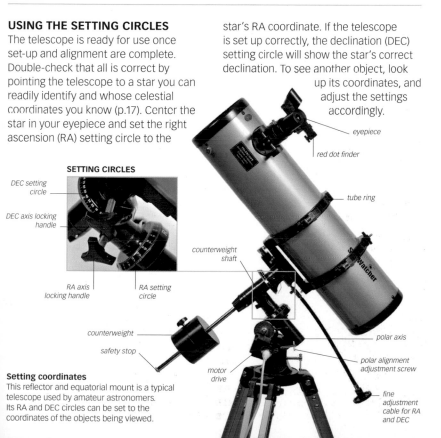

eyepiece

red dot finder

tube ring

SETTING CIRCLES

DEC setting circle

DEC axis locking handle

RA axis locking handle

RA setting circle

counterweight shaft

counterweight

safety stop

motor drive

polar axis

polar alignment adjustment screw

fine adjustment cable for RA and DEC

Setting coordinates
This reflector and equatorial mount is a typical telescope used by amateur astronomers. Its RA and DEC circles can be set to the coordinates of the objects being viewed.

GO-TO TELESCOPES

The beauty of this type of telescope is that it makes celestial objects easy to find. The go-to uses computer technology to find and track night-sky objects. An observer simply chooses what to look at and then waits for the telescope to turn to it.

EASY OBSERVING

Many amateur telescopes are sold with mounts equipped with electric drives and control handsets that allow instruments to track stars across the night sky. The go-to is equipped with motors, but it takes automated observing further still. Once aligned to the night sky, it uses its built-in computer system to automatically locate and move to a multitude of different celestial objects. In this way,

it makes navigating the sky very easy for those with little knowledge of celestial coordinates. Although the go-to can be more costly than a conventional telescope, it has become increasingly popular and affordable.

Tracking planets
The go-to telescope will find and track planets such as Jupiter (above), stars, and deep-sky objects. Once set up correctly, the target object appears in the center of the field of view.

altitude setting circle

finderscope

eyepiece

handset for mount

azimuth setting circle

fork mount

Catadioptric telescope on a go-to mount
This 8 in (200 mm) telescope is connected to a personal computer and controlled by a planetarium software package. A computer can also update some handsets with newly discovered objects.

HANDSETS

The go-to is controlled by a computer handset, sometimes referred to as a paddle. This contains a list of celestial objects, and once the telescope has

Entering coordinates
Choose a target object from the handset's database, or key in the name or coordinates of the object you have chosen to observe.

been set up, it is simply a matter of choosing an object from the menu. The handset then communicates with the telescope mount, instructing it where to turn to find the chosen target. The handset's database contains thousands of objects, and each time one is chosen the telescope moves into the correct position. Some go-to handsets offer a pre-determined tour of the sky. Models with Global Positioning System (GPS) receivers automatically set the date, time, and location each time the instrument is switched on.

SETTING UP A GO-TO TELESCOPE

To set up a go-to telescope it must first be aligned to the night sky. Next, switch on the mount and handset and enter the date, time, and location. Finally, center a chosen star in the telescope's field of view. Once this third stage has been repeated for several widely separated stars, the telescope is ready for use.

1 Mount the telescope
Use a spirit level to check that the tripod is level. Gently lower the mount and telescope onto the tripod head and secure it in position.

2 Set position
Move the telescope into its home position. For an alt-az fork-mounted telescope, this can mean aligning two arrows (left). An equatorial mount will need polar alignment.

3 Prepare the telescope
Connect the mount to the power pack and switch on the mount. Remove the lens cover.

4 Enter start data
Enter the date, time, and location into the handset as prompted. Some go-to telescopes require you to select your location from a menu.

5 Align the telescope
Alignment methods vary. Typically, you select a bright star, such as Sirius, to align with. Select the star on the handset. The telescope will move automatically to where it calculates the star to be.

Jupiter selected on handset

7 Set the destination
The go-to telescope is now ready. Find the name of the object you wish to observe in the handset's menu and press "go-to" or "enter." The telescope will move to center the chosen object in the eyepiece.

6 Adjust alignment
Look through the eyepiece to view the selected star. If it is off-center, use the finder to center it, using the directional buttons on the handset. Repeat steps 4 and 5 to align with two or three more stars.

TELESCOPE OBSERVING

One of the joys of observing the night sky is discovering the Universe for yourself. All you really need is a good telescope, but various accessories are also worth buying. These allow you to observe more detail and in greater comfort.

CHANGING VIEW

A telescope's ability to collect more light than the naked eye or binoculars means that it can transform your view of a celestial object. For example, to the naked eye, Orion's arms stretch out from the kite-shaped pattern of stars that makes up his body. Individual stars mark his shoulders, head, and belt. A dim, fuzzy patch of light beneath Orion's belt marks the location of the hunter's sword. In reality, this patch is a massive star-forming cloud of gas and dust called the Orion Nebula—the largest and brightest star-forming region visible in the sky.

3 Medium 8 in (200 mm) telescope view
Viewed through a medium telescope, the nebula's structure becomes visible. Through the eyepiece it appears gray-green, but a long-exposure image reveals a range of pinks, blues, and white.

1 Naked-eye view
Even in a sky affected by light pollution, the figure of Orion is visible. The red star Betelgeuse is center top in this view; below is a row of three stars marking Orion's belt; below this is the Orion Nebula.

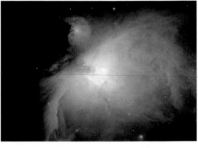

4 Large 87 in (2.2 m) telescope view
This long-exposure image through a wider aperture captures more of the nebula's structure. The Trapezium star system at its core emits radiation that shapes the surrounding volumes of dust and gas.

2 Binocular view
The Orion Nebula is the brightest star-forming region in the night sky. Its shape and form start to become visible when seen through binoculars. Stars not visible to the naked eye now also come into view.

When seen through binoculars, the nebula transforms into a cloud with structure and varied brightness. A medium 8 in (200 mm) telescope changes the view even more; the nebula's finer structure becomes visible, and its colors become apparent in images taken through the telescope. These colors are associated with the nebula's consituent elements—mainly hydrogen, oxygen, and sulfur. Increasing the aperture to 87 in (2.2 m) reveals extraordinary detail; the newborn Trapezium star system can be seen blazing at the nebula's heart.

TELESCOPE VIEW

Telescopes produce images that are upside down and flipped left to right. The view you receive is the one you would get if you were standing on your head. This can be disorienting, especially if you are trying to relate what you are seeing to a chart. A small piece of equipment called a diagonal, or star diagonal, which fits between the telescope and the eyepiece, partly rectifies this problem. It uses a prism or an angled mirror to send the light

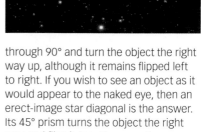

Inverted image
The image below shows how Orion appears to the naked eye. Left is a view of the same constellation seen through a telescope without a diagonal attachment.

through 90° and turn the object the right way up, although it remains flipped left to right. If you wish to see an object as it would appear to the naked eye, then an erect-image star diagonal is the answer. Its 45° prism turns the object the right way and flips it too. However, these corrected views come at a cost, because a diagonal absorbs and disperses some of the light that forms the image. For this reason many astronomers prefer the clearer, although inverted, view.

Viewing through a diagonal
A diagonal fits between the telescope and its eyepiece. Its position allows the viewer to look down rather than up into the telescope.

EYEPIECE FILTERS

A filter is a disk of colored glass that can be fitted to a telescope's eyepiece to prevent certain wavelengths of light, or colors, from reaching the eye. There are many types available, from the simplest colored filter used to enhance the details of planetary features, to specialist filters for taking images of everything from planets to galaxies.

A yellow filter can bring out detail when viewing Mars' polar ice caps; a blue one increases contrast in Jupiter's belts and its Great Red Spot; others subdue the orange glow of light pollution caused by artificial lights. There are also highly specialized narrow-band filters that allow only very specific wavelengths associated with the light emission of different glowing gases to pass through.

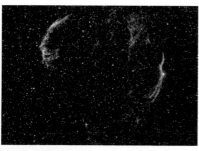

Deep sky filter
This view of the Veil Nebula, a supernova remnant, was achieved using narrow-band filters for deep-sky observing. The filters create greater contrast between the blackness of the background space and the nebula.

Fitting a filter
Filters typically screw into the rear of the eyepiece, which is then attached to the telescope. They enhance certain details, but reduce the amount of light that reaches the eye.

ASTROPHOTOGRAPHY

Taking images of night-sky objects can be a hugely rewarding experience. The equipment needed ranges from a basic point-and-shoot digital camera to sophisticated systems mounted on motorized telescopes and using specialist computer software.

CAMERA AND TRIPOD

A camera can be used to take pictures of the night sky, but without a telescope it is limited to recording little more than naked-eye views of stars, the Moon, bright planets, and aurorae. A tripod holds the camera steady, allowing you to make exposures of several seconds at a time. Experiment by photographing the same view at a range of exposures, such as 5, 10, 20, and 40 seconds. A longer exposure will produce an image in which the stars are trailed, appearing as arcs across the sky as the Earth rotates.

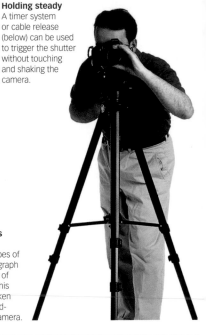

Holding steady
A timer system or cable release (below) can be used to trigger the shutter without touching and shaking the camera.

Star trails
One of the simplest types of astrophotograph is an image of star trails. This one was taken with a tripod-mounted camera.

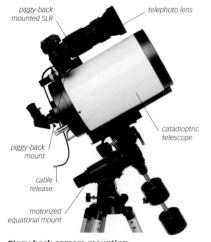

piggy-back mounted SLR

telephoto lens

catadioptric telescope

piggy-back mount

cable release

motorized equatorial mount

Piggy-back camera mounting
A piggy-backed camera can track the stars as the telescope moves to follow them. A telephoto lens can take wide-field images of constellations.

PIGGY-BACKING

To take long-exposure pictures without producing star trails requires that your camera can follow the movement of the stars. This can be achieved by attaching the camera to a telescope connected to a motorized equatorial mount (p.43). The simplest way to attach the camera is to fit it to the telescope in a piggy-back position. This set-up allows you to record the object that is aligned with the telescope, but the image itself is the one seen and captured by the camera. This method is useful for taking long-exposure shots of large expanses of sky and can be effective for capturing nebulae or galaxies that are spread across large areas. A special mount can be bought to attach the camera to the telescope.

PRIME-FOCUS PHOTOGRAPHY

An alternative way of photographing night-sky objects is the prime-focus technique. This allows the image produced by the telescope to be recorded, rather than the image produced by the camera on its own. To achieve this, the camera's lens is removed and the camera body is

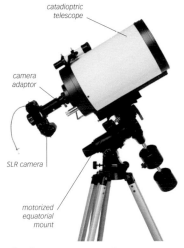

catadioptric telescope

camera adaptor

SLR camera

motorized equatorial mount

Prime-focus camera mounting
The prime-focus technique places the camera in the eyepiece position, with or without the eyepiece present. A motorized equatorial mount is needed to keep the target object in the telescope's field of view.

attached to the telescope eyepiece holder by means of an adaptor. This effectively turns the telescope into a giant telephoto lens, providing greater magnification of a narrower field of view.

The prime-focus technique allows galaxies, nebulae, and star clusters to be captured with long exposures (lasting several minutes). This requires that the telescope is attached to a motorized equatorial mount, which keeps the camera fixed on its target in the sky.

Prime-focus image of the Dumbbell Nebula
This view of the Dumbbell Nebula was made using the prime-focus technique. The image was recorded on a computer, and a software package was then used to sharpen it and enhance the color (p.57).

telescope eyepiece

point-and-shoot digital camera

adaptor

adjustment screw

Aligning camera and telescope
A simple adaptor allows the lens of a point-and-shoot digital camera to be aligned accurately and securely with the eyepiece of the telescope. Experiment with the distance between the camera lens and the telescope to achieve optimum results.

DIGISCOPING

Another way of using a camera and telescope together is a technique known as afocal imaging, or digiscoping. This involves aligning the lens of a point-and-shoot digital camera with the telescope's eyepiece and simply taking a picture. At its most basic, this can be done by mounting the camera on a tripod and pointing it down the telescope eyepiece. The alternative is to use a dedicated adaptor to fix the camera to the eyepiece. Most point-and-shoot cameras take reasonable pictures of the Moon.

If choosing a camera specifically for digiscoping, look for one that has a large LCD screen and allows you to vary the exposure time manually. Use the camera's self-timer to avoid shaking the camera when taking a picture.

»

»DIGITAL SLR PHOTOGRAPHY

Digital SLR (single lens reflex) cameras are the most common type of astro-imaging camera. They are more sensitive to light than point-and-shoot cameras, their shutters can be left open for long periods, and they have interchangeable lenses. An exposure of about 30 seconds by a digital SLR will show the constellations and Milky Way star fields. Exposures of several minutes with the prime-focus technique yield close-ups of galaxies, nebulae, and star clusters but require a mount that tracks the sky with great accuracy. Many SLRs have a remote-control feature; this allows you to take pictures without shaking the camera.

wide-angle lens

digital SLR camera

camera mount

Digital SLR imaging
The digital SLR camera can be used attached to a tripod (left) or mounted on a telescope in either prime-focus or piggy-back style.

CCD IMAGING

A CCD (charge-coupled device) camera is a dedicated astronomy camera, unlike point-and-shoot digital or digital SLR cameras, which are designed primarily for everyday use. Although CCD technology is used in digital cameras, the dedicated CCD camera gives the best results, taking vivid images of faint objects in seconds rather than minutes. A typical CCD camera is a simple device that is attached to the telescope eyepiece and linked to a computer by cable. The camera converts the light of astronomical objects into digital code and shows the objects on screen. It has a cooling system to reduce heat-generated interference.

7 in (175 mm) catadioptric telescope

CCD camera attached to eyepiece

cable connects camera to computer

equatorial fork mount with drive

CCD camera and telescope
The CCD camera is attached to the telescope eyepiece and linked to the computer by a cable. It works by converting the light that falls on its light-sensitive silicon chip into digital data to produce an image.

Combining images
Dedicated software can be used to combine separate images shot through different filters to produce a detailed full-color picture.

WEBCAM ASTRONOMY

Earth's atmosphere can cause problems for the astrophotographer, affecting the seeing (p.33) and blurring the view. An ingenious way around this is to use a webcam to make images. Once attached to your telescope, this small video camera can send images in real time to your computer. A short video recorded by the webcam can be broken down into individual frames by computer software. The best of these images are filtered out by the software and then stacked together to create a final image. In this way, the image produced is much sharper and more detailed than any in the original video.

Setting up a webcam
Dedicated astronomy webcams slot into the telescope in place of the eyepiece. A cable connects the webcam to the computer. Once the telescope is pointing at the target object, the webcam is operated from the computer using the webcam's software.

USING A WEBCAM TO IMAGE THE MOON AND THE PLANETS

1 Set up the telescope. If it has a motor drive, make sure it is tracking the Moon or planet. Set up the computer and the webcam's image-capture software.

2 Center the target in the eyepiece, then swap the eyepiece for your webcam. Check it is connected and communicating with the software.

3 Using the video feed from the software, focus the image so that the view of the planet or the lunar landscape is in sharp focus.

4 Set the camera exposure, using the control software if necessary. Start by setting the recording time to 60 seconds; adjust as needed.

5 If imaging a planet, do not record for much longer than 60 seconds; the rotation of the planet on its axis may blur fine detail.

6 Set the camera running. It will now capture a short video in the form of an AVI (Audio Video Interleave) format video file.

7 Once recorded, load the video into your stacking software. Follow the software instructions to produce a stacked final image.

WEBCAM IMAGE OF MARS

IMAGE PROCESSING

Images recorded directly onto a computer or downloaded onto it can all be improved using a software program such as Adobe Photoshop. This allows you to crop, rotate, and clean up your images, and to adjust properties such as brightness and contrast to bring out detail. Another control to experiment with is the color balance. Changing individual colors is a trick often used by professional astronomers to make a detail stand out. Make sure you save the image once you have finished enhancing it.

Digital enhancement
This screenshot shows an image of Saturn in Photoshop, software that can be used to enhance or manipulate an image's features, such as its contrast, sharpness, and color.

RECORDKEEPING

Astronomers have traditionally kept both written and visual records of what they see. Despite the advent of digital technology, this method is still popular today. It fosters a keen eye for detail and makes for a highly personalized observational record.

KEEPING A LOG

It is good to get into the habit of recording what you see. Make your notes in an exercise book or large-format diary, or buy a specially made logbook. The commercial logs are formatted to make recordkeeping easy; they have room set aside for sketches and provide helpful information. Whatever your choice, your log should be portable and easy to use. Fill it out as you observe or immediately afterward, always recording the key facts. Keeping a log makes you look and think more carefully about what you see. It will become a keepsake and a reference book for the future. As you look through it, you will see the progress you have made as an observer.

RECORDING OBSERVATIONS

Routinely record key facts in your observation log, making sure to list them in the same order each session. By doing this, details will not be forgotten and the logbook will become an easy-to-use information source. For each observation include:

- The observer's name and location
- The date and time of observation
- The name of the object (and constellation)
- The coordinates of the object
- The observing conditions (the seeing, transparency, cloud cover, temperature)
- Details of any equipment used (for example, telescope, eyepiece, and filter)
- A written description of the target object
- The orientation of any drawings or photographs made

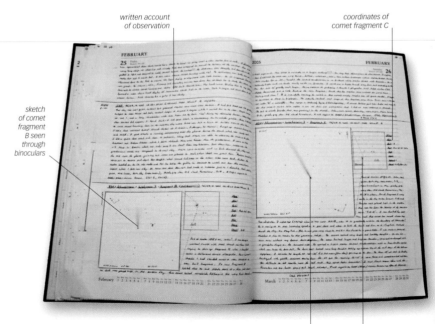

written account of observation

coordinates of comet fragment C

sketch of comet fragment B seen through binoculars

Astronomer's diary, with sketches
This logbook of an astronomer based in Northern Ireland contains records of comet-hunting sessions. The sketches feature fragments of Comet Schwassmann–Wachmann 3, which had broken into pieces.

sketch of comet fragment C seen through binoculars

sketch of naked-eye view of comet fragment C

VISUAL RECORD

Whatever you see, it is worth making a sketch of it. Drawing not only provides a permanent record of your observations, it is also an excellent way of training your eyes to pick up detail. Sketching the Moon's surface, for example, will train your eyes to identify mountains and craters and the shadows they cast. Try recording the prominent markings along

Sketching the Moon
A notebook and soft pencil is used to sketch lunar craters and maria seen through a telescope. If observing with the naked eye or binoculars, record the light and dark regions on the Moon's face and the nightly lunar phase.

the terminator—the boundary between the sunlit and dark parts of the Moon. Sketch several prominent features before filling in detailed lines, shading, and other features. For stellar objects such as clusters, nebulae, and galaxies, the trick is to start with the stars—the brighter the star, the bigger the dot on your paper. By drawing the brightest stars first, you establish a framework for filling in the remainder of your sketch.

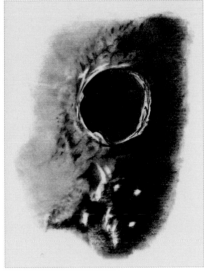

Copernicus Crater, the Moon
The key to sketching surface features of the Moon is to start by lightly sketching the main shapes. The side of a charcoal pencil is particularly good for adding shadows.

RECORD SHARING

It is always interesting to look at someone else's log, not only to see what they have observed, but how they have recorded information. Many astronomers upload their observations onto the internet, making them easy to share with other observers near and far. Such online astronomy forums give you the chance to learn from the work of others. You will see how some sketchers use a pen to add different densities of dots instead of pencil shading. Others sketch in the negative, using dark pencil to represent stars on white paper, and then use computer graphics to turn their images into bright stars on a dark background.

Jupiter
This colored-pencil sketch of Jupiter is based on an image taken by the Voyager 1 spacecraft. Working from an image is a good way to practice your technique.

The Sun
To produce this solar-telescope view of the Sun, the observer made a sketch using white and black pastel pencils. Color was added using Photoshop software.

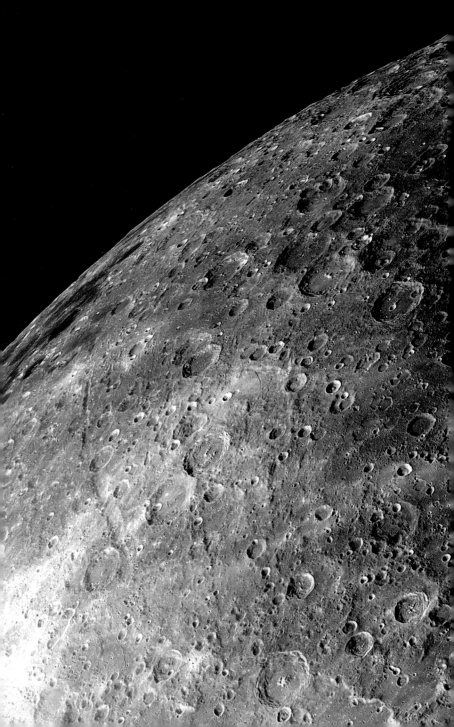

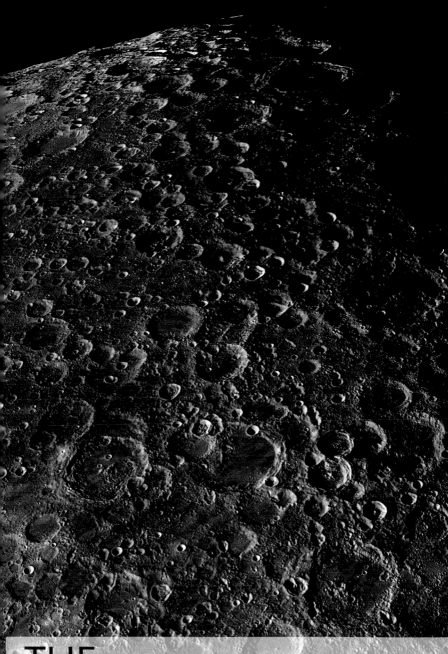

THE
SOLAR SYSTEM

The Moon's south pole, seen in a mosaic of 130 images taken with a 14 in (355 mm) telescope

THE SUN AND PLANETS

Situated in the Milky Way, the Solar System consists of the Sun and the bodies within its gravitational influence. Chief among these are the eight planets, but the Solar System also contains many other objects, including moons, asteroids, and comets.

THE STRUCTURE OF THE SOLAR SYSTEM

At the heart of the Solar System is the star we call the Sun. Its gravity holds the planets in orbit around it, while its heat and light sustain life on Earth. Nearest the Sun are the four inner rocky planets: Mercury, Venus, Earth, and Mars. Lying outside Mars' orbit is the Main Belt of asteroids, consisting of a ring of numerous lumps of rock and metal; it also contains the dwarf planet Ceres. Beyond are the giant outer planets: Jupiter, Saturn, Uranus, and Neptune. The Sun's influence extends beyond Neptune and includes the Kuiper Belt (p.107) and Oort Cloud (p.109), zones of icy bodies from which comets originate. The Kuiper Belt also contains Pluto and other dwarf planets (pp.106–7).

The orbits of the planets
The planets orbit the Sun in the same direction and in roughly the same plane (called the ecliptic). The planets and their orbits are not shown to scale in this diagram.

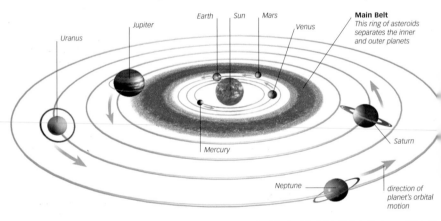

Uranus

Jupiter

Earth   Sun   Mars

Venus

Main Belt
This ring of asteroids separates the inner and outer planets

Mercury

Saturn

Neptune

direction of planet's orbital motion

THE FORMATION OF THE SOLAR SYSTEM

The Solar System is thought to have originated about 4.6 billion years ago as a rotating cloud of gas and dust. Gravity caused the cloud to contract, and its central region became denser and hotter, eventually evolving into the Sun. In the spinning disk (known as a protoplanetary disk) the gas and dust gradually coalesced to form small bodies called planetesimals. These then collided together many times, forming the planets we see today. The debris left over from planet formation can be seen throughout the Solar System as asteroids and comets.

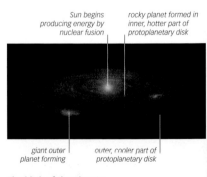

Sun begins producing energy by nuclear fusion

rocky planet formed in inner, hotter part of protoplanetary disk

giant outer planet forming

outer, cooler part of protoplanetary disk

The birth of the planets
Rocky planets formed from accreting planetesimals in the inner, hotter part of the protoplanetary disk. In the outer, cooler region, gas accumulated around rocky cores to form the giant outer planets.

THE INNER ROCKY PLANETS

Closest to the Sun are four relatively small rocky planets: Mercury, Venus, Earth, and Mars. All these planets are thought to possess metallic cores surrounded by rocky mantles. All four also have atmospheres, although Mercury's and Mars' are considerably more tenuous than Earth's and Venus's. Only Earth has liquid water, although there is evidence that Mars may once have had a liquid-water ocean. The inner planets are all visible with the naked eye.

THE OUTER GIANT PLANETS

Beyond the main asteroid belt are the giant planets: Jupiter, Saturn, Uranus, and Neptune. Each consists of an enormous ball of gas (composed mainly of hydrogen and helium) with, it is thought, an inner rocky core. Because the giant planets are so far from the Sun, they are extremely cold. Jupiter and Saturn are clearly visible with the naked eye; Uranus can also be seen with the naked eye but is very faint. Binoculars or a telescope are needed to observe Neptune.

Mercury
The closest planet to the Sun and the smallest in the Solar System, Mercury is heavily cratered and has huge extremes of temperature from day to night.

Venus
Similar in size to Earth, Venus is shrouded in cloud. Its surface is shaped by volcanism and is hotter than any other planet, at about 867°F (464°C).

Jupiter
The largest and most massive planet in the Solar System, Jupiter has an atmosphere with a distinctive banded appearance and swirling weather systems.

Saturn
The second largest but least dense of the planets, Saturn has a pale yellow, banded upper atmosphere and is surrounded by an extensive ring system.

Earth
The largest of the rocky planets, Earth is unique in the Solar System in having abundant liquid water, an oxygen-rich atmosphere, and intelligent life.

Mars
The outermost rocky planet, the Red Planet (as Mars is sometimes called) has a surface littered with tectonic features, rocky plains, and dry riverbeds.

Uranus
A pale blue planet about four times the size of Earth, Uranus is tipped on its side so that its faint ring system and moons encircle it from top to bottom.

Neptune
The most distant planet from the Sun and the smallest of the giant planets, Neptune has a dynamic atmosphere, with winds of over 1,200 mph (2,000 kph).

ASTEROIDS AND COMETS

In addition to the eight planets and their moons, the Solar System also contains thousands of other objects. These include asteroids, most of which orbit the Sun in the Main Belt between Mars and Jupiter. There are also other asteroid groups, such as the Trojans, which accompany Jupiter in its orbit, and near-Earth asteroids, whose orbits bring them close to our planet. The inner Solar System is also visited by comets (p.108), which originate from far beyond Neptune.

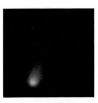

Comet Hyakutake
Comets are chunks of ice and dust that follow looping orbits around the Sun. In 1996, Hyakutake passed near Earth and was easily visible with the naked eye.

Asteroid Eros
Eros is one of the largest near-Earth asteroids, at about 21 miles (34km) long. In 2001, NASA's Near Earth Asteroid Rendezvous probe landed on its surface.

Jupiter in the night sky
The brightest object in this image of the night sky is Jupiter. The orange-colored spot below it is Antares, one of the 20 brightest stars. To the left is the central bulge of the Milky Way.

OBSERVING THE PLANETS

When and where a planet is visible in the sky depends on its position in space relative to Earth and the Sun. Planets can be observed only when they are neither hidden behind the Sun nor between Earth and the Sun, when they are obscured by glare.

OBSERVING WITH THE NAKED EYE

When not obscured by the Sun's glare, all the planets except Neptune are visible to the naked eye. They fall into two groups: inferior and superior planets. Inferior planets have orbits inside Earth's. Mercury and Venus are the two inferior planets. They are always near or fairly near the Sun in the sky and are visible either in the western sky around or after sunset or in the east before or around dawn. Venus is brighter than anything in the night sky except the Moon. Mercury is visible only infrequently, but when present it looks like a star in the evening or morning twilight. Superior planets have orbits beyond Earth's. They are Mars, Jupiter, Saturn, Uranus, and Neptune. When observable, they can be seen for anything from a few hours up to the entire night. Of the four superior planets visible to the naked eye, Mars has a red hue and is of very variable brightness; Jupiter is brighter than any star; Saturn looks like a moderate to bright star; and Uranus is so faint it is only just visible.

OBSERVING WITH BINOCULARS AND TELESCOPES

Through binoculars, Mars and Saturn appear as bright points of light, Jupiter can be seen as a small disk, and the phases of Venus are visible. A small telescope reveals some detail, including the polar caps of Mars, Jupiter's four largest moons, and Saturn's rings. A large telescope reveals more detail and also shows Mercury, Uranus, and Neptune as disks rather than points of light.

The Moon through binoculars
Through binoculars, individual features can be seen on the Moon, such as its larger craters and maria (dark "seas" of basaltic lava).

Jupiter with a small telescope
Through a small telescope, Jupiter appears clearly as a disk, and some or all of its four largest Moons are usually just visible.

Saturn with a large telescope
Through a large amateur telescope it is possible to see a reasonable amount of detail in the spectacular ring system around Saturn.

THE SUPERIOR PLANETS

The superior planets appear brightest at opposition, when they are on the opposite side of Earth from the Sun and closest to Earth. Over the long term, the superior planets normally move west-to-east across the celestial sphere. However, as a superior planet nears opposition, its eastward movement halts and for a time it moves backward in the sky (called retrograde motion) before restarting its normal motion.

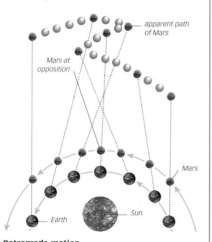

Retrograde motion
As Mars nears opposition, Earth overtakes it on the inside, and for a couple of months Mars appears to go into reverse across the night sky.

THE INFERIOR PLANETS

When an inferior planet is on the far side of the Sun from Earth—a point called superior conjunction—it is hidden by the Sun. When between Earth and the Sun (inferior conjunction), the planet is obscured by the Sun's glare. As a result, the inferior planets are best observed around their times of elongation: the two points on their orbits at which they appear farthest from the Sun as viewed from Earth. But even at elongation, Mercury can be seen only at dusk and dawn, while Venus is visible for up to three hours after dusk or before dawn. When visible, the inferior planets show phases, like the Moon.

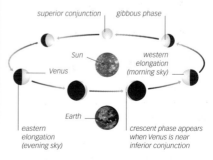

Phases of inferior planets
Inferior planets (here, Venus) appear as phases, because they are always on the far side of the Sun when their full disks would be visible.

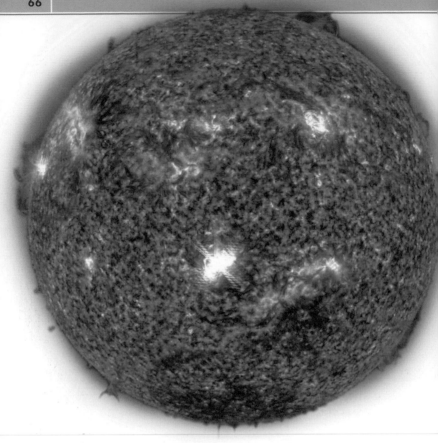

PROFILE

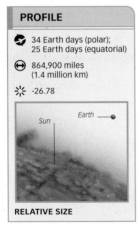

- 34 Earth days (polar); 25 Earth days (equatorial)
- 864,900 miles (1.4 million km)
- -26.78

Sun

Earth

RELATIVE SIZE

THE SUN

Our local star, the Sun, is an incandescent ball of exceedingly hot plasma (ionized gas) with 750 times the mass of all the Solar System's planets combined. At its core, which has a temperature of 28.3 million °F (15.7 million °C), nuclear fusion reactions produce helium from hydrogen and generate huge amounts of energy, which escapes from the Sun as heat, light, and other types of radiation, including X-rays, ultraviolet radiation, and radio waves. In addition, the Sun also emits a continual stream of charged particles, called the solar wind. The Sun's rate of energy production and surface temperature have stayed relatively constant since it formed about 4.6 billion years ago, and will remain so for about another 5 billion years. It will then swell into a red giant star and destroy the inner planets.

VISIBLE SURFACE AND INTERIOR

The Sun's visible surface layer, which has a temperature of about 9,570 °F (5,300 °C), is the photosphere. Beneath it is the convective zone, where hot gas bubbles rise to the surface, cool, then sink again, making the photosphere appear to boil. The next layer down is the radiative zone, where energy is radiated from the core to the convective zone. The core makes up 2 percent of the Sun's volume but about 60 percent of its mass.

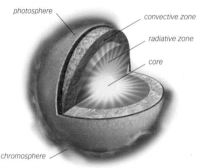

The Sun's layers
The Sun's interior consists of a core, radiative zone, and convective zone. Photons (tiny packets of energy) released in the core take about 100,000 years to reach the photosphere.

THE SOLAR WIND

The Sun's corona is so hot that some of its particles have enough energy to escape from the Sun's gravity and stream away into space. This stream of particles, called the solar wind, consists mainly of electrons and protons and surges through space at over 1 million mph (about 1.6 million kph). When the solar wind reaches Earth, some of the particles collide with atoms in the upper atmosphere, causing aurorae (p.118).

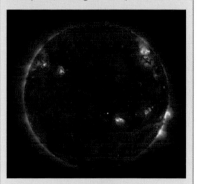

The origin of the solar wind
In this ultraviolet image of the Sun, the large, dark area near the center is a hole in the corona (the outer layer of the Sun's atmosphere). A substantial part of the solar wind streams out of such coronal holes.

ATMOSPHERE

The Sun's atmosphere has three layers. Just above the brilliant photosphere lies the chromosphere, which is about 1,200 miles (2,000 km) deep. Going up through the chromosphere, the temperature gradually rises, but the density drops by a factor of 5 million.

Next is a thin, irregular layer called the transition region, within which the temperature rises from about 36,000 °F (20,000 °C) to 1.8 million °F (1 million °C). The outer layer is the corona, which is a million million times less dense than the photosphere and extends for millions of miles into space.

The chromosphere
The Sun's chromosphere is visible here as the hazy, orange-red area above the sharper-edged photosphere. The temperature in its upper part is about 36,000 °F (20,000 °C).

The corona
During a total solar eclipse, as here, the corona appears as a halo around the Sun. Believed to be heated by a magnetic process, its temperature ranges from 1.8 to 3.6 million °F (1–2 million °C).

»

» SOLAR PHENOMENA

Disturbances in the Sun's magnetic field—due to different parts of the Sun spinning at different speeds—result in various phenomena at the surface, such as sunspots and solar flares. The surface activity varies from a minimum to a maximum over an 11-year cycle.

SUNSPOTS

Sunspots are dark, relatively cool areas on the Sun's surface. They are typically 930–31,000 miles (1,500–50,000 km) across and appear singly or in groups. The darker, cooler central part of a spot is called the umbra; the paler, hotter outer part is the penumbra. Over each 11-year solar cycle the number of spots varies from none during a solar minimum to as many as 100 during a solar maximum.

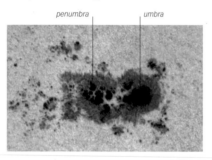

penumbra   umbra

SOLAR FLARES

Violent explosions at the Sun's surface, accompanied by brilliant outbreaks of light and other types of radiation, are known as solar flares. They result in vast additional quantities of charged particles and radiation being blasted off into space, creating disturbances in the solar wind (p.67).

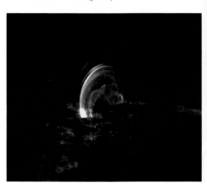

PROMINENCES

A prominence is a huge bright loop or cloud of plasma, up to several hundred thousand miles long, which extends from the photosphere into the corona (upper layer of the Sun's atmosphere). What causes prominences to form is not fully understood. Typically, a prominence lasts about a day, but some may persist for weeks or months. Some of them break apart and give rise to coronal mass ejections.

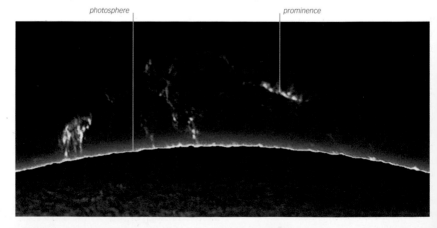

photosphere   prominence

FILAMENTS, FACULAE, AND SPICULES

When a prominence is viewed with the Sun as a backdrop rather than space, it appears dark compared to the photosphere and is called a solar filament. Other features observable at the Sun's surface include faculae (intensely bright, active regions often associated with sunspots) and spicules, which are short-lived jets of gas, some 2,000–6,000 miles (3,000–10,000 km) long, that shoot upward from the photosphere. Granulation refers to the mottled appearance of the photosphere.

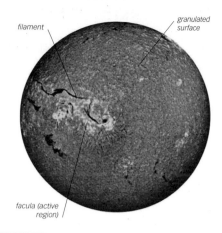

filament

granulated surface

facula (active region)

SOLAR QUAKES

Some solar flares are closely followed by solar quakes—a series of waves that move across the Sun's photosphere. Scientists recorded a solar quake for the first time in July 1996, following a flare that is visible in the sequence below as the white blob with a "tail" to its left. The quake itself is visible as the spreading circular waves in the second and third images. These waves look like ripples on a pond but they were 2 miles (3 km) high and reached a maximum speed of 250,000 mph (400,000 kph). Over the course of an hour, they traveled about 80,000 miles (130,000 km)—a distance equal to 10 Earth diameters—before fading into the background.

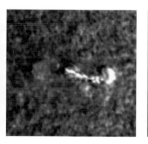
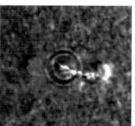
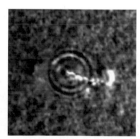

CORONAL MASS EJECTIONS

Coronal mass ejections (CMEs) are huge clumps of plasma that have broken away from the Sun and are heading out into space at speeds as high as 6.8 million mph (11 million kph). During periods of solar maximum, as many as three CMEs can occur per day. They often originate from active regions on the Sun's surface, such as groupings of sunspots associated with flares. On reaching Earth, they can cause particularly strong aurorae (p.118) and geomagnetic storms, with disruption to radio communications and damage to electrical transmission lines.

»

» OBSERVING THE SUN

The centerpiece of the Solar System is a fascinating object to study, but strict safety precautions are essential. Never look directly at the Sun with the naked eye or through any instruments without the use of special filters; the concentration of light can cause blindness.

OBSERVING THE SUN

There are a number of options for safely observing the Sun. Once the province of professional astronomers, solar telescopes—specialized pieces of equipment incorporating filters that enable direct viewing of the Sun—are now available for amateurs. These instruments enable solar features, such as sunspots and solar prominences, to be observed easily. Alternatively, such features can also be observed directly using an ordinary telescope or binoculars fitted with special add-on solar filters. Without special equipment, the safest way to view the Sun with an ordinary small telecope is by projecting the Sun's image through the eyepiece onto a piece of white cardboard. The farther the cardboard is from the eyepiece, the larger but fainter the Sun's image will be.

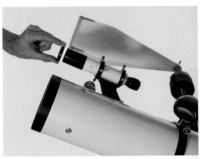

1 Cap the finderscope
Never look for the Sun in the telescope by eye. For safety, place a cap over the finderscope. Also keep a cap over the front of the main telescope tube for the majority of the time.

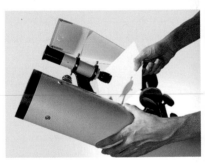

2 Locate the Sun
To aim the telescope at the Sun, move it around until you find a position in which the main telescope tube or finderscope casts the shortest shadow on a piece of cardboard.

solar filter on objective lens

solar filter on finderscope

Refractor with add-on solar filter
Add-on solar filters should be fitted over the front of both the main tube and finderscope of an ordinary amateur telescope. The filters must be used with caution to avoid incidental light.

3 Focus the image
Remove the cap over the front of the main tube and focus the Sun's image through the eyepiece onto a piece of white cardboard. Adjust the alignment and focus to obtain a sharp image.

SOLAR OBSERVATORIES

Telescopes have been used to observe the Sun for about 400 years, but they are hindered by factors such as the weather and the atmosphere, which absorbs X-rays and most ultraviolet radiation. To help counter these, Earth-based solar observatories are usually sited at high altitude, where interference from the atmosphere is at a minimum, and the telescopes are also specialized in various ways. They are usually housed in white towers to minimize hot air currents that could otherwise degrade the image. Sometimes a moving mirror (heliostat) at the top of the tower is used to reflect the solar radiation into the main parts of the telescope, which are often in a cooled underground room. The main optical elements and light paths may also be housed in a vacuum to minimize the effects of air turbulence.

Teide Solar Observatory
At this observatory on Tenerife, the air has been removed from the Vacuum Tower Telescope (the tallest tower) to limit image distortion.

SOLAR SPACE MISSIONS

Because of the problems of studying the Sun from Earth's surface, much solar investigation is now carried out by spacecraft. For example, the Solar and Heliospheric Observatory (SOHO) was launched into orbit around the Sun in 1995 to study the Sun's structure and the solar wind. More recently, the Solar Dynamics Observatory (SDO) was launched into Earth orbit to study the Sun's corona (outer atmospheric layer), magnetic field, seismic activity, and variations in its emission of extreme ultraviolet light.

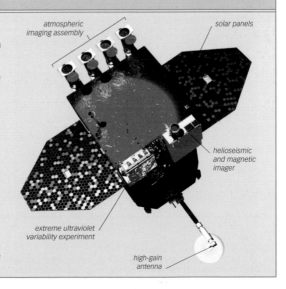

atmospheric imaging assembly

solar panels

helioseismic and magnetic imager

extreme ultraviolet variability experiment

high-gain antenna

Solar Dynamics Observatory
Launched in early 2010, the SDO weighs about 6,615 lb (3,000 kg). Although studying the Sun, it will be in Earth orbit for its entire mission.

- ⦿ 27.32 Earth days
- ↻ 27.32 Earth days
- ⊕ 2,159 miles (3,475 km)
- ☼ -12.7

Earth

Moon

RELATIVE SIZE

THE MOON

Earth's natural satellite, the Moon, is thought to have formed about 4.5 billion years ago as a result of a collision between the young Earth—which was smaller than today's Earth—and an object about the size of Mars. The impact spewed out a huge amount of rocky debris, which eventually coalesced to form the Moon. Even though it has only 1.2 percent of the mass of Earth, the Moon is the fifth-largest planetary satellite in the Solar System. Its gravity influences Earth's oceans, and when full, the Moon is the brightest object in our sky after the Sun. However, it is too small to retain a substantial atmosphere, and geologic activity has long since ceased on it. Twelve men have walked on its surface and have brought back to Earth over 838 lb (380 kg) of lunar rock.

ORBIT

The Moon has an elliptical orbit around Earth, so the distance between the two varies. At its closest to Earth (called perigee), the Moon is 10 percent closer than at its farthest point (apogee). The Moon takes 27.32 Earth days to spin on its axis, which is the same time it takes to orbit Earth. This phenomenon (known as synchronous rotation) keeps one side of the Moon—the near side—permanently facing Earth, although a small oscillation in the Moon's orbit (called libration) allows about 18 percent of the far side to be seen from Earth.

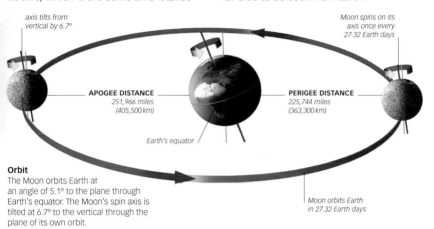

axis tilts from vertical by 6.7°

Moon spins on its axis once every 27.32 Earth days

APOGEE DISTANCE
251,966 miles
(405,500 km)

PERIGEE DISTANCE
225,744 miles
(363,300 km)

Earth's equator

Moon orbits Earth in 27.32 Earth days

Orbit
The Moon orbits Earth at an angle of 5.1° to the plane through Earth's equator. The Moon's spin axis is tilted at 6.7° to the vertical through the plane of its own orbit.

STRUCTURE AND ATMOSPHERE

The lunar crust is about 30 miles (48 km) thick on the near side and about 46 miles (74 km) thick on the far side. The upper part of the rocky mantle is solid but its deeper parts are semimolten. The core consists of an outer part of molten iron around a solid, iron-rich inner part. The Moon's atmosphere is extremely thin, with a total mass of only about 24.6 tons (25,000 kg), and partly because of this, the surface temperature varies widely, from as low as -274°F (-170°C) at night to as high as 257°F (125°C) by day. The Moon's gravity is only one-sixth that of Earth, and the lunar atmosphere is escaping all the time, but it is also replenished by the solar wind and outgassing from lunar rocks.

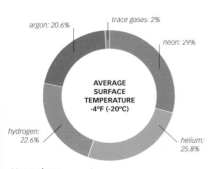

argon: 20.6%

trace gases: 2%

neon: 29%

AVERAGE SURFACE TEMPERATURE
-4°F (-20°C)

hydrogen: 22.6%

helium: 25.8%

crust of granitelike rock

solid, iron-rich inner core surrounded by molten iron outer core

rocky mantle

Atmosphere
The neon and hydrogen have mainly been deposited by the solar wind. The argon and helium are mainly derived from radioactive decay processes in lunar rocks.

Structure
Like Earth, the Moon has a crust, mantle, and core, but the core is relatively small. The Moon's average density is about 60 percent that of Earth—similar to the density of Earth's mantle.

»

⟩⟩ PHASES OF THE MOON

The Moon has no light of its own but is lit up by the Sun. As it orbits Earth, the sunlit portion of its near side goes through phases. A complete phase cycle takes 29.5 days. During the first half of each cycle, the sunlit part grows (waxes) until the whole near side is lit up at full Moon, when the Moon is on the opposite side of Earth from the Sun. During the second half of the cycle, the sunlit portion shrinks (wanes) until the near side is completely dark at new Moon, when the Moon lies between Earth and the Sun.

WAXING CRESCENT · FIRST QUARTER · WAXING GIBBOUS · FULL MOON

WANING GIBBOUS · LAST QUARTER · WANING CRESCENT · NEW MOON

OBSERVING THE MOON

The Moon is a wonderful object to observe. Binoculars alone clearly reveal its broad, dark plains, mountain ranges, and the larger craters, and with a telescope considerably more detail can be seen. However, viewing it at full Moon can be disappointing, because then the Moon's surface is evenly lit by the Sun and its features are relatively indistinct. A better time to observe it is during the weeks on either side of the new Moon, when sunlight strikes the Moon's surface obliquely, producing long shadows and throwing the landscape features into sharp relief. An interesting feature to look at is the terminator—the line that divides the Moon's sunlit and dark sides. Along this line, sunlight strikes the Moon's surface at the narrowest angle, revealing the greatest amount of detail.

Naked eye
Without optical aid, some lighter and darker areas are plainly visible on the lunar surface, and some prominent craters may just be discernible.

Binoculars
The Moon still appears as a whole disk but the image is bigger. Large ray craters, such as Tycho (at bottom), can now easily be seen.

Small telescope
Only part of the Moon's disk is now visible, but extra detail can be seen in its light (highland) regions and dark maria (lava-filled plains).

Large telescope
Many mountain ranges and craters are now visible. The large crater here is Copernicus. Above it is the mountain range called Montes Carpatus.

MAPPING AND EXPLORING THE MOON

The earliest maps did little more than record dark and light areas on the Moon. From the 1600s, the telescope enabled detail to be added, but the major advance in mapping came with spaceflight. In 1959, the Soviet probe Luna 3 returned the first images of the Moon's far side, and In 1966–1967 NASA's five Lunar Orbiters imaged 99 percent of the lunar surface, paving the way for NASA's Apollo manned missions from 1969 to 1972. Since the 1990s, probes from several countries have continued investigations of the Moon's surface.

Luna 9
This Soviet probe made a soft landing on the Moon in 1966, becoming the first craft to make a soft landing on any celestial object. It sent images back to Earth for three days.

LUNAR RECONNAISSANCE ORBITER

The Lunar Reconnaissance Orbiter (LRO) is a NASA robotic spacecraft that began orbiting the Moon in June 2009. The orbiter's objectives are to make a 3D map of the lunar surface and to investigate the lunar environment, including radiation levels, potential resources, and possible water-ice deposits in the polar regions. The LRO mission is considered a pathfinder for possible future manned NASA Moon landings.

The LRO spacecraft
Orbiting about 31 miles (50 km) above the Moon, the LRO has seven instruments on board to map the lunar surface and investigate the lunar environment.

The first manned Moon landing
On July 21, 1969, Neil Armstrong exited Apollo 11's Lunar Module *Eagle* and became the first person to walk on the Moon. This image from that mission shows him in front of the Lunar Module.

>> SURFACE FEATURES

The Moon's surface bears the scars of a long history of bombardment by asteroids and comets as well as volcanic activity. The ancient Greeks thought the dark areas were water, and these areas were later given names such as mare (sea) and oceanus (ocean). It is now known that they are lowland basins blasted out by impacts during the Moon's early existence and later filled with flows of dark lava. Other than the seas and brighter highland areas, the most obvious features are ray craters, formed when impacting objects sent out splashes of molten rock as they hit the surface.

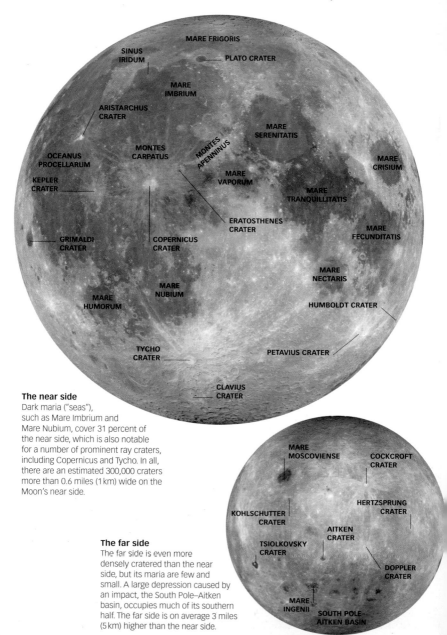

MARE FRIGORIS
SINUS IRIDUM
PLATO CRATER
MARE IMBRIUM
ARISTARCHUS CRATER
MARE SERENITATIS
MONTES CARPATUS
MONTES APENNINUS
OCEANUS PROCELLARUM
MARE CRISIUM
KEPLER CRATER
MARE VAPORUM
MARE TRANQUILLITATIS
ERATOSTHENES CRATER
GRIMALDI CRATER
COPERNICUS CRATER
MARE FECUNDITATIS
MARE NECTARIS
MARE NUBIUM
MARE HUMORUM
HUMBOLDT CRATER
TYCHO CRATER
PETAVIUS CRATER
CLAVIUS CRATER

The near side
Dark maria ("seas"),
such as Mare Imbrium and
Mare Nubium, cover 31 percent of
the near side, which is also notable
for a number of prominent ray craters,
including Copernicus and Tycho. In all,
there are an estimated 300,000 craters
more than 0.6 miles (1 km) wide on the
Moon's near side.

MARE MOSCOVIENSE
COCKCROFT CRATER
HERTZSPRUNG CRATER
KOHLSCHUTTER CRATER
AITKEN CRATER
TSIOLKOVSKY CRATER
DOPPLER CRATER
MARE INGENII
SOUTH POLE–AITKEN BASIN

The far side
The far side is even more
densely cratered than the near
side, but its maria are few and
small. A large depression caused by
an impact, the South Pole–Aitken
basin, occupies much of its southern
half. The far side is on average 3 miles
(5 km) higher than the near side.

LUNAR SEAS

The maria, or "seas," are large, dark plains on the lunar surface. Most formed between 3 and 3.5 billion years ago as volcanically erupted basaltic lava flooded into shallow basins formed by earlier impacts.

MARE CRISIUM

Mare Crisium (Latin for "Sea of Crises") is near-circular, with a diameter of about 370 miles (600 km), and occupies a basin formed by an impact that occurred some 3.9 billion years ago. It has an extremely smooth floor, which varies in height by less than 295 ft (90 m).

MARE IMBRIUM

One of the largest lunar maria, with a diameter of 698 miles (1,123 km), Mare Imbrium (Latin for "Sea of Showers") sits in the Imbrium impact basin, which formed about 3.85 billion years ago. It is surrounded by mountain ranges, such as Montes Carpatus and Montes Apenninus to its south and east, and Montes Jura to its northwest. These were thrown up by the impact that created the basin and rise as high as 3.5 miles (5.5 km).

MARE TRANQUILLITATIS

Mare Tranquillitatis (Latin for "Sea of Tranquillity") has an average diameter of about 542 miles (873 km). The impact basin in which it formed may have been created as long as 4 billion years ago, although it only flooded with lava about 3.6 billion years ago. This mare was the site of the first manned moon landing, NASA's 1969 Apollo 11 mission.

MARE SERENITATIS

Mare Serenitatis (Latin for "Sea of Serenity") is about 435 miles (700 km) in diameter and formed within an impact basin created 3.9 billion years ago. In 1972, Apollo 17 astronauts landed in the Taurus-Littrow valley, on the southeastern edge of Mare Serenitatis, where their discoveries included a boulder that had been ejected from an impact crater in Serenitatis.

»

» IMPACT CRATERS

Impact craters mark where objects such as asteroids hit the Moon. After an impact, part of the lunar surface and some or all of the impactor melt and splash outward then resolidify. Some molten material may flow backward then solidify to form a peak in the crater.

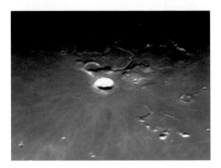

ARISTARCHUS CRATER

Aristarchus, the bright crater at the center of this image, is one of the brightest large formations on the lunar surface. It is a relatively young crater, at only about 450 million years old, and is about 25 miles (40 km) in diameter. Its wall has a series of nested terraces that were produced by concentric slices of rock in the wall slipping downward, both widening the crater and making it shallower.

CLAVIUS CRATER

Clavius is a huge, ancient crater, 140 miles (225 km) in diameter, situated toward the Moon's south pole, in the southern highlands. It was formed by an impact about 4 billion years ago. Clavius has a relatively low rim and is pockmarked with smaller craters. Five of these form a curving chain of craters that arc across Clavius's floor.

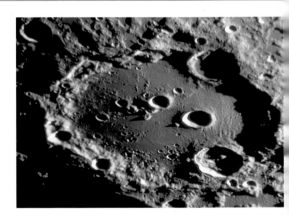

COPERNICUS CRATER

This prominent ray crater is about 66 miles (107 km) across and some 800 million years old. It has high central peaks, massive terraced walls, and a crater floor that lies 2.3 miles (3.7 km) below its rim.

The vicinity of Copernicus is peppered with secondary craters formed by boulders thrown out during the impact that formed it. Fine, light-gray rock particles ejected during the formation process make up the crater's rays.

ERATOSTHENES CRATER

This relatively deep crater lies on the southern edge of Mare Imbrium at the western end of Montes Apenninus and not far from the larger, more prominent ray crater Copernicus. Eratosthenes is 36 miles (58 km) in diameter, 2.2 miles (3.6 km) deep, and is thought to have formed as a result of an impact about 3.2 billion years ago. It has a well-defined circular rim, terracing on its inner wall, an irregular floor, and central mountain peaks. Eratosthenes itself lacks rays but it is crossed by rays from Copernicus.

HUMBOLDT CRATER

This large, ancient crater near the southeastern edge of the Moon's near side is about 120 miles (190 km) in diameter and 3.8 billion years old. It has a low, worn rim and central peaks that form a range on its lava-filled floor. The floor is also criss-crossed with radial and concentric clefts or channels—some of which look like tubes through which lava once flowed—and pockmarked with smaller craters.

TYCHO CRATER

This prominent young crater lies in the southern highlands. Some 63 miles (102 km) in diameter, 3 miles (4.8 km) deep, and only about 100 million years old, Tycho is one of the most sharply defined craters on the Moon. At its center is a mountain peak towering 1.9 miles (3 km) above a rough, infilled inner region. The crater is surrounded by distinctive rays that extend as far as 930 miles (1,500 km) from the crater rim.

»

» OTHER SURFACE FEATURES

In addition to maria ("seas") and impact craters, other surface features on the Moon include montes (mountains and mountain ranges), sinuous ridges called wrinkle-ridges, valleys and rilles (narrow grooves or depressed channels in the surface), and rupes (escarpments).

MONTES CARPATUS

This mountain range runs runs along the southwestern edge of Mare Imbrium (in the foreground of the image) to the north of Copernicus (the large crater near the horizon). The range extends for more than 180 miles (290 km) and was formed about 3.9 billion years ago as a result of the impact that created the Imbrium basin.

MONTES APENNINUS

The Apennine mountains form a sharp, curved southeastern edge to the Mare Imbrium impact basin. Pushed up by the shock wave from the Imbrium impact, the mountain chain stretches for about 375 miles (600 km). Mons Huygens, at 3.5 miles (5.5 km) high the Moon's tallest mountain, forms part of the chain.

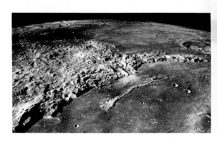

ARIADAEUS RILLE

This linear depression is over 185 miles (300 km) long and runs roughly in a westerly direction from the western edge of Mare Tranquillitatis. The rille is thought to have formed when a section of crust sank between two parallel fault lines. Relatively young, the rille has few craters or other features overlying it.

HADLEY RILLE

A winding valley near the northern end of Montes Apenninus, Hadley Rille averages about 1 mile (1.6 km) wide and 1,300 ft (400 m) deep. It is best known for the fact that the Apollo 15 mission landed close to it. In this image, an Apollo 15 astronaut and lunar rover are at the edge of the rille.

SCHROTER'S VALLEY

This meandering rille runs through elevated terrain, the Aristarchus plateau, to the northwest of Aristarchus crater. The largest sinuous rille on the Moon, Schroter's Valley is approximately 60 miles (100 km) long, 0.6 miles (1 km) deep, and gradually narrows from 6 miles (10 km) wide at one end to less than 0.6 miles (1 km) wide at the other. It is thought to have been created by molten lava, erupted from a volcanic vent, carving a channel in the lunar surface.

THE ALPINE VALLEY

Also called Vallis Alpes, the Alpine Valley is a spindle-shaped depression that cuts through a highland region known as the Montes Alpes on the northeastern edge of Mare Imbrium. The valley is 100 miles (166 km) in length and varies from 4 to 6 miles (7 to 10 km) in width. Running down the middle of the valley is a slender rille (channel) about 2,300 ft (700 m) wide and 400 ft (20 m) deep. The valley probably formed when a slab of lunar crust sank between fault lines.

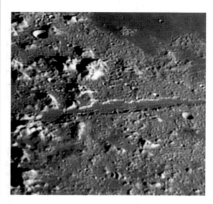

SINUS IRIDUM

Sinus Iridum (Latin for "Bay of Rainbows") is a semicircular extension to the Mare Imbrium. A lava-filled plain, Sinus Iridum is about 147 miles (236 km) wide and is surrounded from the northeast to the southwest by the Montes Jura mountain range. In this image, the large crater at the top is Bianchini, and the large crater at the bottom is Helicon.

THE STRAIGHT WALL

Also known as Rupes Recta, the Straight Wall (the long, dark line in this image) is an escarpment that runs through Mare Nubium. Caused by movement along a fault in the Moon's crust, the escarpment is about 78 miles (125 km) long, some 800–1,000 ft (240–300 m) high, and about 1–2 miles (2–3 km) wide.

Eclipse of the Sun
This multiple-exposure image shows the stages of a total solar eclipse, including the moment of totality (the central point), when the Moon completely covers the Sun's disk.

ECLIPSES

An eclipse is the temporary dimming or cutting off of light from a particular celestial object, either because it passes into the shadow of another astronomical body or because another object passes between it and the observer.

TYPES OF ECLIPSES

The term "eclipse" is most often used to describe either a lunar eclipse, when the Moon moves into Earth's shadow, or a solar eclipse, in which the Moon either partly or completely obscures the Sun's disk as seen from Earth. Taken together, eclipses of these types occur several times a year. However, the term "eclipse" can also refer to similar types of events beyond the Earth–Moon system: for example, a moon passing into the shadow cast by its host planet, or one of a pair of stars in a binary star system (p.127) temporarily obscuring the other star. Two phenomena related to eclipses are occultations and transits. An occultation is an event in which an apparently larger object completely blocks out the light from a more distant, apparently smaller one (for example, the Moon hiding a star). A transit is the opposite—it occurs when a smaller object passes across the face of a larger one and obscures just a small part of it, as when Venus or Mercury pass across the face of the Sun.

HOW LUNAR ECLIPSES OCCUR

Lunar eclipses occur when the Sun, Earth, and Moon are aligned, with Earth in the middle, so that the Moon passes through Earth's shadow. There are three types of lunar eclipses: penumbral eclipses, in which the Moon passes through Earth's penumbra (part shadow), leading to only a slight dimming; partial eclipses, in which part of the Moon passes through Earth's umbra (full shadow); and total eclipses, in which the whole Moon passes through the umbra. In a total eclipse the Moon often appears a dim red color, due to red light from the Sun reaching the Moon after refraction by Earth's atmosphere.

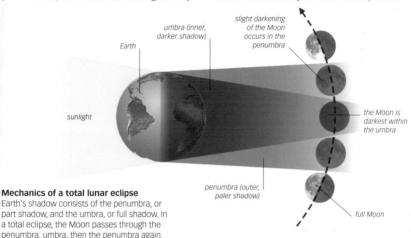

umbra (inner, darker shadow)

Earth

slight darkening of the Moon occurs in the penumbra

sunlight

the Moon is darkest within the umbra

penumbra (outer, paler shadow)

full Moon

Mechanics of a total lunar eclipse
Earth's shadow consists of the penumbra, or part shadow, and the umbra, or full shadow. In a total eclipse, the Moon passes through the penumbra, umbra, then the penumbra again.

1 Full Moon
Lunar eclipses occur at times of full Moon. The first sign of an eclipse is a slight darkening of the Moon as it passes into Earth's penumbra.

2 Moon entering umbral shadow
Once one edge of the Moon passes into Earth's umbra, a much darker region begins to move across the face of the Moon.

3 Moon approaching totality
The umbral shadow slowly creeps across the Moon's face. It may take about an hour before the shadow completely covers the Moon.

4 Moon at totality
At totality, no sunlight falls directly on the Moon, but it looks red because some sunlight is refracted toward it by Earth's atmosphere.

»

›› HOW SOLAR ECLIPSES OCCUR

An eclipse of the Sun occurs when the Moon blocks sunlight from reaching parts of Earth. The Sun's diameter is 400 times greater than the Moon's, but it is also 400 times farther away from Earth. As a result, the two objects look the same size in the sky. There are three main types of solar eclipse. In a partial eclipse, viewers across a broad swathe of Earth's surface see the Sun partly hidden for a period of time. During a total eclipse, viewers in a narrow strip within such a swathe (called the path of totality) see the Sun completely hidden for a short time; in the rest of the swathe, only a partial eclipse is visible. In an annular eclipse, the Moon blocks out all but a narrow ring of the Sun's disk; this occurs when the Moon is farther from Earth than average, so that its disk looks smaller than normal. Solar eclipses happen two or three times a year, but total eclipses occur only about once every 18 months.

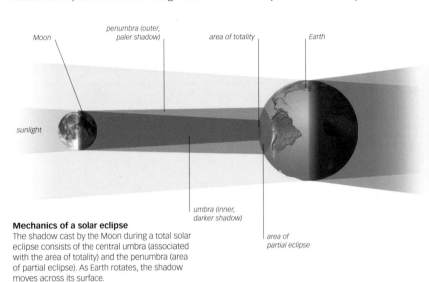

Moon · penumbra (outer, paler shadow) · area of totality · Earth · sunlight · umbra (inner, darker shadow) · area of partial eclipse

Mechanics of a solar eclipse
The shadow cast by the Moon during a total solar eclipse consists of the central umbra (associated with the area of totality) and the penumbra (area of partial eclipse). As Earth rotates, the shadow moves across its surface.

OBSERVING SOLAR ECLIPSES

The Sun should never be viewed directly with the naked eye or through any instrument (such as binoculars, a telescope, or a camera) because its intense light can cause eye damage. During an eclipse, it is safe to observe the Sun directly only during totality. If any part of the bright region is uncovered—even if this is just a thin crescent—it should not be looked at. Safe methods for viewing an eclipse before and after totality include using special eclipse-viewing glasses or by projecting an image of the Sun through a small pinhole in a piece of cardboard onto any flat, light-colored surface. Ordinary sunglasses should never be used to view an eclipse.

Eclipse-viewing glasses
Glasses with special solar filters can be used for viewing an eclipse, but they should first be checked to ensure the filters have no holes.

Leaves as pinholes
Gaps between leaves on a tree act like pinholes: each tiny gap projects an image onto the ground, producing numerous images of a solar eclipse seen here.

SOLAR ECLIPSE PHENOMENA

During a total solar eclipse, it typically takes about an hour and a half from the point when the Moon just starts to cover the Sun's disk to the start of totality, the period when the Sun's bright disk is entirely hidden. Totality itself typically lasts for 1–3 minutes but can last up to 7 minutes 30 seconds. During totality, the viewing location becomes dark and the obscured Sun presents a dramatic sight as its corona becomes visible. Just before totality and at its end, phenomena such as Baily's beads and the diamond ring effect may be seen. An annular eclipse is equally spectacular, producing what looks like a ring of light in the sky.

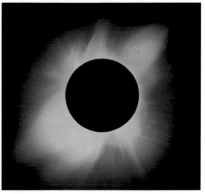

Corona
During the brief period of totality, the Sun's corona—an extremely hot, irregularly shaped part of its atmosphere—becomes visible.

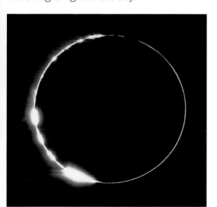

Baily's beads
Just before totality and at its end, the Moon's rough surface allows small patches of sunlight, known as Baily's beads, to break through.

Diamond ring effect
The diamond ring effect occurs when a single patch of light breaks through, together with a circular sliver of light around the edge of the Moon.

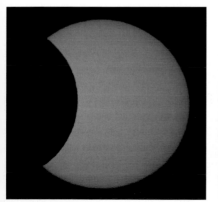

Partial solar eclipse
In a partial eclipse, or during the period before or after totality, the Moon obscures a gradually enlarging or diminishing portion of the Sun's disk.

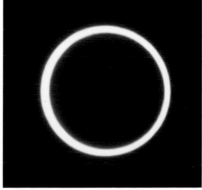

Annular solar eclipse
In an annular eclipse, the Moon's disk is too small to cover the Sun completely. This occurs when the Moon is farther than normal from Earth.

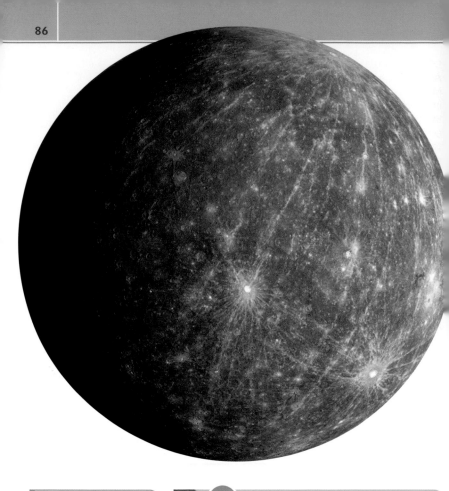

PROFILE

- 36.0 million miles (57.9 million km)
- 88 Earth days
- 58.7 Earth days
- 3,032 miles (4,879 km)
- 0
- -1.9

Earth

Mercury

RELATIVE SIZE

MERCURY

Scorched by the Sun and scarred by eons of asteroid impacts, Mercury is one of the most extreme planets in the Solar System. At 3,032 miles (4,879 km) in diameter, it is also the smallest of the eight planets. During the day, Mercury's surface temperature reaches 806°F (430°C), while at night it falls to -292°F (-180°C)—a wider range than on any other planet. Mercury's surface is heavily cratered, and one of its best-known geologic features is the Caloris Basin, a crater about 960 miles (1,550 km) wide that formed when a large asteroid crashed into the planet long ago. Other notable surface features include extinct volcanoes, huge plains that formed as ancient lava flows solidified, and vast cliffs that are hundreds of miles long.

ORBIT, STRUCTURE, AND ATMOSPHERE

Mercury makes one orbit around the Sun in less time than any other planet—just 88 Earth days. Its orbit is also the most eccentric of any planet: its distance from the Sun varies from only 28.6 million miles (46.0 million km) at perihelion to 43.4 million miles (69.8 million km) at aphelion. Mercury's interior composition is unusual. The planet's remarkably high density suggests that it has a large, iron-rich core, roughly 2,240 miles (3,600 km) in diameter. Mercury's atmosphere is extremely thin, with a total mass of only about 2,200 lb (1,000 kg), and is made up of oxygen (42 percent), sodium (29 percent), hydrogen (22 percent), helium (6 percent), and trace gases (1 percent).

crust of silicate rock

rocky silicate mantle

iron core

Structure
Mercury's core is thought to be composed of iron. It is probably surrounded by a rocky silicate mantle and, on top of that, a silicate crust scarred by asteroid impacts and ancient volcanism.

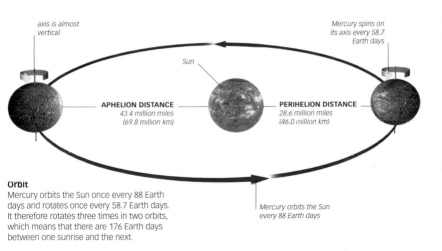

axis is almost vertical

Mercury spins on its axis every 58.7 Earth days

Sun

APHELION DISTANCE
43.4 million miles
(69.8 million km)

PERIHELION DISTANCE
28.6 million miles
(46.0 million km)

Orbit
Mercury orbits the Sun once every 88 Earth days and rotates once every 58.7 Earth days. It therefore rotates three times in two orbits, which means that there are 176 Earth days between one sunrise and the next.

Mercury orbits the Sun every 88 Earth days

OBSERVING MERCURY

Mercury is visible with the naked eye and can be seen low in the twilight sky, either shortly before sunrise or just after sunset. However, because of Mercury's fast orbit, it does not remain visible for long. The best time to observe Mercury is when it is at its largest separation from the Sun in the sky—its greatest elongation—which occurs six or seven times a year. Because Mercury is never far from the Sun—even at greatest elongation—take care to avoid looking at the Sun accidentally, especially if using binoculars or a telescope.

Mercury through a telescope
Due to its proximity to the Sun, Mercury is tricky to observe. Through a large telescope it appears as a fuzzy disk that, like all the inner planets, exhibits phases.

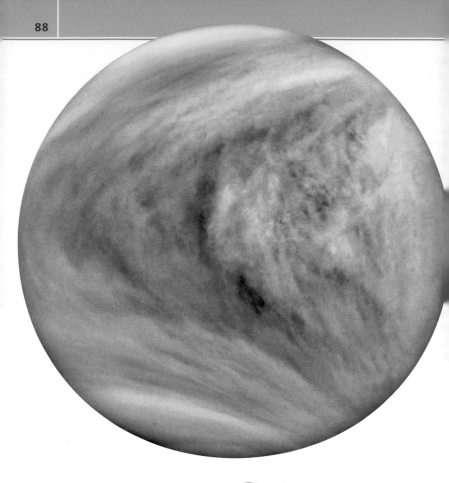

PROFILE

- 67.2 million miles (108.2 million km)
- 224.7 Earth days
- 243 Earth days
- 7,521 miles (12,104 km)
- 0
- -4.7

Earth

Venus

RELATIVE SIZE

VENUS

The second planet from the Sun, Venus is similar in size and structure to Earth. However, it is always covered by thick, unbroken cloud that conceals a hostile environment of extreme temperatures and pressures. Several unmanned probes have landed on the planet's surface and have relayed images that reveal a barren landscape shaped by vast lava flows and enormous volcanoes, some more than 60 miles (100 km) across. The volcanism that created the desolate surface is thought to have occurred in a global flurry of activity around 500 million years ago, although there may still be small regions of volcanic activity on Venus today. In the past, it was suggested that Venus might have oceans under its thick clouds, but it is now known that there is no liquid water there.

ORBIT, STRUCTURE, AND ATMOSPHERE

Venus orbits the Sun in 224.7 Earth days and spins on its axis in 243 Earth days, with the result that a day on Venus lasts longer than its year. Its rotational axis is titled at 177.4°, so the north pole is at the bottom of the planet. It appears to spin in the opposite direction to the other planets (apart from Uranus, whose spin axis is nearly horizontal). Structurally, Venus resembles Earth, with a solid metallic inner core, a molten metallic outer core, a rocky mantle, and a solid crust. The atmosphere is extremely thick, creating a pressure of about 92 Earth atmospheres at the surface. Above Venus's polar regions, the atmosphere forms huge, swirling vortices. The dense clouds contain sulfuric acid and create a powerful greenhouse effect that produces an average surface temperature of 464°C (867°F).

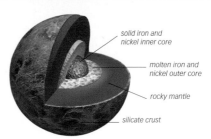

solid iron and nickel inner core

molten iron and nickel outer core

rocky mantle

silicate crust

Structure
Venus has a solid nickle-iron inner core, a molten nickel-iron outer core, a rocky mantle, and a silicate crust—a similar structure to that of Earth

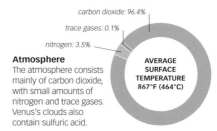

carbon dioxide: 96.4%

trace gases: 0.1%

nitrogen: 3.5%

AVERAGE SURFACE TEMPERATURE 867°F (464°C)

Atmosphere
The atmosphere consists mainly of carbon dioxide, with small amounts of nitrogen and trace gases. Venus's clouds also contain sulfuric acid.

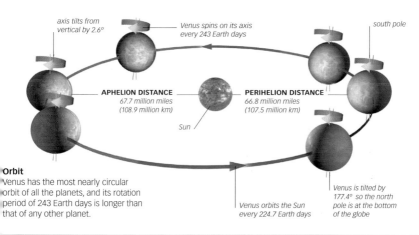

axis tilts from vertical by 2.6°

Venus spins on its axis every 243 Earth days

south pole

APHELION DISTANCE
67.7 million miles
(108.9 million km)

PERIHELION DISTANCE
66.8 million miles
(107.5 million km)

Sun

Orbit
Venus has the most nearly circular orbit of all the planets, and its rotation period of 243 Earth days is longer than that of any other planet.

Venus orbits the Sun every 224.7 Earth days

Venus is tilted by 177.4° so the north pole is at the bottom of the globe

OBSERVING VENUS

Venus is one of the easiest planets to observe due to its brightness—it is brighter than anything else in the sky except the Sun and Moon. Venus is visible at certain times during the year, either before sunrise or after sunset. It appears close to the Sun in the sky, so only look for it when the Sun is below the horizon, preferably around greatest elongation (p.65). Through binoculars or a telescope, Venus shows phases that change as it orbits the Sun.

Venus through a telescope
Venus's phases are easily visible through a telescope. When in its crescent phase, Venus is relatively close to us and at its largest in the sky.

141.6 million miles
(227.9 million km)

687 Earth days

24.6 hours

4,221 miles (6,792 km)

2

-2.9

Earth

Mars

RELATIVE SIZE

MARS

Known as the Red Planet because of its rust-red color, Mars is the fourth planet from the Sun and the outermost of the rocky planets, orbiting at an average distance of 141.6 million miles (227.9 million km) from the Sun. It is roughly half the size of Earth and is now a dry, barren planet with a surface marked by huge extinct volcanoes, large canyon systems, and rock-strewn plains. Vast dust storms whip around the planet, and clouds and falling snow have been spotted by spacecraft sent to explore its surface. Like Earth, Mars has seasons and polar ice caps. In the past, Mars was probably very different—evidence from orbiting space probes and planetary rovers suggests that liquid water flowed across the Martian surface billions of years ago.

ORBIT

Mars orbits the Sun once every 687 Earth days, so a year on Mars is nearly twice as long as a year on Earth. Because Mars is orbiting the Sun at a different speed than Earth, it exhibits what is called retrograde motion: when viewed from Earth, Mars sometimes appears temporarily to reverse the direction of its motion across the night sky (p.65). Mars' axis of rotation is tilted, as a result of which the planet experiences seasonal changes as it orbits the Sun. Mars' orbit is highly elliptical, and the planet receives much more solar radiation at its nearest point to the Sun than at its farthest. Consequently, surface temperatures can range from about 68°F (20°C) to -193°F (-125°C).

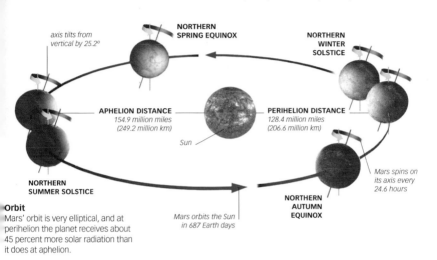

axis tilts from vertical by 25.2°

NORTHERN SPRING EQUINOX

NORTHERN WINTER SOLSTICE

APHELION DISTANCE
154.9 million miles
(249.2 million km)

PERIHELION DISTANCE
128.4 million miles
(206.6 million km)

Sun

Mars spins on its axis every 24.6 hours

NORTHERN SUMMER SOLSTICE

Mars orbits the Sun in 687 Earth days

NORTHERN AUTUMN EQUINOX

Orbit
Mars' orbit is very elliptical, and at perihelion the planet receives about 45 percent more solar radiation than it does at aphelion.

STRUCTURE AND ATMOSPHERE

The structure of Mars is similar to that of Earth, with a distinct crust, mantle, and core. Previously thought to be solid, Mars' metallic core is now believed to be partially molten. The atmosphere of Mars is extremely thin—the surface pressure is about one-hundredth that on Earth—and consists mainly of carbon dioxide. Recent studies have shown that the trace gases include some methane, which may have originated from geological activity or possibly even microbial life. Winds in the atmosphere periodically cause dust storms that cover the entire planet.

trace gases: 0.4%

carbon dioxide: 95.3%

argon: 1.6%

nitrogen: 2.7%

AVERAGE SURFACE TEMPERATURE
-81°F (-63°C)

Atmosphere
The atmosphere of Mars is dominated by carbon dioxide. It also contains small quantities of nitrogen and argon, and various trace gases, including oxygen and methane.

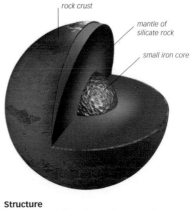

rock crust

mantle of silicate rock

small iron core

Structure
Beneath Mars' thin rock crust lies a mantle of silicate rock. At its heart, the planet is thought to have an iron core that scientists think may be partially molten.

»

» SURFACE FEATURES

Mars' surface is marked by volcanoes, craters, and chasms. Much of the Northern Hemisphere consists of relatively smooth plains, whereas the Southern Hemisphere is typically cratered highland. The planet's major surface features are mostly in a wide band centered around the equator. Mars' most striking feature is the Valles Marineris, a canyon system more than 2,500 miles (4,000 km) long, but its most famous is Olympus Mons, the largest volcano in the Solar System. Mars' surface is also scarred by cratering, which is still occurring today—images taken from orbit reveal small craters that formed in the last decade. Channels and gullies have also been detected, but it is not yet clear whether they were formed by the flow of water or by dry landslides.

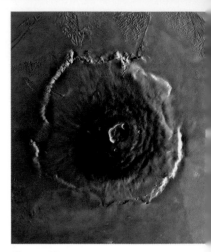

Olympus Mons
This extinct volcano is the largest and tallest in the Solar System: 388 miles (624 km) in diameter and 16 miles (25 km) high. It is so large that it can be seen in images captured with large amateur telescopes.

Bonneville Crater
This image of Bonneville Crater taken by NASA's *Spirit* rover shows large rocks that could have been blasted from the surface by the impact that created the crater.

MOONS OF MARS

Mars has two small, cratered moons: Phobos and Deimos. Both are irregular in shape and appear similar to asteroids; in fact, many astronomers believe that they are actually asteroids that have been captured by Mars' gravity. Phobos is the larger of the two, at about 17 miles (27 km) long and 14 miles (22 km) wide. It is also nearest to Mars, orbiting at an average distance of 5,827 miles (9,378 km). Phobos makes one orbit of Mars every 7 hours and 39 minutes. Deimos is about 9 miles (15 km) long and 7 miles (12 km) wide, and orbits Mars at an average distance of 14,577 miles (23,459 km); it orbits Mars once every 30 hours and 18 minutes.

PHOBOS

DEIMOS

OBSERVING MARS

Mars is visible in the sky for at least some of the year, but it usually looks like an ordinary star. However, every two years and two months it comes into opposition, and then it is close to Earth and at its largest and brightest. During opposition, Mars becomes one of the brightest objects in the night sky, and it appears to the naked eye as a brilliant orange-red point of light that is easily discernible against the background stars. Every 15 or 17 years, Mars' elliptical orbit brings it particularly close to Earth, providing the best view of the planet. However, even when Mars is closest to Earth, a telescope is needed to see surface details; the larger the telescope, the more detail will be visible.

With binoculars

Through binoculars Mars usually looks like a brilliant point of orange-red light. However, with a good pair of powerful binoculars and good viewing conditions, it may be possible to see a tiny disc when Mars is at opposition.

With a small telescope

Through a small telescope some of the largest surface features, such as the polar ice caps, can be seen when viewing conditions are good.

With a large telescope

Through a large amateur telescope more detail will be visible, including some of the larger dark surface features, such as the huge Syrtis Major Planum.

Maps of Mars
The four views of Mars (above) collectively show the complete surface of Mars, labeled to indicate some of the major surface features.

PROFILE

⬛ 483.8 million miles
(778.6 million km)

◎ 11.86 Earth years

🌀 9.93 hours

⊕ 88,846 miles (142,984 km)

● 64

☀ -2.9

Jupiter

Earth

RELATIVE SIZE

JUPITER

Within the Solar System, Jupiter is second only to the Sun in size and mass. It is 11 times the diameter of Earth and almost 2.5 times the combined mass of the other seven planets of the Solar System. Fifth out from the Sun, Jupiter is also about five times the distance of Earth from the Sun. Its vast globe has a distinctively banded atmosphere that is affected by storms and winds of up to 388 mph (625 kph), and its immense gravity has meant it has clung on to or captured a huge collection of moons. The giant planet also has a faint ring system and a powerful magnetic field generated by electric currents in an internal layer of metallic hydrogen. This field is stronger than that of any other planet and dominates a vast, bubblelike volume of space around Jupiter.

ORBIT

Jupiter has quite an eccentric elliptical orbit, with the nearest (perihelion) and farthest (aphelion) distances from the Sun differing by 47.3 million miles (76.1 million km). The planet completes each orbit in a little less than 12 years and rotates once on it own axis in just less than 10 hours, which is the shortest rotational period of all the planets. This rapid spin pushes out Jupiter's equatorial region, making the planet 6.5 percent wider at its equator than at its poles. Jupiter's spin axis is tilted from the vertical by only 3.1°, which means that the planet has no seasons because neither hemisphere ever becomes tilted significantly toward or away from the Sun.

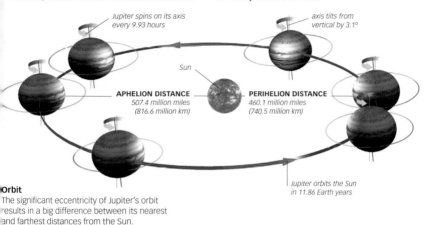

Jupiter spins on its axis every 9.93 hours

axis tilts from vertical by 3.1°

Sun

APHELION DISTANCE
507.4 million miles
(816.6 million km)

PERIHELION DISTANCE
460.1 million miles
(740.5 million km)

Jupiter orbits the Sun in 11.86 Earth years

Orbit
The significant eccentricity of Jupiter's orbit results in a big difference between its nearest and farthest distances from the Sun.

STRUCTURE AND ATMOSPHERE

At Jupiter's center there is thought to be a dense core with a mass several times that of Earth. The core is surrounded by a layer of liquid metallic hydrogen, which, in turn, is surrounded by a layer of liquid hydrogen and helium. This layer is thought to merge gradually into the atmosphere of hydrogen and helium gas. The segregation of the upper atmosphere into bands is due to a combination of factors: heat from the Sun and the planet's interior; the planet's spin; the resulting large-scale atmospheric movements; and condensation of hydrogen compounds to form the different-colored clouds.

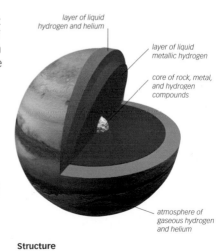

methane and other trace gases: 0.3%

hydrogen: 89.6%

helium: 10.1%

CLOUD-TOP TEMPERATURE
-162°F (-108°C)

Atmosphere
Jupiter's atmosphere consists mainly of hydrogen and helium. Trace amounts of methane and other hydrogen-containing compounds color its upper layers.

layer of liquid hydrogen and helium

layer of liquid metallic hydrogen

core of rock, metal, and hydrogen compounds

atmosphere of gaseous hydrogen and helium

Structure
From the surface downward, the composition of Jupiter's interior gradually changes from gaseous to liquidlike due to the increasing temperature and pressure.

»

⟩⟩ RINGS AND MOONS

Jupiter is surrounded by a thin, faint ring system in four distinct parts. The brightest ring, called the main ring, is accompanied on its inside by a sparse, doughnut-shaped structure called the halo ring and on its outside by a broad, thick, two-part structure called the gossamer ring. Jupiter has at least 64 moons, more than any other Solar System planet. The four largest—Ganymede, Callisto, Europa, and Io—are known as the Galilean moons because they were discovered by the Italian scientist Galileo Galilei in 1610. Of the other moons, 47 have been discovered since January 2000.

Ganymede
The largest moon in the Solar System, Ganymede has an icy surface with distinctive bright and slightly more ancient dark areas. The polar regions are covered by frost (the pale mauve areas in this color-enhanced image).

Callisto
The surface of Jupiter's second largest moon is pockmarked by craters and multi-ringed structures created by asteroid impacts—the bright areas are the impact scars.

Europa
Europa has a smooth, icy crust that reflects light well, making it one of the brightest moons in the Solar System. Beneath the crust lies an ocean of liquid water.

Io
The most volcanically active body in the Solar System, Io is a world of volcanic vents (the dark spots in this color-enhanced image), lava flows, and high-reaching lava plumes.

The main ring
Jupiter's brightest ring, the main ring, is about 4,000 miles (6,500 km) wide but less than 20 miles (30 km) deep. It is probably made of dust ejected from the moons Adrastea and Metis.

THE GREAT RED SPOT

Jupiter's atmosphere is affected by colossal, oval-shaped storms whipped up by the planet's internal heat and its rapid rotation. The largest of these storms is the Great Red Spot—an enormous, anticlockwise-rotating, high-pressure system that has been raging for at least 350 years.

Giant storm
The largest storm complex in the Solar System, Jupiter's Great Red Spot is about twice the size of Earth.

north polar region

north temperate belt

north tropical zone (includes the paler bands above and below)

north equatorial belt

equatorial zone

south equatorial belt

Great Red Spot

south tropical zone

south temperate belt

south polar region

OBSERVING JUPITER

Jupiter is easy to spot with the naked eye, because even at its faintest it outshines Sirius, the brightest star. With its 12-year solar orbit, Jupiter moves fairly slowly across the celestial sphere, passing through about one constellation per year. It comes into opposition (closest proximity to Earth), and thus appears at its brightest, every 13 months. Jupiter appears as a disk through binoculars, and a small telescope clearly reveals some or all of its four largest moons. Other aspects of Jupiter, noticeable through a larger telescope, are the squashed shape of the planet, caused by its rapid rotation, and its surface banding. The dark bands, containing warmer gas, are called belts; the lighter bands, containing cooler gas, are known as zones. Two dark bands on either side of the equator—the north and south equatorial belts—are always prominent.

Jupiter's belts and zones
The top of Jupiter's colorful atmosphere forms its visible surface. The outlines of the belts and zones—and features within them, such as the Great Red Spot—are always changing as a result of vast, swirling atmospheric disturbances.

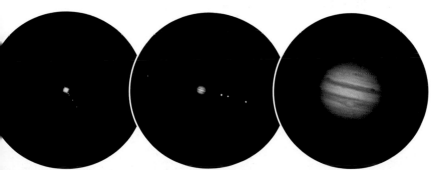

With binoculars
Through binoculars, Jupiter appears as a bright disk. Some of its four largest moons (or even all four, depending on their position) may just be discernible.

With a small telescope
With a small telescope, Jupiter's largest moons (the Galilean moons) are clearly visible, and it may also be possible to see some of Jupiter's surface banding.

With a large telescope
With a large telescope, the surface belts and zones are visible in some detail. Because of Jupiter's fast spin, its surface features move across the planet in a single night.

PROFILE

▣	890.8 million miles (1.43 billion km)
◉	29.46 Earth years
⚡	10.7 hours
⊕	74,898 miles (120,536 km)
●	62
✳	-0.3

Earth

Saturn

RELATIVE SIZE

SATURN

Famous for its impressive rings, Saturn is the sixth planet fom the Sun and the second largest in the Solar System (after Jupiter), with an equatorial diameter of 74,898 miles (120,536 km). It consists almost entirely of the lightest elements, hydrogen and helium, as a result of which it is the least dense planet in the Solar System. In the night sky, Saturn can be spotted easily with the naked eye, while a telescope will show its rings and even some of its many moons, which are remarkable in their own right. For example, Titan, the largest moon, has a thick atmosphere, lakes of liquid methane and ethane, and perhaps even icy volcanoes. Today, a great deal is known about the Saturnian system, largely thanks to the Cassini–Huygens mission, which reached there in 2004.

ORBIT

Saturn takes 29.46 Earth years to make one orbit around the Sun. However, it rotates once on its axis every 10.7 hours, so a day on Saturn is much shorter than on Earth. In addition, Saturn's rapid spin causes its equator to bulge, making it about 10 percent wider at the equator than at the poles. The planet has a marked axial tilt of 26.7°, which means that there are seasonal differences on Saturn as it moves around the Sun. The axial tilt is also the reason why the view of the rings from Earth changes as Saturn orbits the Sun (p.101).

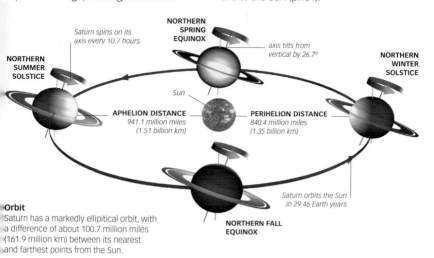

Orbit
Saturn has a markedly ellipitical orbit, with a difference of about 100.7 million miles (161.9 million km) between its nearest and farthest points from the Sun.

STRUCTURE AND ATMOSPHERE

Saturn's atmosphere forms the planet's subtly banded visible surface. Ferocious winds whip through the atmosphere at speeds of about 1,120 mph (1,800 kph), and sometimes huge storms also develop. The Cassini–Huygens mission has even taken images revealing bursts of lightning high in Saturn's atmosphere. The outer gaseous atmosphere gradually merges into a deeper layer, where the hydrogen and helium are liquid, and deeper still the hydrogen and helium become like metals. At the planet's center there is probably a core of rock and ice.

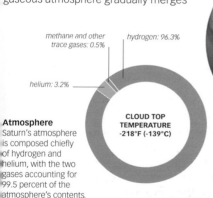

Atmosphere
Saturn's atmosphere is composed chiefly of hydrogen and helium, with the two gases accounting for 99.5 percent of the atmosphere's contents.

Structure
Saturn's gaseous atmosphere envelops shells of hydrogen and helium, in both liquid and liquid metallic forms, around a rock and ice core.

≫ MOONS

Saturn is orbited by 62 moons, which differ greatly in size and distance from the planet. While some are as small as a few miles across, Titan is about 3,200 miles (5,150 km) in diameter and is the second largest moon in the Solar System (after Jupiter's Ganymede).

Some of Saturn's moons orbit in its ring system whereas others are a huge distance away from the planet. Most of the moons are small, frozen worlds, although a few are active, dynamic places. For example, Enceladus has enormous fountains of icy material erupting from its southern hemisphere.

Titan
The largest moon, Titan orbits 758,000 miles (1.22 million km) from Saturn. It has a thick atmosphere, lakes of liquid methane and ethane, and a weather system.

Rhea
The second-largest moon, Rhea is 949 miles (1,528 km) across and orbits 327,000 miles (527,000 km) from Saturn. Unusually, it is thought to have its own rings.

Iapetus
The third-largest moon, Iapetus is 914 miles (1,471 km) across and orbits 2.21 million miles (3.56 million km) from Saturn. One of Iapetus' sides is covered in dark material.

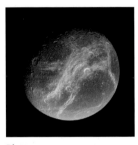

Dione
The fourth-largest moon, Dione is 698 miles (1,123 km) across and orbits 234,260 miles (377,000 km) from Saturn. Parts of its icy surface have long, deep canyons.

Tethys
The fifth-largest moon, Tethys is 662 miles (1,066 km) across and orbits 183,100 miles (294,670 km) from Saturn. It shares its orbit with the moons Telesto and Calypso.

Enceladus
The sixth-largest moon, Enceladus is 318 miles (512 km) across and orbits 147,900 miles (238,020 km) from Saturn. From near its south pole, jets of ice erupt into space.

Mimas
The seventh-largest moon, Mimas is 246 miles (396 km) across and orbits 115,280 miles (185,520 km) from Saturn. It has a large crater, Herschel, 87 miles (140 km) wide.

Hyperion
This moon orbits 919,630 miles (1.48 million km) from Saturn and, at about 230 miles (370 km) long, is one of the largest irregularly shaped bodies in the Solar System.

Phoebe
The largest of Saturn's outer moons, Phoebe is about 143 miles (230 km) in diameter and orbits 8.05 million miles (12.95 million km) from Saturn.

RINGS

Saturn's ring system is the largest of any around the outer giant planets. The rings are composed of lumps of ice, which vary in size from tiny frozen particles to icy boulders about the size of a car. Closest to Saturn is the D ring, the inner edge of which is 41,570 miles (66,900 km) from the planet. The most easily seen are the main rings—the C, B, and A rings. Beyond these are the faint F, G, and E rings. In 2009, NASA's Spitzer Space Telescope found another, diffuse ring 3.7 million miles (6 million km) from Saturn. Although enormous in diameter, Saturn's rings can be as thin as 33 ft (10 m) in places.

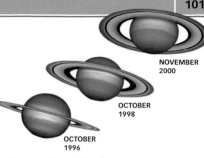

NOVEMBER 2000

OCTOBER 1998

OCTOBER 1996

Changing views of the rings
Saturn's axis of rotation is tilted at 26.7°, so the view of the planet from Earth changes as Saturn orbits around the Sun.

Saturn's main ring system
The main ring system comprises the C, B, and A rings, which together extend from 46,390 miles (74,658 km) to 84,991 miles (136,780 km) from Saturn.

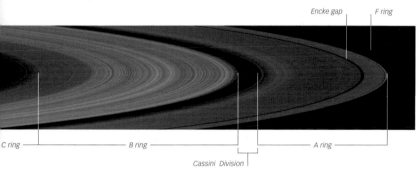

Encke gap | F ring

C ring | B ring | A ring

Cassini Division

OBSERVING SATURN

Saturn is visible to the naked eye for about 10 months of the year, appearing as a bright, yellowish star. However, through even a modest telescope, the view of the planet, with good observing conditions, is impressive. A small telescope will reveal the planet's disk and rings as well as its largest moon, Titan. Larger telescopes may reveal subtle details on the planet's pale disk and in its ring system. The best time to observe Saturn is when it is at opposition, which occurs annually, about two weeks later each year.

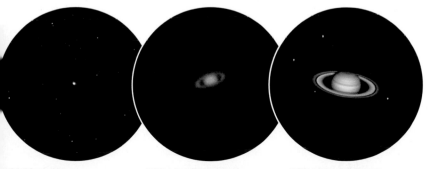

With binoculars
Through binoculars, Saturn appears as a bright, yellowish point of light, but neither the rings nor the moons can be seen.

With a small telescope
Saturn's ring system can be seen through even a small telescope, and it may be possible to spot Saturn's largest moon, Titan.

With a large telescope
A large telescope will reveal details in Saturn's atmosphere and ring system, and will also show some of its moons more clearly.

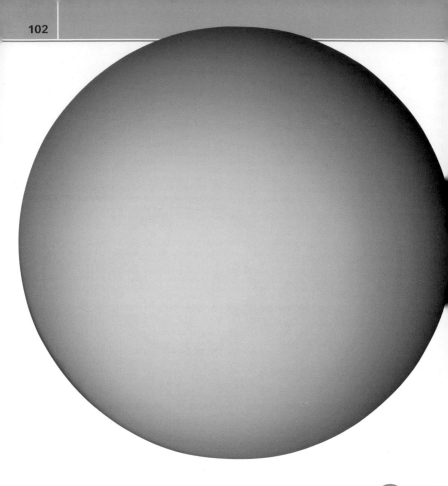

PROFILE

▨	1.78 billion miles (2.87 billion km)
◉	84 Earth years
⟳	17.2 hours
⊖	31,763 miles (51,118 km)
●	27
☀	5.3

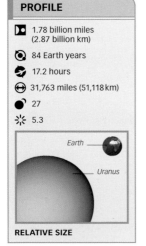

Earth

Uranus

RELATIVE SIZE

URANUS

Uranus is the third-largest planet and lies twice as far from the Sun as its neighbor Saturn. It is pale blue—the color comes from methane gas in its atmosphere—and featureless, with a sparse ring system and an extensive family of moons. Probably because of a collision with a planet-sized body not long after Uranus formed, its spin axis is tipped over by 98°. As a result, Uranus moves along its orbital path on its side, its spin is retrograde (moving in the opposite direction to that of most other planets), and from Earth its moons and rings appear to encircle it from top to bottom. Uranus was the first planet to be discovered with a telescope—in 1781, by the British astronomer William Herschel—but little was known about it until the Voyager 2 spacecraft flyby in 1986.

ORBIT, STRUCTURE, AND ATMOSPHERE

Uranus takes 84 years to orbit the Sun. Due to the length of its orbit and the planet's extreme tilt, its polar regions spend periods of about 21 years pointing either toward the Sun and thus in continuous daylight or away from it in complete darkness. A large part of Uranus is a liquid layer, composed of water, ammonia, and methane, which surrounds a rocky core. Enveloping the liquid layer is a deep gaseous atmosphere.

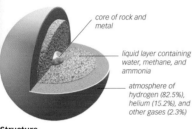

core of rock and metal

liquid layer containing water, methane, and ammonia

atmosphere of hydrogen (82.5%), helium (15.2%), and other gases (2.3%)

Structure
Uranus's visible surface is the outer edge of its atmosphere, which consists mainly of hydrogen. Below the atmosphere is a liquid layer, then a small core of rock and metal.

RINGS AND MOONS

Uranus has 13 rings, many of them extremely narrow, which extend from 23,500 to 61,000 miles (38,000 to 98,000 km) from the planet. They are composed of dark particles, ranging in size from a few inches to feet across, and some dust. Uranus has 27 moons. The five largest (Titania, Oberon, Umbriel, Ariel, and Miranda) were discovered between 1787 and 1948; the remainder have been discovered since the mid-1980s.

Titania
This image from the Voyager 2 spacecraft shows Titania, the largest moon of Uranus at 980 miles (1,578 km) in diameter.

Uranus's rings
This infrared image, taken by the Keck Observatory, Hawaii, in 2004, shows the innermost 11 of Uranus's 13 rings; four of the five main satellites are also visible here.

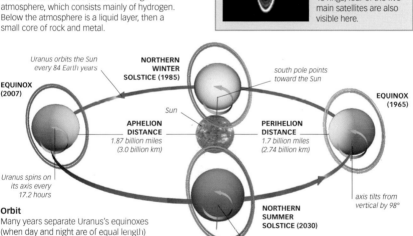

Uranus orbits the Sun every 84 Earth years

NORTHERN WINTER SOLSTICE (1985)

south pole points toward the Sun

EQUINOX (2007)

Sun

EQUINOX (1965)

APHELION DISTANCE
1.87 billion miles (3.0 billion km)

PERIHELION DISTANCE
1.7 billion miles (2.74 billion km)

Uranus spins on its axis every 17.2 hours

axis tilts from vertical by 98°

Orbit
Many years separate Uranus's equinoxes (when day and night are of equal length) and solstices (when the poles experience continuous light or dark).

NORTHERN SUMMER SOLSTICE (2030)

south pole points away from the Sun

OBSERVING URANUS

Uranus is just visible to the naked eye, but a telescope is needed to see it as more than a pinpoint of light. Uranus moves slowly across the celestial sphere, taking some 5 to 10 years to traverse each zodiacal constellation. Since 2010, it has been in or near Pisces, where it will remain until 2019, when it will move into Aries.

Uranus through a telescope
Even through a moderate-sized amateur telescope, Uranus appears as only a small pale disk. The planet is so remote that the amount of sunlight it receives is only 0.25 percent of that reaching Earth.

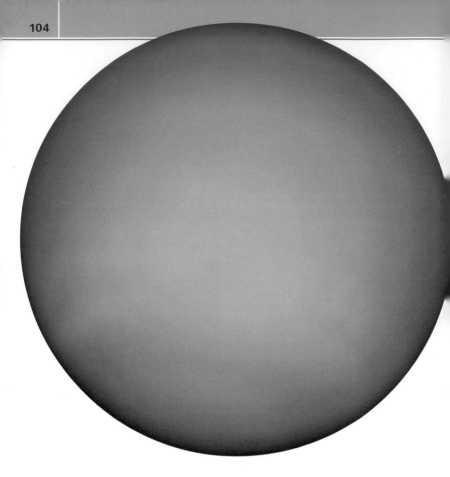

PROFILE

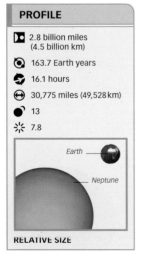

- 2.8 billion miles (4.5 billion km)
- 163.7 Earth years
- 16.1 hours
- 30,775 miles (49,528 km)
- 13
- 7.8

Earth

Neptune

RELATIVE SIZE

NEPTUNE

Neptune is the smallest and coldest of the four giant outer planets, as well as the most distant from the Sun. Just one spacecraft, Voyager 2, has been to investigate this remote world. When the probe flew by in 1989, it provided the first close-up view of Neptune, revealing it to be a translucent, crisp blue globe with wispy white clouds of frozen methane in its atmosphere and a ring system. Neptune has 13 moons; the largest, Triton, was discovered in 1846 and a second, Nereid, in 1949. Six further moons were found by Voyager 2, and the remaining five were discovered after 2000. Only Triton is of substantial size—1,700 miles (2,700 km) in diameter—and spheroidal. The remaining moons are small and irregularly shaped; some orbit in gaps within Neptune's ring system.

ORBIT, STRUCTURE, AND ATMOSPHERE

Neptune takes 163.7 Earth years to orbit the Sun, so it has completed only one circuit since its discovery in 1846. Structurally, Neptune is similar to Uranus, with a core of rock and metal surrounded by a liquid layer of water, ammonia, and methane. Above this is a hydrogen-dominated atmosphere affected by greater wind speeds—up to 1,240 mph (2,000 kph)—than those found on any other planet. The boundaries between its layers are not clearly defined.

core of rock and metal

liquid layer containing water, methane, and ammonia

atmosphere of hydrogen (80%), helium (19%), and methane and trace gases (1%)

Structure
Neptune's visible surface is the outer edge of its atmosphere. Below this is a liquid layer, which surrounds a small core of rock and metal.

RINGS OF NEPTUNE

Neptune's ring system extends from 25,500 to 40,000 miles (41,000 to 64,000 km) from the planet. There are five main rings; from outermost to innermost, these are Adams, Arago, Lassell, Le Verrier, and Galle. There is also an extremely faint unnamed sixth ring inside Adams. The rings are made of small, dark particles of unknown composition and are so faint that their existence was not confirmed until Voyager 2 flew past in 1989.

Rings from Voyager 2
The two bright rings in this image are Adams, on the outside, and Le Verrier. Inside Le Verrier is the fainter Galle ring.

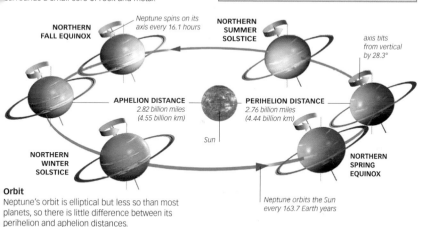

NORTHERN FALL EQUINOX

Neptune spins on its axis every 16.1 hours

NORTHERN SUMMER SOLSTICE

axis tilts from vertical by 28.3°

APHELION DISTANCE
2.82 billion miles
(4.55 billion km)

PERIHELION DISTANCE
2.76 billion miles
(4.44 billion km)

Sun

NORTHERN WINTER SOLSTICE

NORTHERN SPRING EQUINOX

Orbit
Neptune's orbit is elliptical but less so than most planets, so there is little difference between its perihelion and aphelion distances.

Neptune orbits the Sun every 163.7 Earth years

OBSERVING NEPTUNE

With an apparent magnitude of 7.8, Neptune can be observed only through binoculars or a telescope. Through a small telescope it appears as a tiny blue disk. Its long orbit means it takes many years to move through each zodiacal constellation. Since early 2011 it has been in Aquarius, where it will remain until 2023.

Neptune through a telescope
The vivid blue of Neptune's atmosphere makes it stand out strikingly against the blackness of space when viewed through a telescope. With binoculars, Neptune appears as only a faint star.

Pluto and its moons
The dwarf planet Pluto (the central white disk) is now known to have four moons: Hydra (top left), Charon (bottom left), Nix (bottom right), and P4 (top right). P4 was discovered in 2011.

DWARF PLANETS AND THE KUIPER BELT

Beyond the most distant planet, Neptune, lies the cold and dark outer region of the Solar System. Numerous small, icy bodies populate this region. It also contains several larger objects classified as dwarf planets, including Pluto.

BEYOND NEPTUNE

The planetary wilderness that lies outside Neptune's orbit is arguably one of the most enigmatic and least known regions of the Solar System. However, it is known that beyond Neptune there is a huge, flattened disk known as the Kuiper Belt. This contains mainly small bodies but also has a few larger objects, including dwarf planets such as Pluto, Haumea, and Makemake. It may also be the source of some comets that visit the inner Solar System. Most comets, however, originate from the Oort Cloud (p.109), a vast spherical region lying outside the Kuiper Belt. In 2006, NASA launched its New Horizons probe to explore the region beyond Neptune and specifically to study Pluto and other objects in the Kuiper Belt. The probe, which passed Jupiter in 2007, is due to make its closest approach to Pluto in 2015.

DWARF PLANETS

As well as the eight main planets of the Solar System, five other Solar System objects are recognized by astronomers as dwarf planets. These include the former planet Pluto, which was reclassified as a dwarf planet in 2006, partly as a result of the discovery of similar-sized objects orbiting beyond Neptune. The largest dwarf planets are Pluto and Eris, each of which is about 1,430 miles (2,300 km) in diameter. These, together with Haumea and Makemake (which were discovered in 2004 and 2005, respectively), orbit beyond Neptune. Ceres, the other dwarf planet, orbits much closer to the Sun, in the Main Belt of asteroids.

Ceres
Unlike the other dwarf planets, Ceres is also an asteroid and orbits the Sun in the Main Belt between Mars and Jupiter. At about 960 km (597 miles) across, Ceres is the largest object in this region.

Surface of Pluto
This Hubble Space Telescope image of Pluto's mottled surface reveals dark areas that may be carbon-rich deposits resulting from methane breakdown.

THE KUIPER BELT

The Kuiper Belt is an enormous disk of cold bodies, known as Kuiper Belt Objects (or KBOs), that lies beyond the orbit of Neptune. About 1,100 KBOs have been found—including the dwarf planets Pluto, Haumea, and Makemake—and it is thought that there may be hundreds of thousands in all. The Kuiper Belt is also thought to be the source of some of the comets that occasionally visit the inner Solar System.

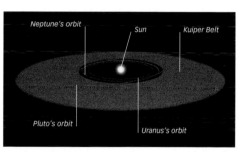

Location of the Kuiper Belt
Situated beyond the orbit of Neptune, the Kuiper Belt stretches between about 3.7 and 7.4 billion miles (6 and 12 billion km) from the Sun.

THE DISCOVERY OF PLUTO

Pluto was discovered in 1930 by the American astronomer Clyde Tombaugh (1906–97) while working at the Lowell Observatory in Arizona. He had been capturing images of small patches of the night sky, to see if anything moved against the background stars. Eventually he spotted a point of light that appeared to change position between frames—it was Pluto.

Moving star
These two images were taken by Clyde Tombaugh in 1930. When the images were compared, one "star" (shown by red arrows) appeared to change position. This was Pluto.

Comet Hale–Bopp
This large comet was visible to the naked eye for 18 months, reaching peak brightness in April 1997. Here, its spectacular tails of dust (white) and gas (blue) can be seen.

COMETS

A comet is a chunk of ice, rock particles, and dust that orbits the Sun. Comets that get close enough to the Sun develop bright clouds of material around themselves, and some also produce spectacular long, glowing tails.

WHAT COMETS ARE

The heart, or nucleus, of a comet is an irregularly shaped, spinning chunk of dirty ice, from a few hundred feet to tens of miles across. The "dirt" consists of dust and rock particles. The ice is mainly water but also contains carbon dioxide and other substances. While in orbit beyond Jupiter, a comet is in a cold, dormant state. If it moves toward the inner Solar System, heat from the Sun causes some ice near its surface to sublimate (change directly from solid to gas), releasing dust, ice particles, and gas that disperse to form a bright, fuzzy cloud around the nucleus called a coma. This reaches its maximum size when the comet is closest to the Sun. The solar wind and solar radiation push the gas and dust particles away from the coma, creating two tails. The dust particles eventually form a ring of dust along the comet's orbital path. When, as it orbits the Sun, Earth intersects one of these rings, the dust enters the atmosphere and produces meteors (pp.112–13).

THE ORIGIN OF COMETS

Comets are debris from the formation of the Solar System. Once the planets formed, vast numbers of leftover chunks of material were ejected beyond the orbits of the giant outer planets and became what are now comets and other icy objects. Most comets are thought to originate either from the Kuiper Belt (p.107) or from the Oort Cloud. It is theorized that comets from the Kuiper Belt are pulled into orbits that periodically bring them near the Sun by the gravity of the giant planets, whereas comets from the Oort Cloud are nudged into solar orbit by the gravity of passing stars.

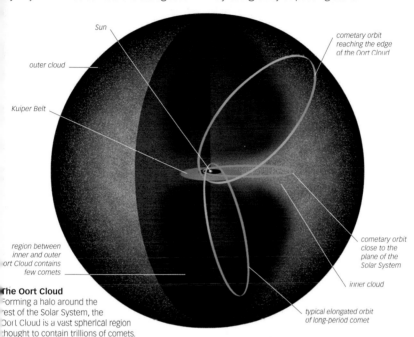

Sun

outer cloud

Kuiper Belt

cometary orbit reaching the edge of the Oort Cloud

region between inner and outer Oort Cloud contains few comets

cometary orbit close to the plane of the Solar System

inner cloud

typical elongated orbit of long-period comet

The Oort Cloud
Forming a halo around the rest of the Solar System, the Oort Cloud is a vast spherical region thought to contain trillions of comets.

COMETARY ORBITS AND TAILS

Comets that visit the inner Solar System have elliptical orbits around the Sun. As a comet approaches the Sun, it develops two tails, which grow as the comet nears the Sun and shrink as it moves away. Some comets reappear regularly—those arriving more than once every 200 years are called short-period comets, others are known as long-period comets—while a few pass only once through the inner Solar System. The Kuiper Belt is believed to be the origin of some short-period comets and the Oort Cloud to be the source of all other comets.

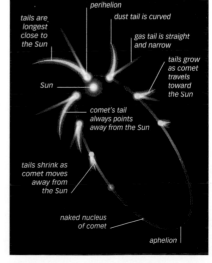

perihelion

tails are longest close to the Sun

dust tail is curved

gas tail is straight and narrow

tails grow as comet travels toward the Sun

Sun

comet's tail always points away from the Sun

tails shrink as comet moves away from the Sun

naked nucleus of comet

aphelion

The tails of comets
A comet's dust tail consists of dust pushed away from the comet's coma by solar radiation. The gas tail is caused by the solar wind interacting with gas in the coma.

»

» NOTABLE COMETS

The Solar System is thought to contain trillions of comets, most of which never travel close to the Sun. Astronomers have catalogued about 480 short-period comets and more than 4,000 overall, but only a few have been visited by spacecraft or created a spectacle visible from Earth.

HYAKUTAKE

This long-period comet came to notice in March 1996, when it passed within only 9 million miles (15 million km) of Earth. In May 1996, the Ulysses spacecraft detected Hyakutake's gas tail as far as 355 million miles (570 million km) from its nucleus—the longest comet tail ever detected. Hyakutake was also the first comet observed to emit X-rays.

HALLEY'S COMET

In 1696, the English astronomer Edmond Halley proposed that comets recorded in 1531, 1607, and 1682 were actually the same one returning every 75–76 years. He was correct, and this comet became known as Halley's Comet. On its last appearance, in 1986, five spacecraft visited the comet. Of these, Giotto took the first ever pictures of a comet's nucleus and revealed that Halley's has an average diameter of 7 miles (11 km).

SHOEMAKER–LEVY 9

Unusually, this comet was discovered orbiting Jupiter, not the Sun, in 1993, having probably been captured by the giant planet 20–30 years earlier. Even more remarkably, it was in 21 pieces (many of them shown below), up to 1.2 miles (2 km) in diameter, having been ripped apart in July 1992, when it passed too close to Jupiter. The 21 fragments subsequently slammed one by one into Jupiter in July 1994.

WILD 2

Wild 2 is a short-period comet that first began orbiting in the inner Solar System in September 1974, after a close encounter with Jupiter. Its nucleus is only 3.4 miles (5.5 km) long. In January 1994, NASA's Stardust spacecraft flew by the comet, capturing dust from its nucleus.

HOLMES

In October 2007, this short-period comet temporarily brightened by a factor of about 500,000, making it visible to the naked eye. This was the most vigorous outburst by a comet ever recorded. Briefly, the comet's coma became larger than the Sun.

MCNAUGHT

McNaught is an example of a single-apparition comet: one that visits the inner Solar System once and is never expected to return. In early 2007, it became the brightest comet since 1965, easily visible with the naked eye—even, for a few days, during daylight. By mid-January 2007, it had also developed a spectacularly long, curved dust tail.

HARTLEY 2

This short-period comet was the target of a flyby by the Deep Impact spacecraft in November 2010. It obtained images of the comet's nucleus, which is peanut-shaped and about 1.2 miles (2 km) long. The nucleus was seen to have jets spewing out ice, dust, and rocks.

SWIFT–TUTTLE

Swift–Tuttle is a short-period comet with an orbit that intersects Earth's. The stream of dust the comet leaves in its wake causes the Perseid meteor shower, which is visible every August. The nucleus of the comet is large, at about 17 miles (27 km) in diameter.

The Leonid meteor shower
Leonid meteors are visible in November every year, appearing to come from the constellation Leo.

METEORS

More commonly known as shooting stars, meteors are among the most beautiful and exciting sights in the night sky. Each of these brief streaks of light marks the entry of a tiny particle of debris into our atmosphere.

METEOROIDS, METEORS, AND METEORITES

Scattered throughout the Solar System are small pieces of debris from comets or asteroids. These tiny particles are known as meteoroids. As Earth orbits around the Sun, it encounters some of these meteoroids, which enter the atmosphere and heat up, creating a streak of light across the night sky—a meteor. Most meteoroids are only about the size of a grain of sand, yet they enter Earth's atmosphere at incredible speed—typically at about 45,000 mph (72,000 kph). This tremendous speed causes the air ahead of the meteoroid to compress and heat up. The meteoroid begins to vaporize, and it is at this point we see a shooting star flash across the sky. The larger the meteoroid, the brighter the resulting meteor will be. Very bright meteors are known as meteor fireballs. If the meteoroid is large enough, it may not be completely destroyed as it passes through the atmosphere and may reach Earth's surface; these objects are known as meteorites.

METEOR SHOWERS

Single meteors can appear anywhere in the sky at any time. These occasional meteors are known as sporadics. However, there are predictable dates in the year when showers of meteors occur. These showers result from Earth crossing debris left in the orbit of a comet or asteroid. For example, the Perseid meteor shower, which peaks in August, is due to dust from the comet Swift–Tuttle. The meteors in a shower all appear to originate from the same point in the sky, known as the radiant, and the constellation in which the radiant is located gives the name to the meteor shower. For example, the Orionid meteor shower (visible in October) has its radiant in the constellation Orion.

Dust trail in space
Comets leave trails of dust debris along their orbital paths around the Sun. When Earth passes through a dust trail, the dust particles enter the atmosphere and heat up, producing a meteor shower.

METEORITES

Meteoroids and asteroids with a mass of over 70 lb (30 kg) are not totally vaporized in Earth's atmosphere. Those that hit the ground are called meteorites. They are classified according to composition: stony meteorites are made of rock; iron ones consist mainly of an iron-nickel alloy; and stony-iron ones are a mixture of rock and an iron-nickel alloy.

STONY-IRON METEORITE

STONY METEORITE

IRON METEORITE

OBSERVING METEOR SHOWERS

Meteor showers are easily visible with the naked eye, and the dates when they appear are predictable; information about them is given in the table on p.330. To get the best view of a meteor shower, find a site that is as dark as possible and has a wide, unobstructed sky. Moonlight or light pollution will reduce the number of meteors you can see. Look towards the radiant constellation but avoid looking straight at the radiant point; instead, scan to either side of it, where the meteor streaks will be longer.

Meteor fireball
Extremely bright meteors are known as fireballs. The one shown here was photographed in California, USA, in 2009 and was so large it left a brief glowing line in the sky, known as a meteor train.

Ida and its moon
Orbiting in the Main Belt, Ida is about 36 miles (58 km) long and has a tiny satellite, Dactyl—seen here on the right—which is just 0.9 miles (1.4 km) in diameter.

ASTEROIDS

Composed mainly of rock and metal, asteroids are small bodies that in nearly all cases orbit the Sun, either within or close to the orbit of Jupiter. The majority are found in a region known as the Main Belt, between the orbits of Mars and Jupiter.

WHAT ASTEROIDS ARE

Asteroids are remnants of a failed process of planet formation that would have resulted in a rocky planet about four times the size of Earth. They are dry, dusty objects, too small to have atmospheres. Except for a few of the largest (which are roughly spherical), asteroids have irregular shapes. They come in a wide range of sizes, with large asteroids greatly outnumbered by smaller ones. Thus, within the Main Belt there are a few dozen with an average diameter greater than 125 miles (200 km), but tens of thousands larger than 12.5 miles (20 km) across, and millions over 1.25 miles (2 km) in diameter. Because asteroids occasionally collide and fragment, their numbers are gradually increasing but their average size is decreasing. In 2006, the largest and most massive object in the Main Belt—a body called Ceres—was also classified as a dwarf planet (pp.106–107). Consequently, Ceres is the only object that is both an asteroid and a dwarf planet.

ORBITS OF ASTEROIDS

All asteroids orbit the Sun in the same direction as the planets. Asteroids in the Main Belt have orbits that are usually close to being circular. Their orbital periods are typically between four and five years. Outside the Main Belt are other groups of asteroids with a variety of orbits. Trojan asteroids move along the same orbit as Jupiter, either about 60° in front of or behind the giant planet. Other groups, called the Amor, Apollo, and Aten asteroids, have orbits that are usually markedly elliptical and either cross or pass close to Earth's orbit. Together, these groups are classed as near-Earth asteroids. Some have the potential for future impact with Earth.

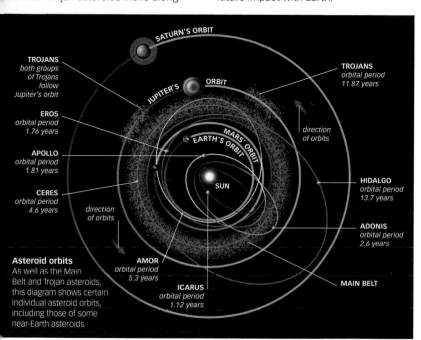

TROJANS
both groups of Trojans follow Jupiter's orbit

EROS
orbital period 1.76 years

APOLLO
orbital period 1.81 years

CERES
orbital period 4.6 years

direction of orbits

AMOR
orbital period 5.3 years

ICARUS
orbital period 1.12 years

SATURN'S ORBIT

JUPITER'S ORBIT

MARS' ORBIT

EARTH'S ORBIT

SUN

direction of orbits

TROJANS
orbital period 11.87 years

HIDALGO
orbital period 13.7 years

ADONIS
orbital period 2.6 years

MAIN BELT

Asteroid orbits
As well as the Main Belt and Trojan asteroids, this diagram shows certain individual asteroid orbits, including those of some near-Earth asteroids.

OBSERVING ASTEROIDS

Vesta (p.116) is the only asteroid visible with the naked eye, but a few dozen others can be seen through binoculars or a small telescope when they are at their brightest, including Ceres, Pallas, Juno, Eros, and Toutatis. The paths of brighter asteroids during specific periods can be found from astronomy publications and internet sites or by using astronomy software. Asteroids look like stars, and to identify a particular one it is usually necessary to make observations and photographs or sketches of the relevant part of the sky over successive nights. In this way, the asteroid can be identified as a point of light that shifts against the background star field.

Path of an asteroid
To photograph asteroids, the background stars must be kept stationary. An asteroid then forms a trail (the curved blue line) in the image as it moves relative to the star field.

»

» NOTABLE ASTEROIDS

VESTA

Situated in the Main Belt, Vesta is the brightest asteroid. After Ceres, It is also the second most massive and second largest asteroid, with an average diameter of about 330 miles (530km). Vesta is one of very few asteroids thought to have a layered internal structure, consisting of a crust, mantle, and metallic core. Occupying much of its south pole (shown below) is a huge depression, about 285 miles (460km) across, caused by an ancient impact. Fragments of Vesta's crust from this impact have landed on Earth.

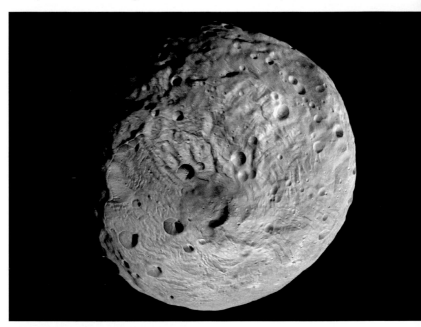

MATHILDE

Mathilde is a large Main Belt asteroid, with an average diameter of about 30 miles (50km). Its surface is darker than that of most asteroids and its average density is much lower, which suggests that it probably consists of loosely packed rock fragments.

GASPRA

About 11 miles (18km) long and orbiting close to the inner edge of the Main Belt, Gaspra has a gray surface, with some of the recently exposed crater edges being bluish and some of the older, low-lying areas appearing slightly red.

EROS

About 21 miles (34 km) long and roughly peanut-shaped, Eros is a near-Earth asteroid with an unstable orbit that crosses the orbit of Mars and approaches Earth's. In 2000, it was the first asteroid to be orbited by a space probe—the NEAR Shoemaker probe—and the following year, the probe made the first ever landing on an asteroid.

IDA

At about 36 miles (58 km) long, Ida is a large asteroid in the Main Belt. It was imaged by the Galileo probe during a flyby in 1993. The probe photographed most of Ida's heavily cratered surface and also discovered that Ida has a tiny moon, called Dactyl, which was the first confirmed satellite of an asteroid.

LUTETIA

Lutetia is a large Main Belt asteroid, with an average diameter of about 60 miles (100 km). The Rosetta probe, which passed within 1,965 miles (3,162 km) of the asteroid in 2010, discovered that its surface is covered by an unusually thick deposit of regolith (loose dust and rock fragments) up to 2,000 ft (600 m) deep.

ITOKAWA

Just 2,065 ft (630 m) long and 820 ft (250 m) wide, this near-Earth asteroid is the first from which material has been collected and returned to Earth. In 2005, the probe Hyabusa landed on Itokawa and collected surface samples. These were sent back to Earth in a capsule, which arrived in 2010. Itokawa has a rough surface studded with boulders. Scientists suspect the asteroid is made up of rock fragments that have become loosely stuck together.

TOUTATIS

A small, near-Earth asteroid, Toutatis is about 2.8 miles (4.5 km) long and sweeps past our planet every four years. In September 2004, it came as close as four times the Earth–Moon distance, and in December 2012 it will again pass close to Earth, at about 18 lunar distances.

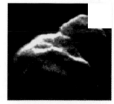

Aurora Borealis
Here, bands and ribbons of green, red, and blue can be seen at different altitudes. The colors stem from light emitted by different gases.

AURORAE

An aurora is a colorful display of light in the sky, caused by the collision of energetic particles from the Sun—the solar wind—with gas particles in Earth's atmosphere. Aurorae are most often seen in the Arctic and Antarctic regions in winter.

NORTHERN AND SOUTHERN LIGHTS

An aurora in the Northern Hemisphere is known as the Northern Lights or Aurora Borealis, whereas one in the Southern Hemisphere is called the Southern Lights or Aurora Australis. Because aurorae occur when charged particles from the Sun are funneled down toward Earth's geomagnetic poles (see opposite), they are most often seen at high latitudes, in zones situated about 10–20° from the poles. The two geomagnetic poles gradually move over time but are currently located northwest of Greenland and in eastern Antarctica. Occasionally, aurorae are seen closer to the equator during disturbances in the solar wind, such as after mass ejections from the Sun (p.69). When an aurora occurs, the light show can tower 190 miles (300 km) or more in the atmosphere and can last for hours. Often, it starts with a low, soft green arc, growing in brightness and size before red, green, blue, and violet rays appear. These then ripple out into ribbons that stretch across the sky.

THE CAUSE OF AURORAE

Aurorae appear when energetic particles from the Sun, known as the solar wind (p.67), become trapped by Earth's magnetic field. The particles are accelerated into regions above the geomagnetic poles, where they collide with gas particles in the atmosphere. As a result, the gas particles become excited, meaning their energy state is raised. As the particles revert to their normal state, they emit energy as light. The color of the light depends on which atmospheric gas is excited, the amount of energy involved, and the altitude of the excited particles. Thus, at heights below about 125 miles (200 km), excited oxygen atoms emit mainly green light; at higher altitudes they emit red light.

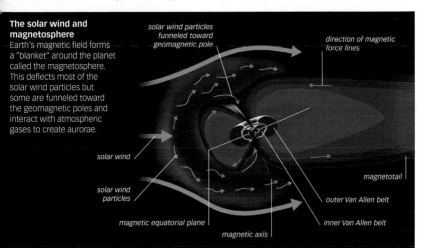

The solar wind and magnetosphere
Earth's magnetic field forms a "blanket" around the planet called the magnetosphere. This deflects most of the solar wind particles but some are funneled toward the geomagnetic poles and interact with atmospheric gases to create aurorae.

solar wind particles funneled toward geomagnetic pole

direction of magnetic force lines

solar wind

magnetotail

solar wind particles

outer Van Allen belt

magnetic equatorial plane

inner Van Allen belt

magnetic axis

OBSERVING AURORAE

The Aurora Borealis is most often seen north of 60°N, usually only between September and April because from May to August much of this region experiences periods of continuous daylight. Rarely, the Aurora Borealis is visible as far south as 35°N. The Aurora Australis is best seen between March and October and from south of 60°S, although it is also occasionally visible in, for example, New Zealand's South Island.

Where aurorae can be seen
The aurorae are most likely to be seen north of 60°N or south of 60°S, in the darker-shaded areas, although they may occasionally be visible outside these regions.

AURORAE ON OTHER PLANETS

Aurorae have been observed on most other planets, including Mars, Jupiter, Saturn, Uranus, and Neptune. A similar phenomenon (called "night glow") has also been seen on Venus, but as the planet has no detectable magnetic field, the causative mechanism of this phenomenon is not known.

Jupiter's aurorae
This composite X-ray and optical image shows Jupiter's polar aurorae (in purple). These are driven by particles from its moon Io, not from the Sun.

Sun dogs
The term "sun dogs" is given to bright patches that appear on either side of the Sun. They are caused by light being refracted by ice in the atmosphere.

ATMOSPHERIC PHENOMENA

As well as astronomical objects and aurorae, other lights may appear in the sky. Many of these are caused by sunlight reaching Earth in various indirect ways, although some are produced by manmade objects or weather phenomena.

TYPES OF LIGHTS IN THE SKY

It is helpful to be aware of unusual sources of lights in the sky, to avoid possible confusion with astronomical objects. Stationary patches or symmetrical patterns of light, seen by day or at night, are most often ice haloes or related phenomena caused by light refracted through ice crystals in the atmosphere. Silvery strands of light that shine in the sky after sunset are usually noctilucent clouds. A triangular glow of light near the horizon often turns out to be a nonatmospheric phenomenon called zodiacal light. Other phenomena that can cause moving lights and flashes across the sky include meteors or shooting stars (p.112), aircraft, satellites, orbiting spacecraft, and the International Space Station, while large light flashes are usually electrical discharges due to thunderstorms.

ICE HALOES

Atmospheric haloes are caused by ice crystals high in Earth's atmosphere refracting light. Light from either the Sun or Moon (that is, sunlight reflected by the Moon) can cause such haloes. The most common form of halo is a circle of light with a radius of 22° that is visible around the Moon or Sun. Also present may be patches of light, called moon dogs or sun dogs (also known as parhelia), arcs, and circles of light that seem to pass directly through the Sun or Moon.

Moon halo and Moon dogs
This photograph taken in Arctic Canada shows several refraction phenomena. These include a halo around the Moon and two patches of light just outside the halo, called moon dogs.

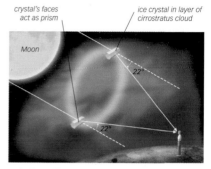

Halo formation
A halo forms when ice crystals in the atmosphere refract moonlight or sunlight through an angle of 22° to the observer. This occurs when light rays pass through two faces of an ice crystal.

NOCTILUCENT CLOUDS

Clouds at extremely high altitudes—at around 50 miles (80 km) high—can shine at night by reflecting sunlight long after the Sun has set or before it has risen. These noctilucent (night-shining) clouds are most often seen between latitudes 50° and 65° north and south, from May to August in the Northern Hemisphere and from November to February in the Southern Hemisphere.

Night-shining clouds
Noctilucent clouds are silvery-blue and usually appear as interwoven streaks. They are only ever seen near the horizon against the background of a partly lit sky.

ZODIACAL LIGHT

The zodiacal light is a faint glow that is sometimes visible in the eastern sky before dawn or in the west after sunset. A related phenomenon, called the gegenschein (German for "counterglow"), is sometimes visible on a dark night as a spot on the celestial sphere directly opposite the Sun's position in the sky. Both phenomena are caused by sunlight reflected off interplanetary dust.

Zodiacal light
The zodiacal light is most distinct just before dawn in fall, far from any light pollution. It forms a rough triangle near the eastern horizon.

Gegenschein
This faint, circular glow, 10° across, is most often spotted at midnight, in an area above the southern horizon (for Northern Hemisphere viewers).

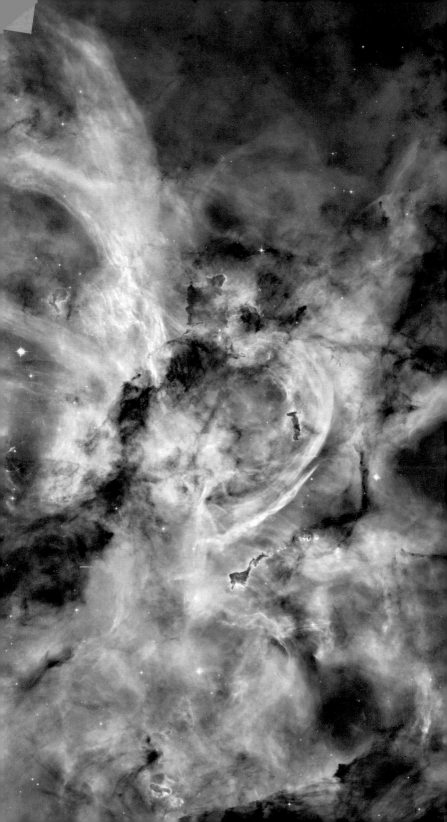

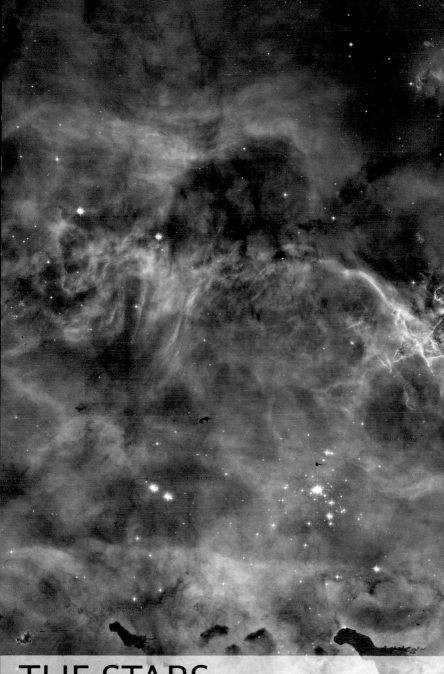

THE STARS
AND BEYOND

The Carina Nebula, a region where new stars are forming and massive old stars are dying

WHAT ARE STARS?

Look up on a clear, dark night and the first thing you see is a sky full of sparkling stars. Each one is a vast sphere of gas, blazing brightly due to the enormous amounts of energy being released by nuclear reactions deep within its heart.

STAR FORMATION

Stars form within enormous clouds of gas and dust strewn throughout space. If a nearby star disturbs these clouds, or if the blast from an exploding star ripples through them, they may collapse, triggering the process of star formation. As part of a gas cloud begins to collapse, a clump forms and grows gradually as more material falls onto it. When this clump has grown sufficiently massive, the temperature at its center is so great that hydrogen nuclei fuse together, beginning the nuclear reactions that cause stars to shine.

The Eagle Nebula
This beautiful region is a birthing ground for stars. Hot young stars at the heart of the Eagle Nebula cause the surrounding gas to glow.

STAR SIZES

When you look up at the night sky, the stars all look like tiny pinpoints of light. In fact, stars have a huge range of sizes—they only appear the same size because of their immense distance from us. Our Sun is a staggering 0.87 million miles (1.4 million km) in diameter, yet it is decidedly small compared to some other stars. For example, the bright star Betelgeuse, easily visible to the naked eye in the constellation Orion, is 800 times larger than our Sun.

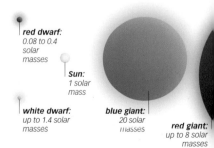

red dwarf:
0.08 to 0.4 solar masses

Sun:
1 solar mass

white dwarf:
up to 1.4 solar masses

blue giant:
20 solar masses

red giant:
up to 8 solar masses

Stellar masses
The least massive stars may only contain 1/100 the mass of our own Sun, while the most massive stars may possess well over 100 solar masses.

THE LIVES OF STARS

Once a star has emerged from its maternal cloud of gas and dust, it will spend the majority of its life converting hydrogen to helium through the process of nuclear fusion. Astronomers call stars in this relatively stable part of their lives "main-sequence" stars. Our Sun is currently a main-sequence star.

Low-mass stars, such as red dwarfs, use up their hydrogen fuel slowly, so they live for extraordinarily long periods. High-mass stars, such as the enormous supergiant stars, use up their energy rapidly, meaning that they live for much shorter periods. When a star has converted all its hydrogen to helium, it can begin to enter the final stages of its life.

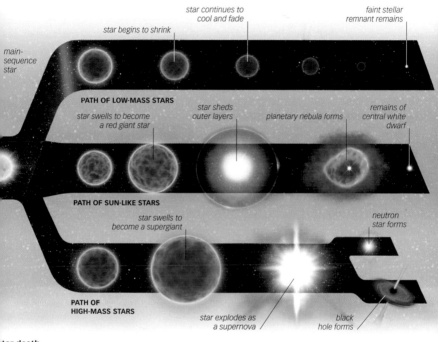

main-sequence star

star begins to shrink

star continues to cool and fade

faint stellar remnant remains

PATH OF LOW-MASS STARS

star swells to become a red giant star

star sheds outer layers

planetary nebula forms

remains of central white dwarf

PATH OF SUN-LIKE STARS

star swells to become a supergiant

neutron star forms

PATH OF HIGH-MASS STARS

star explodes as a supernova

black hole forms

Star death

Not all stars die in the same way. Their fate is linked to how massive they are when they run out of their nuclear fuel. The smallest stars tend to fade away relatively slowly, whereas the most massive stars explode violently as supernovae.

SPECTRAL CLASSES

The light from each star is imprinted with information about its composition. This information can be studied by examining chemical "fingerprints" in the spectrum of colors that make up a star's light. By identifying these fingerprints in the light from different stars, astronomers are able to classify stars into types. Some of these "spectral classes" are listed here.

SPECTRAL CLASSIFICATION OF STARS

Type	Color		Average temperature
O	●	Blue	80,000°F (45,000°C)
B	●	Bluish white	55,000°F (30,000°C)
A	●	White	22,000°F (12,000°C)
F	○	Yellowish white	14,000°F (8,000°C)
G	○	Yellow	12,000°F (6,500°C)
K	●	Orange	9,000°F (5,000°C)
M	●	Red	6,500°F (3,500°C)

MULTIPLE, VARIABLE, AND NEARBY STARS

All of the stars in the night sky that we can see with the naked eye are part of the Milky Way. Some are alone, like our Sun, but others have companions. The brightness of some stars fluctuates over time—these are known as variable stars.

MULTIPLE STARS

Most stars in our galaxy have one or more companion stars. These multiple star systems may consist of a simple pair of stars—a binary star—or several sets of stars, all orbiting a common center of mass. Some stars simply look as if they are very close to one another in the night sky because they lie along the same line of sight from Earth. These are known as optical double stars.

Epsilon Lyrae
The multiple star system Epsilon Lyrae is best observed with a small telescope. It is a double binary system, with two pairs of stars orbiting each other.

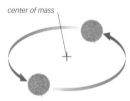

center of mass

Equal mass
In binaries with stars of equal mass, the common center of mass lies midway between the stars.

Unequal mass
If one star in a binary system is more massive than the other, the center of mass lies closer to the higher-mass star.

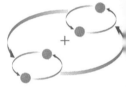

Double binary
In a double binary system, each star orbits its companion, and the two pairs orbit a single center of mass.

VARIABLE STARS

Many of the stars visible in the night sky do not have a steady brightness. Some occasionally dip in brightness while others slowly pulsate. These stars are known as "variable" stars. Some variable stars—such as the star Algol in the constellation Perseus—have a companion that from our perspective moves behind and in front of them, causing their overall brightness to change (see eclipsing binaries). Pulsating variable stars, such as Cepheid variables, are stars that repeatedly puff up and then contract, causing their brightness to go up and down.

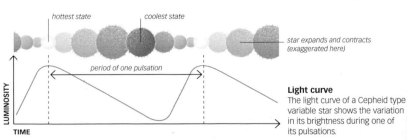

hottest state   coolest state

star expands and contracts
(exaggerated here)

period of one pulsation

LUMINOSITY

TIME

Light curve
The light curve of a Cepheid type variable star shows the variation in its brightness during one of its pulsations.

ECLIPSING BINARIES

Some stars belong to multiple star systems, whose total brightness appears to dip intermittently. These stars are known as eclipsing binaries. The fluctuation in the brightness of an eclipsing binary occurs when one star in the system moves in front of another, creating an eclipse. There is a slight dimming when the fainter star is eclipsed and a more significant dimming when the brighter star is eclipsed.

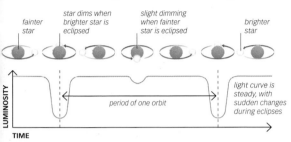

Multiple and variable
Eclipsing binaries are generally both multiple and variable stars—they are multiple because the system contains more than one star and variable because of their changing brightness.

NEARBY STARS

Look up on a clear, dark night and your eyes are met with the sight of hundreds of sparkling stars. These are the stars that inhabit our part of the Milky Way galaxy. Some stars are so bright that they appear close to us, when in reality they are relatively far away. Rigel, for example, is a bright star easily visible to the naked eye in Orion, yet it lies over 700 light-years away. Other bright stars in the night sky really are nearby; Procyon, for example, is just 11 light-years away.

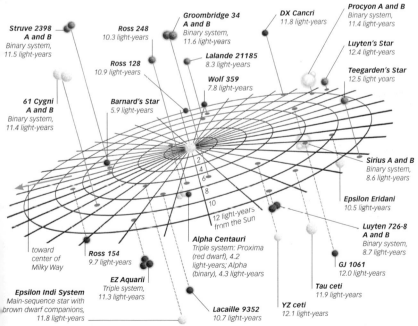

Our stellar neighborhood
This map shows the positions of some of the nearest stars to the Sun. For example, Sirius, the brightest star in the night sky, is 8.6 light-years away, while the closest star to the Sun is Proxima Centauri, just 4.2 light-years distant.

Key to star types
- ● Red dwarf
- ● Yellow main
- ○ White main
- ○ White dwarf

NEBULAE

In most galaxies, the space between stars is permeated by gas and dust. This material is more dense in some places than in others, creating vast gas and dust clouds known to astronomers as nebulae. There are several types of nebulae.

EMISSION NEBULAE

Stars are born within these vast clouds of dust and gas. Often the bright, young stars that are created generate strong stellar winds, which carve out bubbles and caverns in the nebula. At the same time, the radiation from these stars excites the interstellar gas around them. As the gas is excited, it begins to emit light, creating what astronomers call an emission nebula. Some emission nebulae—such as the Lagoon Nebula and the Orion Nebula—can be seen with just a pair of binoculars. A small telescope will begin to reveal their beautiful swirling shapes, as well as the clusters of newborn stars that cause them to shine.

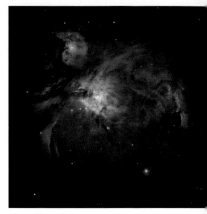

The Orion Nebula
The vast glowing form of M42, commonly known as the Orion Nebula, is arguably one of the most spectacular emission nebulae in the night sky.

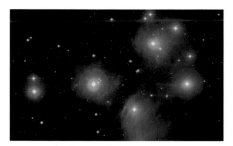

The Pleiades
This open star cluster is visible with the naked eye. However, the reflection nebula that accompanies it is tricky to see.

REFLECTION NEBULAE

Not all nebulae shine due to gas glowing within them. Some scatter and reflect the light from nearby stars back toward Earth. These ethereal nebulae are known as reflection nebulae. The dust spread throughout a reflection nebula is particularly good at scattering bluer wavelengths of starlight, which is why these nebulae typically look blue in long-exposure images.

OBSERVING NEBULAE THROUGH A TELESCOPE

Some bright nebulae can be glimpsed as misty patches of light with the naked eye, while other nebulae require careful observation from a dark-sky site using a telescope. However, most celestial objects, including nebulae, look very different through an eyepiece from how they appear in the beautiful images on these pages. Our eyes are not sensitive enough to see the detail and colors that cameras pick up by gathering light for long periods of time.

A celestial sketch
This beautiful sketch shows how the glowing clouds of the Orion Nebula appear to the eye through a large amateur telescope.

DARK NEBULAE

Many of the dense, billowing clouds of dust and gas in our galaxy do not give off light that we can see—these are dark nebulae. Such nebulae appear silhouetted against bright backdrops, such as sparkling star fields (as in the case of the Coalsack Nebula) or an emission nebula (as in the case of the Horsehead Nebula).

The Horsehead Nebula
The unmistakable Horsehead Nebula is seen here in front of a glowing patch of nebulosity in the constellation Orion.

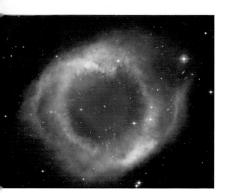

The Helix Nebula
The Helix Nebula in the constellation Aquarius clearly shows shells of glowing material—roughly 2.5 light-years in width—ejected by a dying star.

PLANETARY NEBULAE

When a star like the Sun starts to near the end of its life, it begins to shed shells of material into space. Eventually, it ejects its atmosphere into space, creating a beautiful planetary nebula. At the heart of the nebula a hot, energetic stellar core remains, which causes the surrounding material to shine. The term planetary nebula is slightly misleading; the name was adopted long ago by astronomers who thought that these nebulae looked like the disks of planets through their early telescopes.

SUPERNOVA REMNANTS

Supernova remnants are formed when a massive star dies, producing a spectacularly violent supernova explosion (p.132). These energetic events shred what is left of the star, blasting it out into space at tremendous speeds. Huge glowing filaments of material are created as the explosion crashes violently into the surrounding gas. The supernova remnant then expands over time, forming vast tendrils of gas that stretch far across space.

The Veil Nebula
With a good telescope and dark skies, you can spot some of the brighter patches of gas within the supernova remnant known as the Veil Nebula.

»

» NEBULAE TO OBSERVE

THE ORION NEBULA

Located within the constellation Orion, the Hunter, the Orion Nebula (M42) is one of the easiest and most interesting nebulae to observe. It is a star-forming region located around 1,350 light-years away. A cluster of young stars at its center, known as the Trapezium, causes the nebula's gas to glow. You can see the Orion Nebula with binoculars, but it is best viewed through a small telescope from a dark-sky site.

Swirling streams of gas
The bright center of the Orion Nebula and the sweeping streams of gas around it are easily visible through a small telescope.

THE CARINA NEBULA

Situated roughly 7,500 light-years away, in the constellation Carina, the Carina Nebula is a spectacular star-formation site. Stars are being born throughout this vast, billowing cloud of gas and dust and the bright young stars that already inhabit it make it glow. This nebula stretches approximately 100 light-years across space. It appears in our night sky within a particularly rich region of the Milky Way.

Dark dust
Dark lanes of dust and gas cross the brighter portions of the Carina Nebula in this long-exposure image.

THE CRAB NEBULA

The Crab Nebula (M1) is a supernova remnant located in the constellation Taurus, the Bull. The supernova that created this stellar skeleton occurred in 1054 CE, and it is thought that the supernova was observed by astronomers around the world at that time. To locate the Crab Nebula, first find the Hyades star cluster that marks the "head" of Taurus. Then follow the Bull's "horn" from the star Aldebaran to the star Zeta (ζ) Tauri. The Crab Nebula is just over a degree away from Zeta Tauri.

A colorful crab
The Crab Nebula is a target that is best suited to larger telescopes, which will show it as an elliptical smudge of light.

THE LAGOON NEBULA

The Lagoon Nebula (M8) is a region of star formation roughly 4,000 light-years away in the constellation Sagittarius. Within the nebula is an open cluster of stars, catalogued as NGC 6530. Radiation from bright stars close to the nebula causes it to shine. The Lagoon Nebula can be seen using a telescope or a good pair of binoculars. A small telescope will show it as a misty glow nestled within the Milky Way star fields.

Pink clouds
Long-exposure images of the Lagoon Nebula show its beautiful glowing pink gas clouds and the sparkling collection of stars that accompany it.

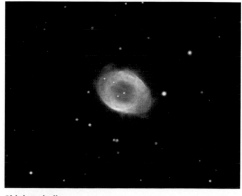

Infrared lagoon
This detailed infrared image reveals vast wisps of dust and gas in the Lagoon Nebula. It was captured by NASA's Wide-Field Infrared Survey Explorer.

THE RING NEBULA

Alongside the Dumbbell Nebula (M27) in the constellation Vulpecula, the Ring Nebula (M57) is arguably one of the finest planetary nebulae in the northern celestial hemisphere. This small glowing ring of light—about 2,300 light-years away—is the remains of a star similar to our own Sun. The nebula is located in the constellation Lyra, the Lyre, between the stars Gamma (γ) and Beta (β) Lyrae. It is an excellent target for a small telescope. The gas that forms the nebula is roughly one light-year across.

Shining shells
This image reveals intricate shells of gas surrounding the Ring Nebula. Through a telescope the nebula looks like a faint gray ring of light.

EXOTIC CELESTIAL OBJECTS

The Universe can be an unbelievably violent and extreme place. Nowhere is this seen more clearly than with the many exotic objects that can be found scattered throughout the cosmos—from transient stellar explosions to all-consuming black holes.

SUPERNOVAE

The deaths of massive stars are marked by powerful explosions known as "type II" supernovae. When a supergiant star runs out of hydrogen fuel, it starts to produce successively heavier elements in its core. At first, the increasingly massive core is held up by internal pressure. But once iron forms, the core can no longer be supported and it collapses. As the surrounding material comes crashing down, it rebounds off the collapsed core, creating a vast shock wave that bursts up through the star. The resultant explosion is tremendously violent, releasing enormous amounts of energy, including light.

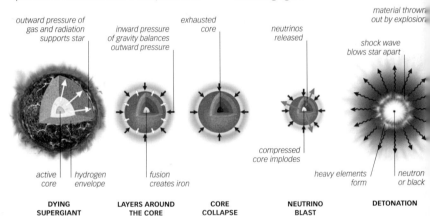

outward pressure of gas and radiation supports star

inward pressure of gravity balances outward pressure

exhausted core

neutrinos released

material thrown out by explosion

shock wave blows star apart

active core *hydrogen envelope*

fusion creates iron

compressed core implodes

heavy elements form *neutron or black*

DYING SUPERGIANT **LAYERS AROUND THE CORE** **CORE COLLAPSE** **NEUTRINO BLAST** **DETONATION**

Supernova formation
The formation of a supernova occurs in several stages at the end of a massive star's life. These vast blasts are responsible for enriching space with the heavy elements that are needed to create much of the everyday world around us.

SUPERNOVA SPOTTING

Astronomers search for supernovae by constantly monitoring distant galaxies for any sign of a bright, new point of light. Many supernovae have been discovered by amateur astronomers carrying out routine searches. The images on the right show the brightening caused by a recent supernova, SN2011DH, in the Whirlpool Galaxy (M51).

BEFORE SN2011dh

DURING SN2011dh

GAMMA-RAY BURSTS

Some of the most distant objects observed by modern space telescopes have been extraordinarily energetic events known as gamma-ray bursts. These come in two distinct types: some are long, while others are over in a flash. Astronomers currently think that short gamma-ray bursts probably form when two neutron stars—or perhaps even a neutron star and a black hole—collide. On the other hand, long gamma-ray bursts are likely to be the result of the catastrophic explosion of a truly enormous star.

Brilliant blasts
The powerful blast of radiation from a gamma-ray burst (shown here in an artwork) can be seen across vast distances. Space telescopes, such as NASA's Swift satellite, constantly monitor the cosmos for signs of these violent events.

PULSARS AND NEUTRON STARS

As a massive star's core collapses in a supernova explosion, protons and electrons are crushed together so strongly that they make neutrons. In the ensuing maelstrom, a ball of neutrons known as a neutron star is born. This bizarre object may only be a few tens of miles across. Neutron stars have extremely strong magnetic fields and may become pulsars (below).

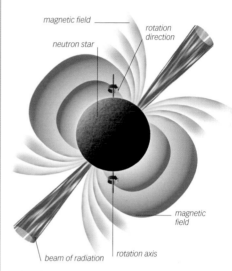

magnetic field

rotation direction

neutron star

magnetic field

beam of radiation

rotation axis

Pulsars
At the heart of a pulsar is a spinning neutron star with a powerful magnetic field. Beams of radiation shoot out from the magnetic poles of the neutron star; if they sweep in Earth's direction, pulses of radiation can be detected.

ACTIVE GALAXIES

At the center of many galaxies is a supermassive black hole. In active galaxies this black hole consumes the surrounding material and is encircled by a vast disk of gas and dust that is very hot. Together with energetic jets emitted from the black hole, this means that active galaxies emit colossal amounts of radiation.

Active galactic nucleus
The orientation of some active galaxies is such that we can see the glowing, energetic region that surrounds the black hole, known as the active galactic nucleus. Such an active galaxy is called a quasar. Quasars are very distant and are probably the centers of older active galaxies.

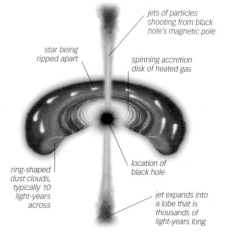

jets of particles shooting from black hole's magnetic pole

star being ripped apart

spinning accretion disk of heated gas

ring-shaped dust clouds, typically 10 light-years across

location of black hole

jet expands into a lobe that is thousands of light-years long

STAR CLUSTERS

Star clusters of all shapes and sizes can be found throughout the Milky Way galaxy. Some, known as globular clusters, are dense balls of thousands of stars; others, called open globular clusters, are looser groupings of stars.

GLOBULAR CLUSTERS

Globular clusters are vast spheres of thousands, or even millions, of stars. They are among the oldest objects in our galaxy—some contain stars that are 12 billion years old. It is thought that there are roughly 150 globular clusters scattered around the disk of the Milky Way galaxy. If it was possible to stand on a planet within a globular cluster, the night sky would be filled with the light from thousands of bright stars.

Omega Centauri
Home to 10 million stars, this is one of the most spectacular globular clusters in the night sky. It is big and bright enough to be seen with the naked eye, but a telescope will show it more clearly. Some astronomers think Omega Centauri may be the remains of a dwarf galaxy.

OPEN CLUSTERS

Open clusters are loose star clusters formed in vast, nebulous clouds of dust and gas. Once the stars materialize out of a star-forming nebula, they drive away the surrounding dust and gas, creating a sparkling open cluster. Open clusters are generally found within the disk and spiral arms of our galaxy, so they often appear in our night skies set against the rich star fields of the Milky Way.

The Butterfly Cluster
The beautiful open cluster M6, the Butterfly Cluster, sits in the constellation Scorpius. It is composed of about 80 stars and is a magnificent sight through a pair of binoculars or a small telescope.

OBSERVING STAR CLUSTERS

Star clusters are arguably among the most rewarding deep-sky objects to observe in the night sky. Some of the brighter star clusters are even visible to the naked eye. A good pair of binoculars or a small telescope will be suitable for tracking down many star clusters. Through the eyepiece of a small telescope, open clusters, such as the Beehive Cluster or the Pleiades, look like glittering diamonds scattered against the blackness of space. Globular clusters tend to appear as fuzzy balls of stars.

Binoculars
Many brighter globular clusters will appear as small, round smudges of light when viewed through a good pair of binoculars.

Small telescope
Away from light pollution and with clear skies, even a small telescope will show the most prominent globular clusters as misty balls of light.

Naked-eye clusters
A few of the biggest and brightest open clusters in the night sky—such as M44 (pictured)—can be seen with just the naked eye.

Large telescope
Viewed through a large amateur telescope, globular clusters can appear as sparkling, apparently three-dimensional, spheres of many stars.

CLUSTERS IN THE MILKY WAY

Open star clusters are typically found nestled within the disk and spiral arms of the Milky Way, where there are large amounts of the dust and gas that are needed for star formation. On the other hand, globular clusters occupy the region around the disk of the galaxy, which astronomers call the galactic halo.

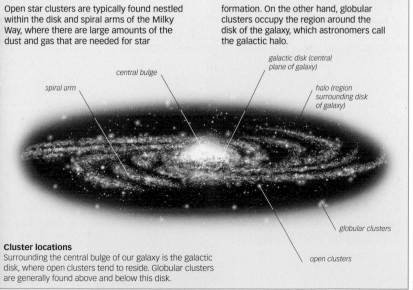

central bulge

spiral arm

galactic disk (central plane of galaxy)

halo (region surrounding disk of galaxy)

globular clusters

open clusters

Cluster locations
Surrounding the central bulge of our galaxy is the galactic disk, where open clusters tend to reside. Globular clusters are generally found above and below this disk.

»

» STAR CLUSTERS TO OBSERVE

M22

This is one of the most beautiful globular clusters to observe in the sky, partly due to its location. It is found in the constellation Sagittarius, among the rich star fields of the Milky Way. M22 is not difficult to find. First, draw an imaginary line between the stars Sigma (σ) and Mu (μ) Sagitarii. You can then find M22 about halfway along this line.

A ball of stars
M22 has a magnitude of about 5, which means that it can be seen with the naked eye. However, it is much more suited to telescopic observation.

THE BEEHIVE CLUSTER

Also known as Praesepe or M44, this is an open cluster in the constellation Cancer, the Crab. With a magnitude of 3.1, it is an easy naked-eye target from a dark sky site. To the naked eye, the Beehive Cluster appears as a small, faint patch of light. However, if you use binoculars or a small telescope it will be revealed as a loose group of many glittering stars.

A stellar swarm
The Beehive Cluster can be spotted with the naked eye from a dark-sky site, but it is particularly beautiful through a pair of binoculars.

47 TUCANAE

The globular cluster 47 Tucanae is one of the jewels of the southern celestial hemisphere. It has a magnitude of 4 and is roughly 200 light-years across. This cluster is easy to find in the night sky, as it is very close to the Small Magellanic Cloud. To find 47 Tucanae, imagine a line between the stars Zeta (ζ) Tucanae and Beta (β) Hydri. You will find 47 Tucane a little over halfway along this line.

A glittering globe
This beautiful, long-exposure image shows the entirety of the magnificent globular cluster 47 Tucanae.

Up close
This image reveals the tightly packed stars located at the heart of the globular cluster 47 Tucanae.

THE PLEIADES

This famous open star cluster is also known as M45 or the Seven Sisters. It is located about 440 light-years away, in the constellation Taurus. The brightest stars in the Pleiades are easily visible with the naked eye, while the cluster itself measure more than one degree across on the sky. Long-exposure images of M45 reveal wisps of nebulosity around several of the cluster's bright stars—this is a reflection nebula, which is unrelated to the stars.

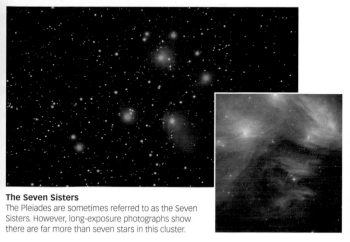

The Seven Sisters
The Pleiades are sometimes referred to as the Seven Sisters. However, long-exposure photographs show there are far more than seven stars in this cluster.

Dusty cloud
This infrared image from NASA's Spitzer Space Telescope shows enormous wisps of dust around the stars within the Pleiades star cluster.

OMEGA CENTAURI

Omega Centauri is the grandest globular cluster in the night sky. With a magnitude of 3.7, it can be seen easily with the naked eye. It sits in the constellation Centaurus, the Centaur, and is composed of around 10 million stars. The cluster is thought to reside about 17,000 light-years from our Solar System and measures an impressive 150 light-years in diameter.

A sparkling sphere
The many stars of the globular cluster Omega Centauri are an incredible sight to behold through a large amateur telescope.

THE DOUBLE CLUSTER

The Double Cluster is an interesting deep sky target made of two open star clusters, catalogued as NGC 869 and NGC 884. It sits in the constellation Perseus, within a bright patch of the Milky Way that passes through this region of the sky. It is visible with the naked eye just over halfway between the bright stars Mirphak (in Perseus) and Ruchbah (in Cassiopeia).

Twinkling twins
The Double Cluster is a wonderful target for a small telescope, which will show the two clusters sitting side by side.

THE MILKY WAY

Travel away from towns and cities and look at the sky on a clear, dark night, and the faint, misty glow of the Milky Way becomes visible in the sky. Amid a Universe of other galaxies, this galaxy is our home in space. The Milky Way contains several hundred billion stars besides the Sun, around which the planets of our Solar System orbit.

The structure of the Milky Way

We may never get a view of our galaxy from far enough away to see its overall structure. However, by studying the distribution of stars in the Milky Way, astronomers are beginning to map out its exact shape. This diagram shows our current understanding of the Milky Way's geography.

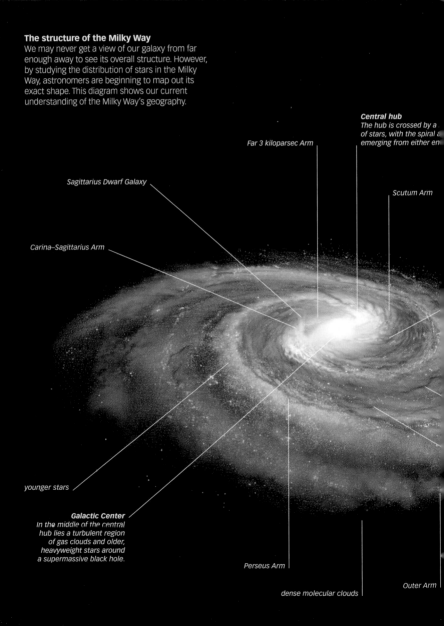

Central hub
The hub is crossed by a of stars, with the spiral a emerging from either en

Far 3 kiloparsec Arm

Sagittarius Dwarf Galaxy

Scutum Arm

Carina–Sagittarius Arm

younger stars

Galactic Center
In the middle of the central hub lies a turbulent region of gas clouds and older, heavyweight stars around a supermassive black hole.

Perseus Arm

dense molecular clouds

Outer Arm

OUR GALAXY

The Milky Way is the name given to the galaxy we live in. Astronomers think that there are somewhere between 200 and 400 billion stars in the Milky Way, one of which is the star we call the Sun. The Milky Way is a spiral galaxy with a pair of sweeping arms extending from a central bulging bar of older stars. The distance from one side of the galaxy to the other is around 100,000 light-years, while the disk of the Galaxy (containing the spiral arms) is roughly 1,000 light-years thick. The spiral arms themselves contain many glowing gas clouds, where stars are being born. Scattered around the Milky Way is a swarm of globular clusters—vast balls of ancient stars.

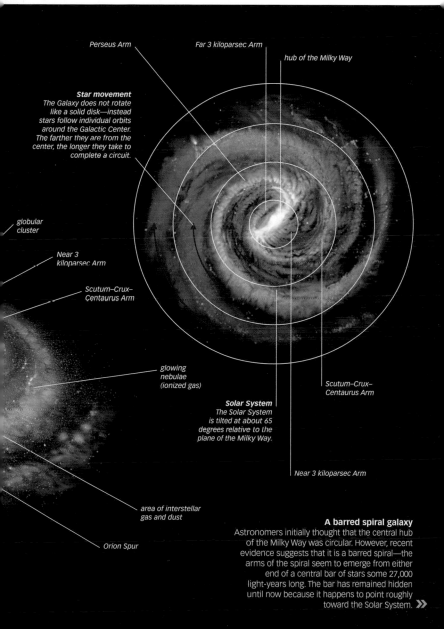

Perseus Arm

Far 3 kiloparsec Arm

hub of the Milky Way

Star movement
The Galaxy does not rotate like a solid disk—instead stars follow individual orbits around the Galactic Center. The farther they are from the center, the longer they take to complete a circuit.

globular cluster

Near 3 kiloparsec Arm

Scutum–Crux– Centaurus Arm

glowing nebulae (ionized gas)

Scutum–Crux– Centaurus Arm

Solar System
The Solar System is tilted at about 65 degrees relative to the plane of the Milky Way.

Near 3 kiloparsec Arm

area of interstellar gas and dust

Orion Spur

A barred spiral galaxy
Astronomers initially thought that the central hub of the Milky Way was circular. However, recent evidence suggests that it is a barred spiral—the arms of the spiral seem to emerge from either end of a central bar of stars some 27,000 light-years long. The bar has remained hidden until now because it happens to point roughly toward the Solar System. »

>> OBSERVING THE MILKY WAY

The Milky Way is best observed on a clear night, away from the light pollution created by towns and cities. Visible to the naked eye as a long swathe of faint light with numerous dark patches scattered across it, the Milky Way is a dramatic sight, arching across the night sky. The dark regions within the Milky Way are vast clouds of dust and gas, which are silhouetted against the combined light of the Milky Way's many stars. A good pair of binoculars or a small telescope are ideal for scanning the spectacularly rich star fields of the Milky Way and the many interesting deep sky objects they contain.

The Milky Way from Earth
We look at the Milky Way from a vantage point within one of its spiral arms. What we see of the Milky Way are its disk, spiral arms, and central bulge.

The Cygnus Rift
This is a cloud of dust and gas within the Milky Way. It is visible from dark sky sites, silhouetted against the star fields of Cygnus.

THE LOCAL BUBBLE

The Solar System lies within a region of hot gas, known as the Local Bubble, which is bounded by colder, denser gas. Many other stars are also found within this region, which is thought to be about 300 light-years wide, though the area is relatively empty compared to other areas of space. The Local Bubble was probably created several million years ago by the powerful blast from a supernova (p.132).

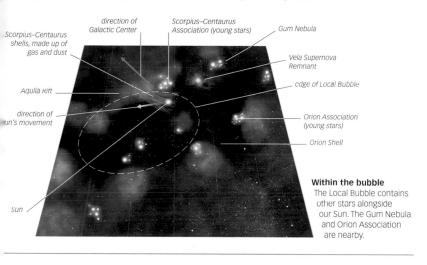

Scorpius–Centaurus shells, made up of gas and dust

direction of Galactic Center

Scorpius–Centaurus Association (young stars)

Gum Nebula

Vela Supernova Remnant

Aquila Rift

edge of Local Bubble

direction of Sun's movement

Orion Association (young stars)

Orion Shell

Sun

Within the bubble
The Local Bubble contains other stars alongside our Sun. The Gum Nebula and Orion Association are nearby.

EXTRASOLAR PLANETS

The family of planets orbiting the Sun, including our Earth, are not the only worlds to exist in the Milky Way. Since the early 1990s, astronomers have been discovering many so-called extrasolar planets orbiting other stars in our galaxy. Many of these planets are large, gaseous worlds but some are rocky. More than 600 of these alien worlds have been found, and that number is rapidly increasing thanks to studies made by missions such as the Kepler space telescope.

The Kepler Mission
The Kepler space telescope studies a specific swathe of our galaxy, looking for any stars that show a slight dimming in brightness caused by a planet passing in front of them.

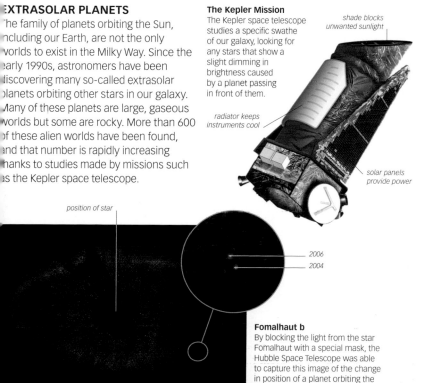

shade blocks unwanted sunlight

radiator keeps instruments cool

solar panels provide power

position of star

2006
2004

Fomalhaut b
By blocking the light from the star Fomalhaut with a special mask, the Hubble Space Telescope was able to capture this image of the change in position of a planet orbiting the star, between 2004 and 2006.

GALAXIES

When we look into the night sky we do so from within our own galaxy, the Milky Way, which is home to several hundred billion stars. There are countless other galaxies of various sizes and shapes scattered throughout the Universe. Some are similar to ours, but others defy even the wildest imaginations.

GALAXY FORMATION

Current theories suggest that the first galaxies to form were irregular, relatively small groupings of stars. Over time, these smaller galaxies merged together to create larger ones, including our own Milky Way. It is thought that further collisions and mergers between these larger galaxies produced some of the vast elliptical galaxies we see scattered throughout space today.

Early galaxies
Tiny "blobs" of light can be seen in this detailed image from the Hubble Space Telescope. These are some of the earliest galaxies known.

Galactic building blocks
Early galaxies such as this one may have merged to create larger galaxies. Some are as small as one-thousandth the size of the Milky Way.

ELLIPTICAL GALAXIES

Elliptical galaxies do not have spiral arms and are largely devoid of the gas and dust seen in other types of galaxies. Ellipticals are generally home to old stars and, as their name suggests, are elliptical in shape.

ESO 325-G004
This enormous elliptical galaxy appears as a blurred ball of light. What we see is the combined glow of millions of stars from over 460 million light-years away.

LENTICULAR GALAXIES

Lenticular galaxies are roughly spherical but with a disk of stars and gas around the nucleus, which is made up of older stars. They have disks, like spiral galaxies, but do not have spiral arms or regions of star formation.

NGC 5866
This edge-on view of NGC 5866 reveals the classic shape of a lenticular galaxy. As well as a clear disk of stars, the galaxy boasts a swirling lane of dust.

COLLIDING GALAXIES

Some galaxies are in the process of intermingling and sometimes merging with each other. These interacting, colliding galaxies often have bizarre shapes, created as their gravity twists and deforms them. They may also have long tails of stars thrown out from the collision.

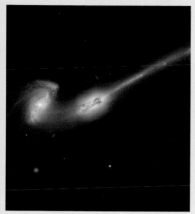

The Mice galaxies
Seemingly frozen in time, these two galaxies are in the process of merging. This detailed image from the Hubble Space Telescope shows vast streams of stars drawn out by the gravitational interactions between the two galaxies.

IRREGULAR GALAXIES

Galaxies that don't have a clear form (spiral, elliptical, or lenticular) are called irregular galaxies. They are often rich in star clusters and regions of star formation. By their very nature, irregular galaxies come in many different shapes and sizes. Their unusual features are often the result of interactions with other galaxies.

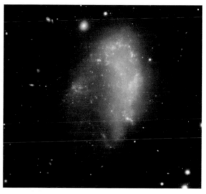

NGC 1427A
This irregular galaxy is littered with star-forming regions (shown in red). The birth of new stars is probably a result of NGC 1427A interacting with nearby gas and other galaxies.

SPIRAL GALAXIES

Of all the different types of galaxy spirals are arguably the most recognizable. These beautiful celestial swirls are typically composed of a central bulge (or nucleus) of older stars surrounded by a flat disk of stars, gas, and dust. The sweeping spiral arms are contained in the disk of a spiral galaxy. The spiral arms contain bright young stars and star-forming nebulae. It is thought that the spiral shape is caused by a density wave that travels slowly around the disk of dust and gas and triggers the birth of new stars.

huge lanes of dust weave through the spiral arms

spiral arms contain young stars and bright star clusters

nucleus or central bulge of galaxy

The Pinwheel Galaxy
The Pinwheel Galaxy is a spectacular example of a spiral galaxy located 25 million light-years away. It is much larger than the Milky Way.

»

» GALAXIES TO OBSERVE

THE ANDROMEDA GALAXY

Of all the galaxies in the night sky the Andromeda Galaxy, M31, is the easiest and most impressive to observe. This spiral galaxy is thought to be slightly larger than the Milky Way and lies 2.5 million light-years away, meaning that it is one of the most distant objects visible to the naked eye. From a dark-sky site, the Andromeda Galaxy is easy to see without a telescope or binoculars. A telescope will show it as an ellipse of light accompanied by two smaller patches of light—the galaxies M110 and M32.

M31 and friends
This image shows the Andromeda Galaxy with its two small satellite galaxies, M110 and M32.

Rings of dust
This unusual view of the dust lanes (seen in pink) within the Andromeda Galaxy was made by observing M31 at infrared wavelengths.

THE WHIRLPOOL GALAXY

M51, more commonly known as the Whirlpool Galaxy, is a spiral galaxy lying well over 23 million light-years away in the constellation Canes Venatici. It has a magnitude of 8.4, which makes it a good target for small telescopes. The Whirlpool Galaxy can be found about 3.5 degrees away from the bright star Alkaid, in Ursa Major. It has a companion galaxy, NGC 5195, which is seen on the left here.

Celestial swirl
The Whirlpool Galaxy sits face-on to us, so that we see its central bulge and swirling spiral arms from above.

THE CIGAR GALAXY

M82, sometimes referred to as the Cigar Galaxy, is a spiral galaxy 12 million light-years away in the constellation Ursa Major. Through a small telescope M82 and its companion M81 can be seen in the same field of view. To find them, first locate the famous asterism the Big Dipper, in Ursa Major. Next, picture a line from the star Phi (φ) Ursae Majoris to the star Upsilon (υ) Ursae Majoris. If you extend this line by roughly 10.5 degrees you will find these two galaxies.

Titanic tendrils
This image shows the distinctive shape of M82. It also reveals tendrils of gas and dust (pink in this image) emanating from the galaxy.

THE LARGE MAGELLANIC CLOUD

The Large Magellanic Cloud is the larger of two notable irregular galaxies, the Magellanic Clouds, that lie close to the Milky Way. Its companion is the Small Magellanic Cloud. The Magellanic Clouds are easy to find in the night sky, because they are visible to the naked eye. The Large Magellanic Cloud sits near the border between the constellations Dorado and Mensa, while the Small Magellanic Cloud is located in Tucana.

Cloud of stars
Through a large telescope the Tarantula Nebula (below center) is visible within the Large Magellanic Cloud.

THE TRIANGULUM GALAXY

Not far from the Andromeda Galaxy lies M33 (the Triangulum Galaxy), in the constellation Triangulum. Like its neighbor, this galaxy can be seen with the naked eye from an observing site free from light pollution. It is a spiral galaxy about 2.8 million light-years away, and through a small telescope it appears as a round patch of light. The Triangulum Galaxy can be found just under two-thirds of the way between the stars Hamal (in Aries) and Mirach (in Andromeda).

Sweeping arms and stellar nurseries
Long-exposure images reveal the spiral arms of M33, as well as numerous star forming regions glowing red within it.

GALAXY CLUSTERS

Many galaxies swarm together in vast groups known as galaxy clusters. Our galaxy, the Milky Way, is part of a small gathering of galaxies called the Local Group. Galaxy clusters can themselves form groups, known as superclusters.

THE LOCAL GROUP

The Local Group is a small collection of roughly 50 galaxies, which includes the Milky Way. Several Local Group galaxies can be observed with amateur telescopes. The Andromeda Galaxy is probably the most famous Local Group galaxy in the northern hemisphere. The Large and Small Magellanic Clouds are two nearby irregular, Local Group, galaxies that can be seen with the naked eye from the southern hemisphere.

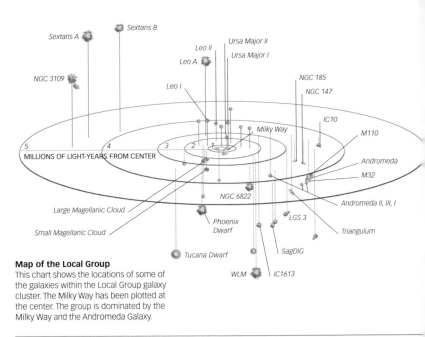

Map of the Local Group
This chart shows the locations of some of the galaxies within the Local Group galaxy cluster. The Milky Way has been plotted at the center. The group is dominated by the Milky Way and the Andromeda Galaxy.

OTHER GALAXY CLUSTERS

The Universe is filled with galaxy clusters, which range from small groups of a few galaxies to huge gatherings of thousands of galaxies. Some may be scattered over a large area while others are known to be extremely tightly packed. One of the closest big galaxy clusters to us is the Virgo Galaxy Cluster (right).

The Virgo Galxy Cluster
Some members of the Virgo Galaxy Cluster can be spotted with an amateur telescope, including the galaxies M84, M86, and M87.

DARK MATTER

Dark matter may be the greatest mystery of modern astrophysics. No one knows exactly what it is, but its presence has been detected throughout the Universe. Its mass influences the rotation of galaxies like ours, and its gravity is known to play a role in holding together huge galaxy clusters. However, dark matter doesn't emit radiation like normal matter and so cannot be "seen." One way astronomers have been able to locate it is by studying how its gravity bends light traveling through space.

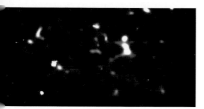

Visible matter
By observing the radiation from stars, galaxies, and gases, astronomers can build a picture of the location of ordinary visible matter (shown here in red).

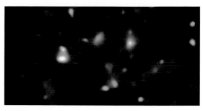

Dark matter
Astronomers chart the distribution of hidden dark matter (shown here in blue, in the same area as the image to the left) by studying how light travelling vast distances is deflected.

LARGE-SCALE STRUCTURE

By carefully measuring the distances to distant galaxies and galaxy clusters, astronomers have been able to piece together charts of the nearby Universe. To make these charts they needed to survey huge swathes of space using powerful telescopes. The results show a spectacular cosmic network of galaxies and galaxy clusters, with vast filaments of galaxies weaving around patches of empty space.

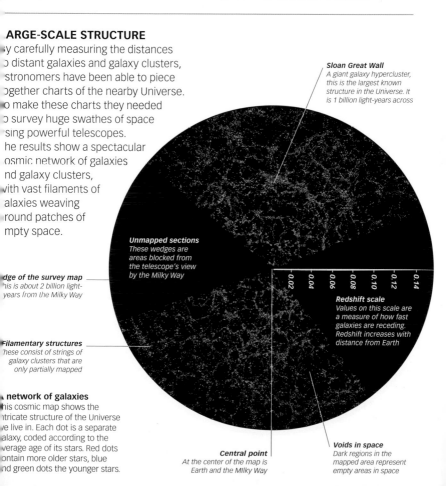

Sloan Great Wall
A giant galaxy hypercluster, this is the largest known structure in the Universe. It is 1 billion light-years across

Unmapped sections
These wedges are areas blocked from the telescope's view by the Milky Way

Edge of the survey map
This is about 2 billion light-years from the Milky Way

0.02 0.04 0.06 0.08 0.10 0.12 0.14

Redshift scale
Values on this scale are a measure of how fast galaxies are receding. Redshift increases with distance from Earth

Filamentary structures
These consist of strings of galaxy clusters that are only partially mapped

A network of galaxies
This cosmic map shows the intricate structure of the Universe we live in. Each dot is a separate galaxy, coded according to the average age of its stars. Red dots contain more older stars, blue and green dots the younger stars.

Central point
At the center of the map is Earth and the Milky Way

Voids in space
Dark regions in the mapped area represent empty areas in space

UNDERSTANDING THE UNIVERSE

Ever since humans first gazed up into the night sky, we have tried to understand the vast Universe we live in. Today, modern telescopes and space probes allow us to examine and explain some of the ways the cosmos works.

RADIATION FROM THE BIG BANG

The Universe we see around us today began 13.7 billion years ago, with a maelstrom of energy that astronomers call the Big Bang. We don't know what was before it (if there was anything) or even what triggered it, but we can find clues that point to its existence. One crucial piece of evidence for the Big Bang is the Cosmic Microwave Background (CMB)—radiation that is visible across the whole sky, lingering from early in the Universe's history.

blue areas are coldest

red areas are warmest

The CMB
This map shows the strength of background radiation over the whole sky. Variations in colour reveal minute temperature differences. The pattern is consistent with a hot gas expanding billions of times over, as predicted by Big Bang theory.

STUDYING FROM AFAR

Astronomers can study a celestial object by examining the features present in the radiation (spectrum) it emits. For example, stars emit many wavelengths. However, as their light passes through their atmospheres, specific wavelengths are absorbed by certain chemicals. This creates dark lines in their spectra known as absorption lines. Conversely, glowing emission lines in a spectrum are created when certain chemicals emit specific wavelengths of light.

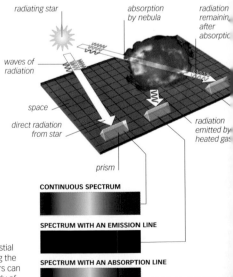

radiating star

absorption by nebula

radiation remaining after absorption

waves of radiation

space

direct radiation from star

radiation emitted by heated gas

prism

CONTINUOUS SPECTRUM

SPECTRUM WITH AN EMISSION LINE

SPECTRUM WITH AN ABSORPTION LINE

Fingerprints in light
Light holds the key to examining a variety of celestial objects across vast cosmic distances. By studying the various lines in an object's spectrum, astronomers can piece together its chemical composition. The study of spectra is called spectroscopy.

REVEALING RADIATION

Our eyes see only a small section of the electromagnetic spectrum, which we call "visible light." Astronomers study the Universe at various wavelengths to build a picture of the processes at work in the cosmos and to observe phenomena that might be invisible or difficult to observe at other wavelengths.

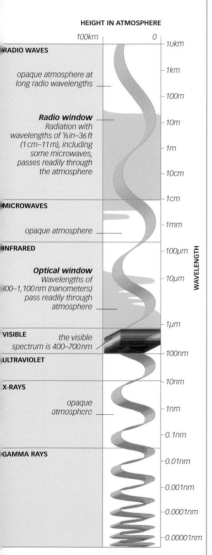

HEIGHT IN ATMOSPHERE

100km   0

10km
1km
100m
10m
1m
10cm
1cm
1mm
100μm
10μm
1μm
100nm
10nm
1nm
0.1nm
0.01nm
0.001nm
0.0001nm
0.00001nm

WAVELENGTH

RADIO WAVES

opaque atmosphere at long radio wavelengths

Radio window
Radiation with wavelengths of ⅓in–36 ft (1 cm–11 m), including some microwaves, passes readily through the atmosphere

MICROWAVES

opaque atmosphere

INFRARED

Optical window
Wavelengths of 300–1,100 nm (nanometers) pass readily through atmosphere

VISIBLE   *the visible spectrum is 400–700 nm*

ULTRAVIOLET

X-RAYS

opaque atmosphere

GAMMA RAYS

The electromagnetic spectrum
This diagram shows the wavelengths of radiation that make up the electromagnetic spectrum. Astronomers use specialist satellites and orbiting space telescopes to look at the wavelengths that can't be seen from Earth's surface.

THE FORMATION OF STRUCTURE IN THE UNIVERSE

If you look closely at the all-sky view of the Cosmic Microwave Background on the opposite page, you will notice that it is not completely smooth—there are small irregularities. Astronomers think that these slight irregularities in the distribution of matter, present at the time the CMB formed, were the seeds of the immense galactic structures we see today. Modern computer simulations have allowed cosmologists to study how these structures might have formed from those early fluctuations.

Emerging structures
These images from a computer simulation show sections of the Universe at different stages in its life, finishing with the present day.

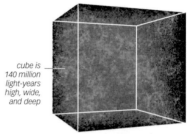

cube is 140 million light-years high, wide, and deep

500,000 YEARS OLD

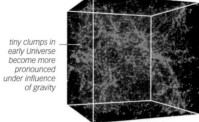

tiny clumps in early Universe become more pronounced under influence of gravity

2.3 BILLION YEARS OLD

further clumping of matter into knots (galaxy superclusters)

13.7 BILLION YEARS OLD

MONTHLY
SKY GUIDE

The Leonid meteors, which appear in the constellation Leo in mid-November each year

HOW TO USE THIS SECTION

This month-by-month guide shows the whole sky as it appears from most places on Earth throughout the year. For each month, a double-page introduction is followed by whole-sky charts for observers in the Northern and Southern Hemispheres. The charts in this section complement THE CONSTELLATIONS section, in which detailed charts show smaller areas of sky.

MONTHLY HIGHLIGHTS AND PLANET LOCATORS

The text and images on these double-page introductions highlight stars, deep-sky objects, and meteor showers that feature prominently in the month. They are intended to be used with the whole-sky charts that follow. These pages also feature a chart for locating the planets.

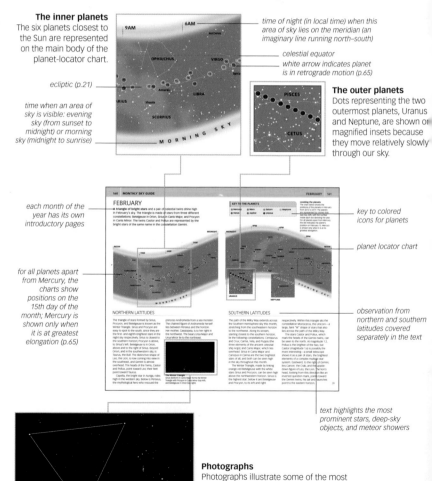

The inner planets
The six planets closest to the Sun are represented on the main body of the planet-locator chart.

ecliptic (p.21)

time when an area of sky is visible: evening sky (from sunset to midnight) or morning sky (midnight to sunrise)

time of night (in local time) when this area of sky lies on the meridian (an imaginary line running north–south)

celestial equator

white arrow indicates planet is in retrograde motion (p.65)

The outer planets
Dots representing the two outermost planets, Uranus and Neptune, are shown on magnified insets because they move relatively slowly through our sky.

each month of the year has its own introductory pages

for all planets apart from Mercury, the charts show positions on the 15th day of the month; Mercury is shown only when it is at greatest elongation (p.65)

key to colored icons for planets

planet locator chart

observation from northern and southern latitudes covered separately in the text

text highlights the most prominent stars, deep-sky objects, and meteor showers

Photographs
Photographs illustrate some of the most interesting objects and phenomena to look out for. On some photographs, linking lines have been added to draw attention to well-known patterns of stars.

THE WHOLE-SKY CHARTS

The introduction to each month is followed by two whole-sky charts.
To use a chart, determine the color-coded horizon lines and
zenith markers for your location, turn to the chart for the
appropriate month, and position yourself and the chart.

60°N
40°N
20°N
0°
20°S
40°S

Lines of latitude
The charts include color-coded lines of reference
so that they can be used various latitudes. Check
your position on Earth on the map above and
identify the latitude closest to your location.

Orientation
To view the sky to the north, face
north and hold the map up, with
the NORTH label closest to your
body. One of the color-coded
lines near the edge of the map
will relate to the horizon in front
of you. To view the south, turn
around and reposition the map.

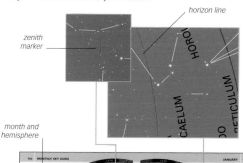

zenith
marker

horizon line

month and
hemisphere

Horizons and zeniths
The stars located near the center of each
chart can be seen near the zenith (the
point directly overhead), while the stars
near the chart's edge appear close to the
horizon. Color-coded lines and crosses
are used to identify the horizon and zenith
respectively—although a difference of up
to 10° of latitude has little effect on the
stars that can be seen.

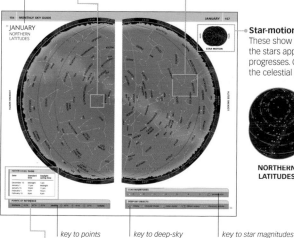

key to points
of reference

key to deep-sky
objects

key to star magnitudes

Star-motion diagrams
These show the direction in which
the stars appear to move as the night
progresses. Circumpolar stars circle
the celestial poles without setting.

**NORTHERN
LATITUDES**

**SOUTHERN
LATITUDES**

Observation times
Each chart shows the sky as it looks at 10pm local
standard time on the 15th of the month. The same
view be seen at other times of the month, as well as
one hour later when daylight-saving time is in use.
To view the sky at a different time, you might need
to consult another chart. Dates and times when each
chart applies are listed in the table next to each chart.

OBSERVATION TIMES		
Date	Standard time	Daylight-saving time
December 15	Midnight	1am
January 1	11pm	Midnight
January 15	10pm	11pm
February 1	9pm	10pm
February 15	8pm	9pm

JANUARY

The constellation of Orion, the Hunter, dominates the evening skies of both Northern and Southern Hemispheres throughout this month. The hunter's figure, with club raised in one hand and lion pelt in the other, is easy to spot, while his sword is marked by the celebrated Orion Nebula. His two hunting dogs, Canis Major and Canis Minor, stand close by, both marked by bright stars.

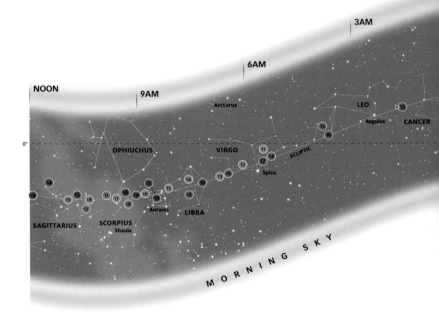

NORTHERN LATITUDES

Orion stands proud above the southern horizon with his dogs to the left of his feet. The dogs are located by their bright stars: Sirius in Canis Major and Procyon in Canis Minor. To either side of Orion's ill-defined head are the constellations Gemini and Taurus.

Gemini is identified by the two bright stars Castor and Pollux at one end, while the head of Taurus, the Bull, charges toward the lion pelt in Orion's hand. The Hyades star cluster, easily visible to the naked eye, locates the bull's face, and the bright star Aldebaran marks its eye. Aldebaran is a red giant, its coloring clear to the naked eye.

Directly overhead and within the path of the Milky Way lies the constellation of Auriga, found by its bright star, Capella.

From here, the Milky Way flows on into Cassiopeia, whose stars make an easily recognizable "W" or "M" shape in the sky, depending on whether they are above or below the north celestial pole. Cassiopeia and the other circumpolar constellations are above the northern horizon. Cassiopeia lies to the left of Polaris, the Pole Star; to the right of Polaris lie the two bears, Ursa Minor and Ursa Major.

METEOR SHOWERS

The Quadrantid meteor shower occurs during the first week of January. The meteors are usually faint and radiate from a point within Boötes, close to where the constellation borders the tail of Ursa Major. The peak of about 100 meteors an hour occurs around January 3–4.

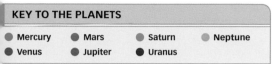

KEY TO THE PLANETS

- ● Mercury
- ● Venus
- ● Mars
- ● Jupiter
- ● Saturn
- ● Uranus
- ● Neptune

Locating the planets

The chart below shows the positions of the planets in January from 2011 to 2019. The planets are represented by colored dots (see key, left), with the number inside each dot denoting the year. For all planets apart from Mercury, the dot indicates the planet's position on January 15. Mercury is shown only when it is at its greatest elongation.

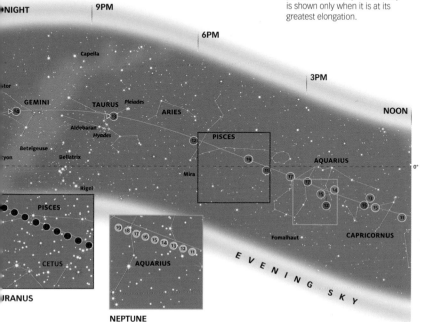

SOUTHERN LATITUDES

Orion stands high in the sky, his head pointing down to the north. The three stars of his belt lie overhead, along with the chain of stars and nebulae, including the Orion Nebula, marking his sword.

Auriga and its bright star Capella lie between Orion and the northern horizon. Procyon, the bright star of Canis Minor, is to the east of Betelgeuse, the orange star marking the hunter's shoulder. To the east of Orion's feet is Sirius, the bright star that marks the head of Orion's larger dog. Just south of Sirius lies M41, an open cluster visible to the naked eye. To the northwest and below Orion is Taurus; to the lower right of Orion, beyond his club, is Gemini, the twins, marked by the bright stars Castor and Pollux close to the horizon.

The view looking south is one of contrasts. To the left, the Milky Way flows from the southeastern horizon up toward Sirius, crossing the rich starfields of Centaurus, Crux, and Carina. In the southwest, however, just one bright star shines out; this is Achernar, at the southern tip of the constellation Eridanus.

Messier 41
M41 is an open cluster in the Canis Major constellation. It contains about 100 stars, and is estimated to be between 190 million and 240 million years old.

»

» JANUARY
NORTHERN LATITUDES

LOOKING NORTH

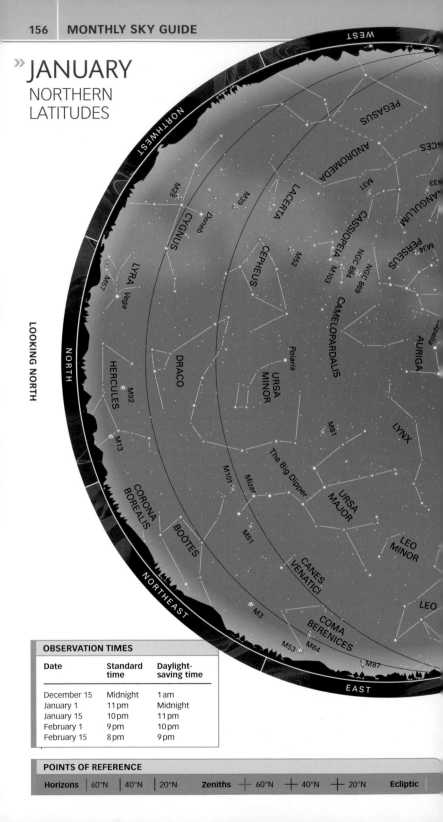

OBSERVATION TIMES

Date	Standard time	Daylight-saving time
December 15	Midnight	1 am
January 1	11 pm	Midnight
January 15	10 pm	11 pm
February 1	9 pm	10 pm
February 15	8 pm	9 pm

POINTS OF REFERENCE

Horizons	60°N	40°N	20°N	Zeniths	+ 60°N	+ 40°N	+ 20°N	Ecliptic

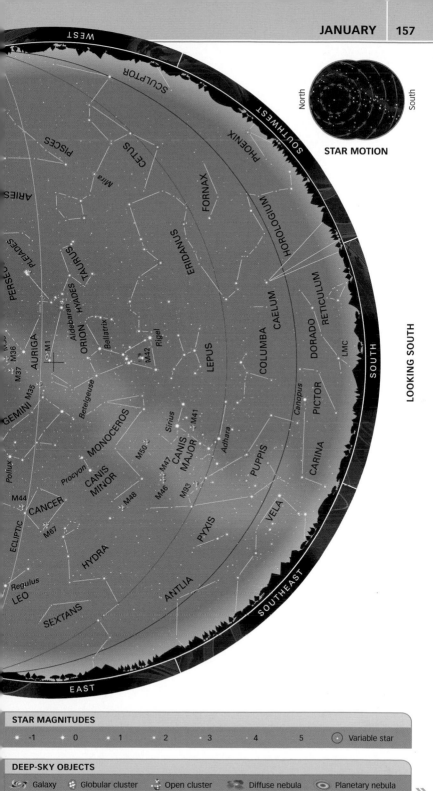

STAR MOTION

North

South

LOOKING SOUTH

STAR MAGNITUDES

-1   0   1   2   3   4   5   ⊙ Variable star

DEEP-SKY OBJECTS

🌀 Galaxy   ⊕ Globular cluster   ⁛ Open cluster   ☁ Diffuse nebula   ⊙ Planetary nebula

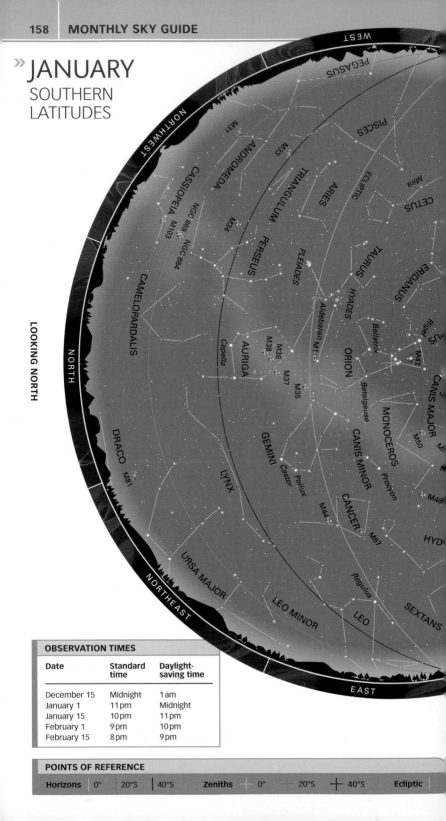

» JANUARY
SOUTHERN LATITUDES

LOOKING NORTH

OBSERVATION TIMES		
Date	Standard time	Daylight-saving time
December 15	Midnight	1 am
January 1	11 pm	Midnight
January 15	10 pm	11 pm
February 1	9 pm	10 pm
February 15	8 pm	9 pm

POINTS OF REFERENCE							
Horizons	0°	20°S	40°S	Zeniths	0° 20°S 40°S		Ecliptic

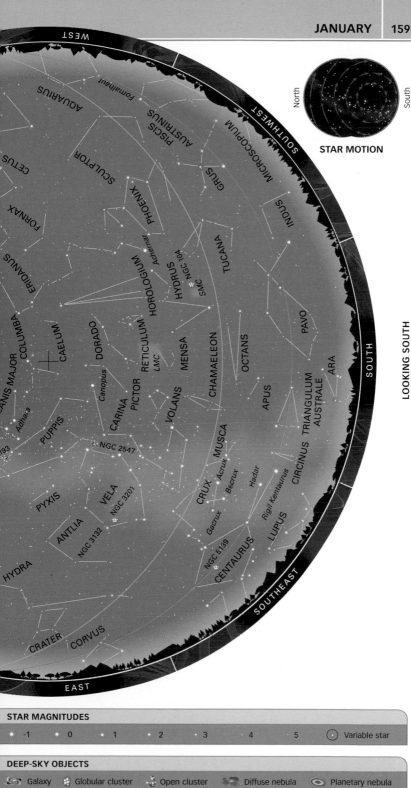

LOOKING SOUTH

STAR MOTION

North

South

STAR MAGNITUDES

-1   0   1   2   3   4   5   ⊙ Variable star

DEEP-SKY OBJECTS

Galaxy   Globular cluster   Open cluster   Diffuse nebula   ⊙ Planetary nebula

FEBRUARY

A triangle of bright stars and a pair of celestial twins shine high in February's sky. The triangle is made of stars from three different constellations: Betelgeuse in Orion, Sirius in Canis Major, and Procyon in Canis Minor. The twins Castor and Pollux are represented by the bright stars of the same name in the constellation Gemini.

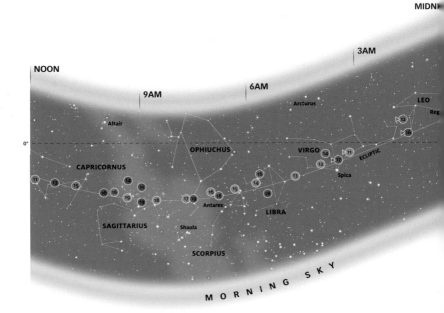

NORTHERN LATITUDES

The triangle of stars formed by Sirius, Procyon, and Betelgeuse is known as the Winter Triangle. Sirius and Procyon are easy to spot to the south, since they are the first- and eighth-brightest stars in the night sky respectively. Sirius is closest to the southern horizon; Procyon is above, to Sirius's left. Betelgeuse is in Orion, above and to the right of Sirius. Beyond Orion, and in the southwestern sky, is Taurus, the Bull. The distinctive shape of Leo, the Lion, is now coming into view in the southeast, and Gemini is almost overhead. The heads of the Twins, Castor and Pollux, point toward Leo; their feet point toward Taurus.

Capella, the bright star in Auriga, rides high in the western sky. Below is Perseus, the mythological hero who rescued the princess Andromeda from a sea monster. The chained figure of Andromeda herself lies between Perseus and the horizon. Her mother, Cassiopeia, is to her right in the northwest. The bears Ursa Major and Ursa Minor lie to the northeast.

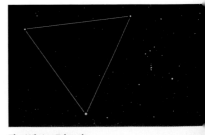

The Winter Triangle
Sirius (bottom) in Canis Major forms the Winter Triangle with Procyon in Canis Minor (top left), and Betelgeuse in Orion (top right).

Locating the planets
The chart below shows the positions of the planets in February from 2011 to 2019. The planets are represented by colored dots (see key, left), with the number inside each dot denoting the year. For all planets apart from Mercury, the dot indicates the planet's position on February 15. Mercury is shown only when it is at its greatest elongation.

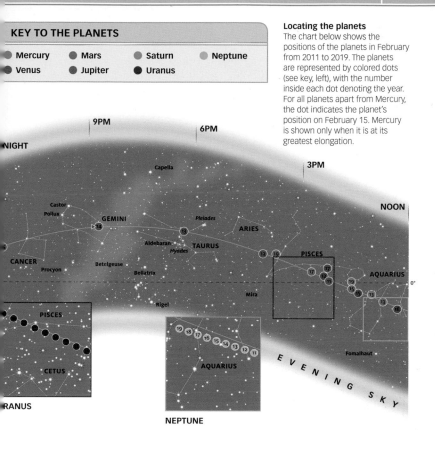

NEPTUNE

SOUTHERN LATITUDES

The path of the Milky Way extends across the Southern Hemisphere sky this month, stretching from the southeastern horizon to the northwest. Along its stream, starting closest to the southern horizon, are the following constellations: Centaurus and Crux; Carina, Vela, and Puppis (the three elements of the ancient celestial ship Argo); and Canis Major, which lies overhead. Sirius in Canis Major and Canopus in Carina are the two brightest stars of all, and both can be seen high in the sky throughout this month.

The Winter Triangle, made by linking orange-red Betelgeuse with the white stars Sirius and Procyon, can be seen high above the northwestern horizon. Sirius is the highest star; below it are Betelgeuse and Procyon, to its left and right

respectively. Within this triangle sits the constellation Monoceros, the Unicorn—a large, faint "W" shape of stars that also lies across the path of the Milky Way.

The stars Castor and Pollux, which mark the heads of the Gemini twins, can be seen to the north. At magnitude 1.2, Pollux is the brighter of the two, but Castor (magnitude 1.6) is possibly the more interesting—a small telescope shows it as a pair of stars, the brightest elements of a complex multiple star system. Eastward, to the right of Gemini, lies Cancer, the Crab, and the upside-down figure of Leo, the Lion. The lion's head, looking from this direction like an inverted question mark, points toward the Gemini twins; his tail and haunches point to the eastern horizon.

»

» FEBRUARY
NORTHERN LATITUDES

LOOKING NORTH

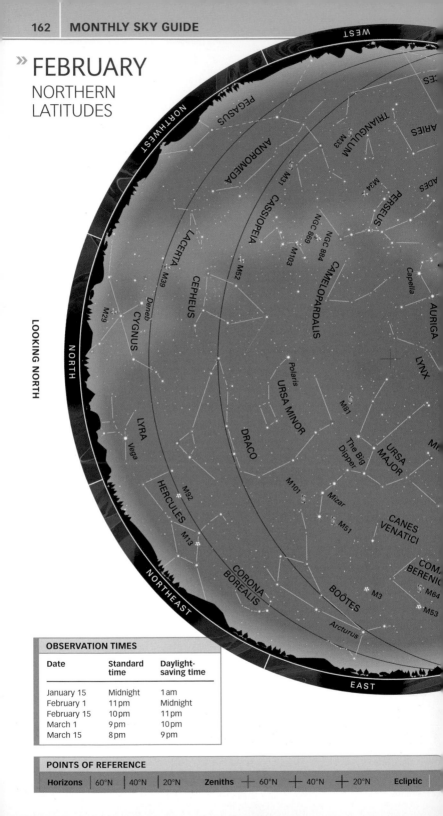

OBSERVATION TIMES

Date	Standard time	Daylight-saving time
January 15	Midnight	1 am
February 1	11 pm	Midnight
February 15	10 pm	11 pm
March 1	9 pm	10 pm
March 15	8 pm	9 pm

POINTS OF REFERENCE

| Horizons | 60°N | 40°N | 20°N | Zeniths | + 60°N | + 40°N | + 20°N | Ecliptic |

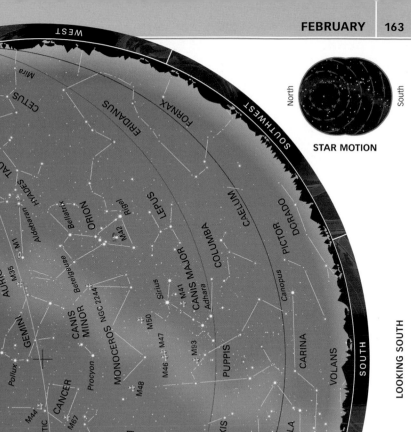

STAR MOTION

North / South

LOOKING SOUTH

WEST

CETUS
Mira
TAUR.
Aldebaran
HYADES
M1
AURIGA
M35
GEMINI
Castor
Pollux
CANCER
M44
M67
ECLIPTIC
LEO
Regulus
HYDRA
SEXTANS
CRATER
CORVUS
M104
VIRGO
87

ERIDANUS
FORNAX
Bellatrix
ORION
M42
Rigel
LEPUS
CANIS MINOR
Procyon
Betelgeuse
NGC 2244
MONOCEROS
M50
M46 M47
M48
M93
Sirius
M41
CANIS MAJOR
Adhara
CAELUM
COLUMBA
PICTOR
DORADO
Canopus
PUPPIS
PYXIS
ANTLIA
CARINA
VELA
VOLANS
SOUTH
SOUTHWEST
SOUTHEAST
EAST

STAR MAGNITUDES

-1   0   1   2   3   4   5   ⊙ Variable star

DEEP-SKY OBJECTS

🌀 Galaxy   ✦ Globular cluster   ✦ Open cluster   ▨ Diffuse nebula   ⊙ Planetary nebula

» FEBRUARY
SOUTHERN
LATITUDES

LOOKING NORTH

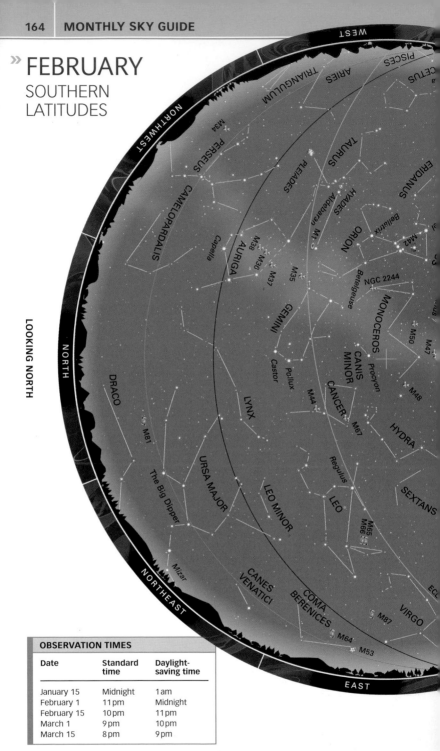

WEST

PISCES

CETUS

TRIANGULUM

ARIES

NORTHWEST

PERSEUS

M34

TAURUS

ERIDANUS

PLEIADES

Aldebaran

HYADES

M1

CAMELOPARDALIS

Capella

AURIGA

M38

M36

M37

M35

ORION

Bellatrix

M42

Betelgeuse

NGC 2244

MONOCEROS

M50

M47

NORTH

GEMINI

Castor

Pollux

CANIS
MINOR

Procyon

M48

DRACO

LYNX

CANCER

M44

M67

HYDRA

M81

The Big Dipper

URSA MAJOR

LEO MINOR

Regulus

LEO

SEXTANS

Mizar

M65
M66

Regulus

CANES
VENATICI

COMA
BERENICES

M87

VIRGO

NORTHEAST

M64

M53

ECL

EAST

OBSERVATION TIMES

Date	Standard time	Daylight-saving time
January 15	Midnight	1 am
February 1	11 pm	Midnight
February 15	10 pm	11 pm
March 1	9 pm	10 pm
March 15	8 pm	9 pm

POINTS OF REFERENCE

Horizons	0°	20°S	40°S	Zeniths	0°	20°S	40°S	Ecliptic

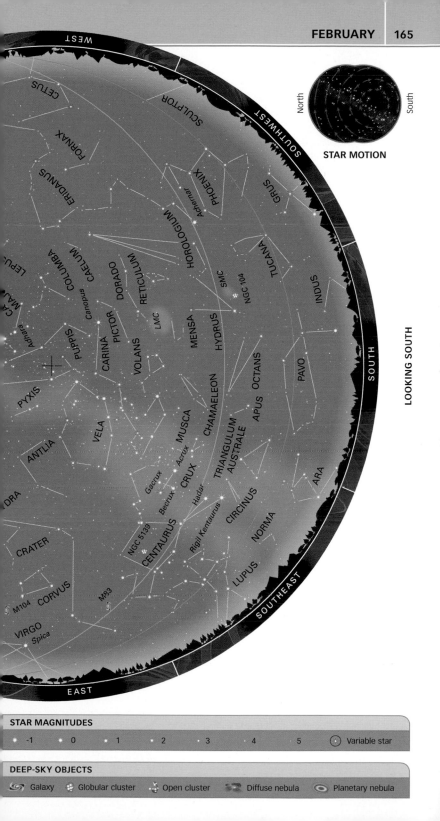

WEST

CETUS

SCULPTOR

SOUTHWEST

North

South

STAR MOTION

FORNAX

ERIDANUS

PHOENIX

Achernar

GRUS

CAELUM

HOROLOGIUM

TUCANA

INDUS

LOOKING SOUTH

LEPUS

COLUMBA

DORADO

RETICULUM

SMC

NGC 104

CA MA

Adhara

Canopus

PICTOR

CARINA

VOLANS

LMC

MENSA

HYDRUS

OCTANS

PAVO

SOUTH

PUPPIS

PYXIS

VELA

CHAMAELEON

APUS

ANTLIA

MUSCA

Acrux

TRIANGULUM
AUSTRALE

ARA

DRA

Gacrux

CRUX

Becrux

Hadar

CIRCINUS

NORMA

CRATER

NGC 5139

CENTAURUS

Rigil Kentaurus

LUPUS

CORVUS

M104

M93

SOUTHEAST

VIRGO

Spica

EAST

STAR MAGNITUDES

-1 · 0 · 1 · 2 · 3 · 4 · 5 · ⊙ Variable star

DEEP-SKY OBJECTS

🌀 Galaxy · 🔵 Globular cluster · ✦ Open cluster · ☁ Diffuse nebula · ◉ Planetary nebula

MARCH

Leo and Virgo take the place of Orion and Gemini as the changing skies herald the start of northern spring and southern fall. On the 20th, or occasionally the 21st, of the month, the Sun crosses the celestial equator as it moves from the southern to the northern sky. Briefly, day and night are of equal length before the Northern Hemisphere nights grow shorter and the southern nights grow longer.

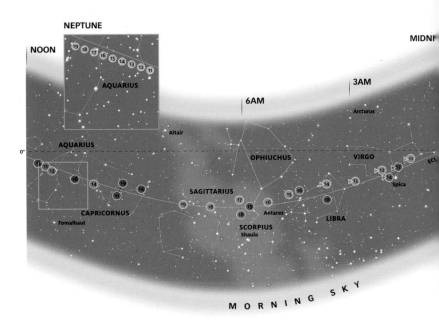

NORTHERN LATITUDES

Due south and high in the sky is the distinctive constellation Leo. Its linked stars really do resemble a crouching lion, and the curve of its head and the shape of its body are easy to pick out in the sky. Leo's brightest star, Regulus, is a blue-white star of magnitude 1.4 that marks the start of an outstretched front leg. Through binoculars, a dimmer, white companion can also be seen. Just below the lion's body are five galaxies, all visible with binoculars in good observing conditions.

Leo looks toward the west, past the faint constellation of Cancer, and on toward the winter constellations of Orion and his retinue, now disappearing from view shortly after sunset. To the left (east) of Leo, the less well-defined constellation Virgo, the Maiden, rises over the eastern

horizon, although it is still easily spotted thanks to the brilliant white star Spica, traditionally said to represent an ear of wheat in the maiden's hand.

Ursa Major, the Great Bear, is situated high in the sky above the northern horizon. The bowl of its saucepan-shaped asterism called the Big Dipper (or the Plough) open "downward" toward Polaris, the Pole Star. Among the stars near the bear's head lies M81, one of the easiest galaxies to find with binoculars. About one Moon-width away lies the smaller and fainter spiral galaxy, M82 (also called Bode's Galaxy), which is easily spotted through a small telescope. Cassiopeia, on the opposite side of Polaris, lies almost directly between the north celestial pole and the horizon, forming a "W" shape.

Locating the planets

The chart below shows the positions of the planets in March from 2011 to 2019. The planets are represented by colored dots (see key, right), with the number inside each dot denoting the year. For all planets apart from Mercury, the dot indicates the planet's position on March 15. Mercury is shown only when it is at its greatest elongation.

KEY TO THE PLANETS

- Mercury
- Venus
- Mars
- Jupiter
- Saturn
- Uranus
- Neptune

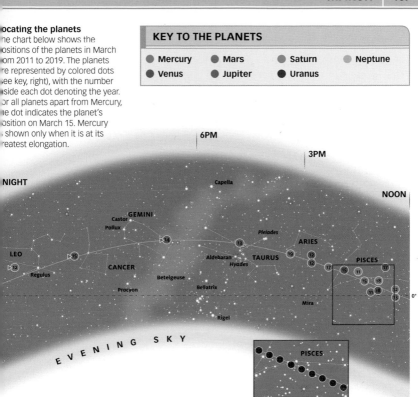

URANUS

SOUTHERN LATITUDES

The view looking south this month is packed with bright stars. Due south are the constellations Carina and Vela. The white supergiant Canopus marks Carina's western tip, and at magnitude -0.62 it is outshone only by Sirius. Sirius itself lies above and west of Canopus.

The southeastern sky is dominated by the two bright stars Alpha (α) and Beta (β) Centauri, which point to the tiny constellation of Crux. Farther east, the first of the southern winter constellations are rising above the eastern horizon. Virgo is leading the way, followed by Scorpius.

Due north is Leo. Its stars draw out a crouching lion, which observers in the Southern Hemisphere see upside-down. The lion is apparently lying on its back, with its head facing west.

Almost equidistant to either side of Leo's brightest star Regulus are two equally bright stars: Procyon in Canis Minor to the west, and Spica in Virgo to the east. Closer to the eastern horizon sits the red giant Arcturus in the constellation Boötes. On the opposite side of the sky, Orion is disappearing over the western horizon.

The Sickle
The head and chest of Leo, the Lion, form an asterism called the Sickle. In the Southern Hemisphere, where the lion appears upside down, the handle of the sickle points upward.

»

MARCH
NORTHERN LATITUDES

LOOKING NORTH

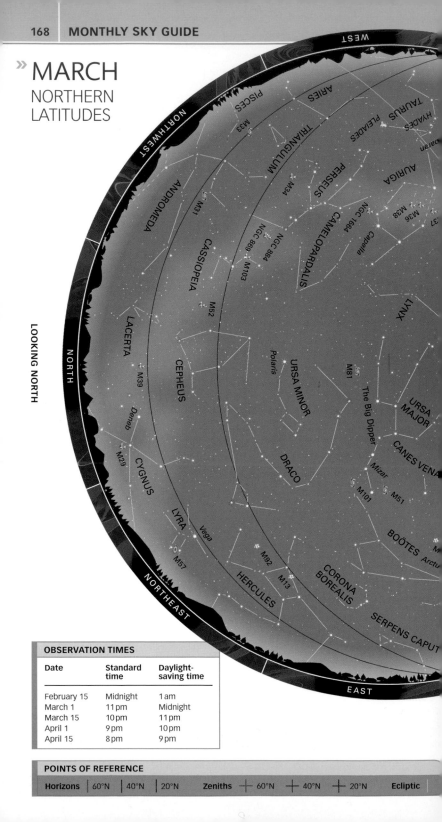

WEST

NORTHWEST

PISCES
M33
ARIES
TRIANGULUM
TAURUS
PLEIADES
HYADES
PERSEUS
M34
NGC 869
NGC 884
NGC 1664
CAMELOPARDALIS
AURIGA
Capella
M36
M38
M37
ANDROMEDA
CASSIOPEIA
M31
M103
M52
LYNX
NORTH
LACERTA
CEPHEUS
Polaris
URSA MINOR
M81
URSA MAJOR
The Big Dipper
M39
Deneb
DRACO
CANES VENA
CYGNUS
M29
M101
Mizar
M51
LYRA
Vega
M57
M92
M13
BOÖTES
Arctu
CORONA BOREALIS
HERCULES
SERPENS CAPUT
NORTHEAST
EAST

OBSERVATION TIMES

Date	Standard time	Daylight-saving time
February 15	Midnight	1 am
March 1	11 pm	Midnight
March 15	10 pm	11 pm
April 1	9 pm	10 pm
April 15	8 pm	9 pm

POINTS OF REFERENCE

Horizons	60°N	40°N	20°N	Zeniths + 60°N + 40°N + 20°N	Ecliptic

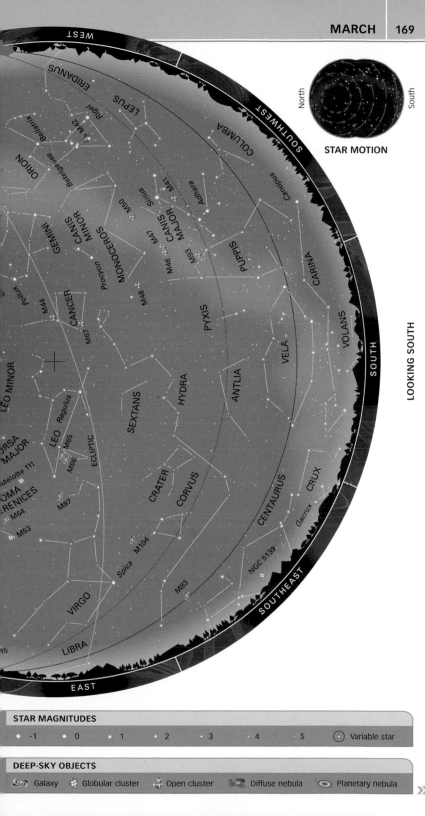

STAR MOTION

North

South

LOOKING SOUTH

STAR MAGNITUDES

-1   0   1   2   3   4   5   Variable star

DEEP-SKY OBJECTS

Galaxy   Globular cluster   Open cluster   Diffuse nebula   Planetary nebula

MARCH
SOUTHERN LATITUDES

LOOKING NORTH

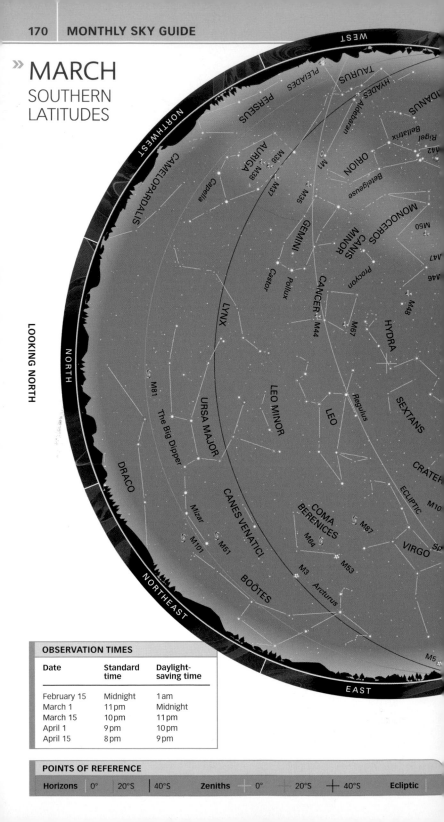

OBSERVATION TIMES

Date	Standard time	Daylight-saving time
February 15	Midnight	1 am
March 1	11 pm	Midnight
March 15	10 pm	11 pm
April 1	9 pm	10 pm
April 15	8 pm	9 pm

POINTS OF REFERENCE

Horizons	0°	20°S	40°S	Zeniths	0°	20°S	40°S	Ecliptic

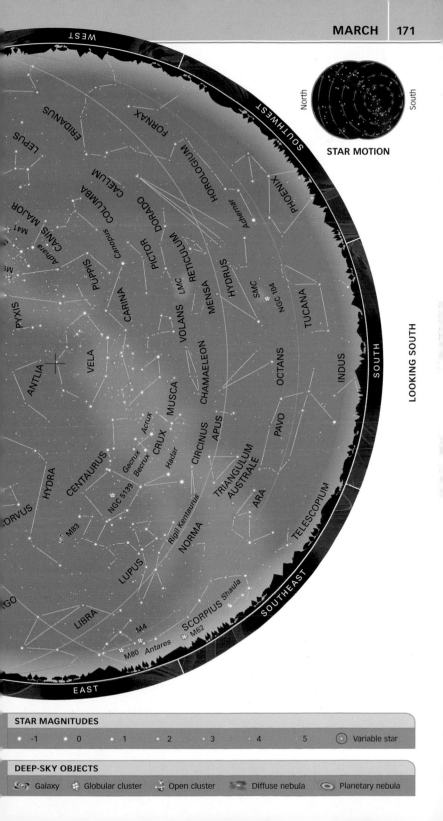

WEST

SOUTHWEST

STAR MOTION

North

South

LOOKING SOUTH

SOUTH

LEPUS

ERIDANUS

FORNAX

CAELUM

COLUMBA

CANIS MAJOR

M41

Adhara

M

PYXIS

PUPPIS

Canopus

PICTOR

DORADO

HOROLOGIUM

Achernar

PHOENIX

RETICULUM

LMC

VOLANS

MENSA

HYDRUS

SMC

NGC 104

TUCANA

ANTLIA

VELA

CARINA

CHAMAELEON

OCTANS

INDUS

HYDRA

CENTAURUS

Gacrux Acrux

Becrux

CRUX

Hadar

MUSCA

CIRCINUS

APUS

TRIANGULUM
AUSTRALE

PAVO

ORVUS

NGC 5139

M83

Rigil Kentaurus

NORMA

ARA

TELESCOPIUM

GO

LIBRA

LUPUS

M4

M80 Antares

SCORPIUS Shaula

M62

SOUTHEAST

EAST

STAR MAGNITUDES

| -1 | 0 | 1 | 2 | 3 | 4 | 5 | ⊙ Variable star |

DEEP-SKY OBJECTS

Galaxy   Globular cluster   Open cluster   Diffuse nebula   Planetary nebula

APRIL

Leo remains high in the sky this month but starts to make room for another zodiac constellation, Virgo, the Maiden. Meanwhile a different mythical figure—Boötes, the Herdsman—is making his presence felt. The third-largest constellation, Ursa Major, the Great Bear, is almost overhead for those in the Northern Hemisphere, while in the Southern Hemisphere Crux is prominently placed.

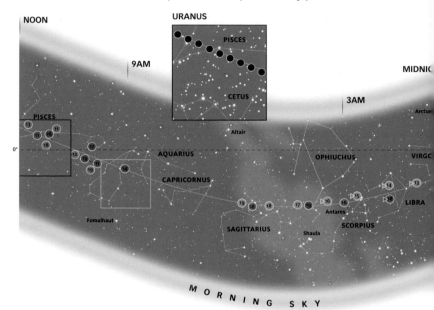

NORTHERN LATITUDES

Leo is high in the southwest sky; its distinctive shape is easily picked out. The lion is looking toward the heads of the Twins, Gemini, who are marked by the bright stars Castor and Pollux. The sky above and below these constellations, however, is relatively barren. The long figure of Hydra, the Water Snake, straggles between Leo and the southern horizon, and while it is the largest constellation, it is far from prominent. To the right, Virgo follows Leo across the sky, with its bright star Spica shining in the southeast. Above and to the east is Arcturus, which, at magnitude 0.05, is the fourth-brightest star in the entire sky.

Looking north, Ursa Major is directly above Polaris, the Pole Star, and almost overhead. Seven stars that form the tail

and rump of the bear make a saucepan shape in profile. Known as the Big Dipper, or the Plough, this is one of the most familiar star patterns in the northern sky.

Cassiopeia sits below the Pole Star and close to the horizon. To the northwest lies yellow-white Capella, the brightest and by far the most conspicuous star in Auriga, the Charioteer, and at magnitude 0.08 the sixth-brightest star in the sky. To the east, Vega in Lyra heralds the arrival of the first of the summer constellations.

METEOR SHOWERS

Lyra is host to the Lyrids meteor shower, which reaches its peak around April 21–22. At this time, a dozen or so meteors can be seen every hour radiating from a point near the star Vega.

Locating the planets

The chart below shows the positions of the planets in April from 2011 to 2019. The planets are represented by colored dots (see key, right), with the number inside each dot denoting the year. For all planets apart from Mercury, the dot indicates the planet's position on April 15. Mercury is shown only when it is at its greatest elongation.

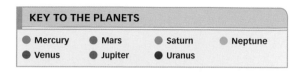

KEY TO THE PLANETS

- Mercury
- Venus
- Mars
- Jupiter
- Saturn
- Uranus
- Neptune

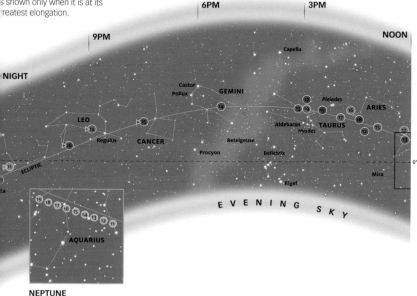

NEPTUNE

SOUTHERN LATITUDES

The Milky Way makes a glorious sight in the southern sky. It extends from the western horizon and climbs high in the sky to the south, before descending to the east. It passes Sirius and incorporates Carina, Crux, Centaurus, and the tail of Scorpius. Crux lies almost due south, the skies above it covered by the faint, meandering body of Hydra, the Water Snake.

Acrux, the brightest star in Crux, marks the base of the Southern Cross. Beta (β) Crucis marks the constellation's left arm, and the sparkling Jewel Box Cluster lies between this star and the prominent dark nebula known as the Coalsack.

Looking to the north, Leo is still well placed but is heading for the western horizon. To the right of its bright star Regulus lie five galaxies that are

visible through binoculars in dark skies. The bright stars Spica in Virgo and Arcturus in Boötes shine prominently in the northeastern sky. Procyon is still visible in the northwest, but Gemini is now sinking below the horizon.

Carina Nebula
This huge nebula in the southern Milky Way is visible to the naked eye. It contains newborn star clusters and the variable star Eta Carinae.

»

» APRIL
NORTHERN LATITUDES

LOOKING NORTH

OBSERVATION TIMES

Date	Standard time	Daylight-saving time
March 15	Midnight	1 am
April 1	11 pm	Midnight
April 15	10 pm	11 pm
May 1	9 pm	10 pm
May 15	8 pm	9 pm

POINTS OF REFERENCE

Horizons	60°N	40°N	20°N	Zeniths	+ 60°N	+ 40°N	+ 20°N	Ecliptic

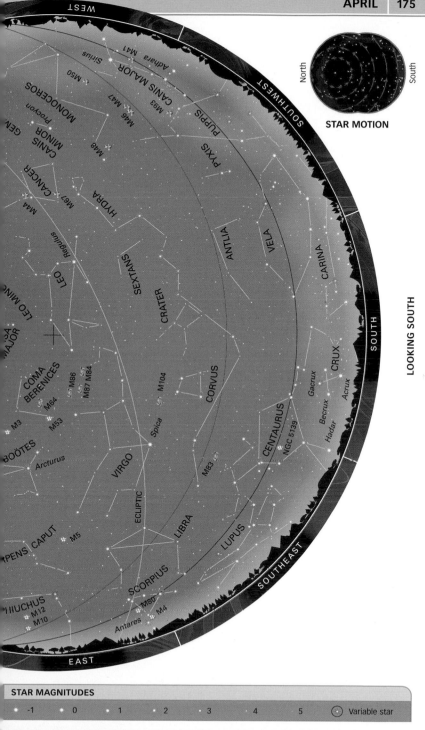

LOOKING SOUTH

STAR MOTION

North / South

STAR MAGNITUDES

-1   0   1   2   3   4   5   ⊙ Variable star

DEEP-SKY OBJECTS

🌀 Galaxy   ✦ Globular cluster   ✦ Open cluster   ▨ Diffuse nebula   ⊙ Planetary nebula

APRIL
SOUTHERN LATITUDES

LOOKING NORTH

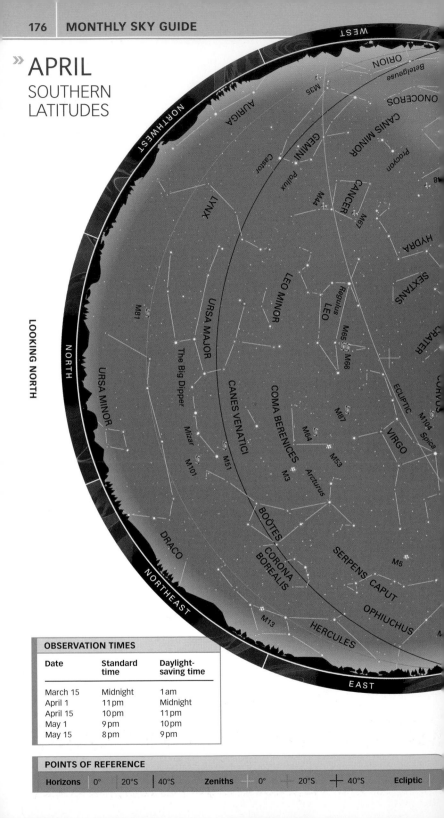

OBSERVATION TIMES

Date	Standard time	Daylight-saving time
March 15	Midnight	1 am
April 1	11 pm	Midnight
April 15	10 pm	11 pm
May 1	9 pm	10 pm
May 15	8 pm	9 pm

POINTS OF REFERENCE

Horizons	0°	20°S	40°S	Zeniths	0°	20°S	40°S	Ecliptic

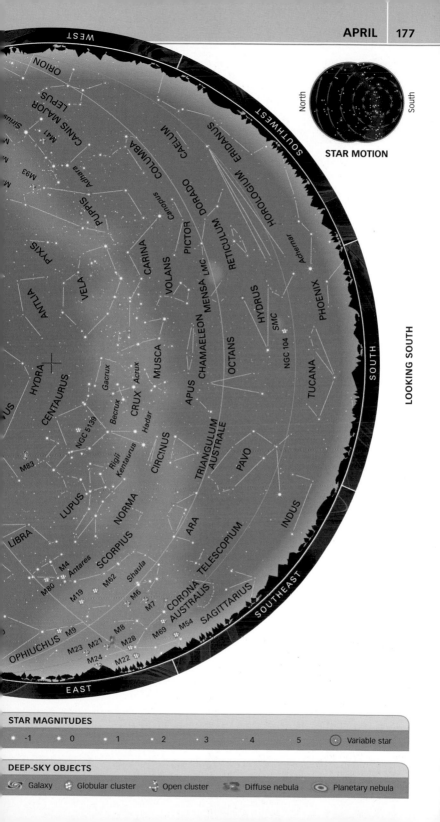

STAR MOTION

LOOKING SOUTH

STAR MAGNITUDES

-1   0   1   2   3   4   5   ⊙ Variable star

DEEP-SKY OBJECTS

🌀 Galaxy   ✦ Globular cluster   ✦ Open cluster   ☁ Diffuse nebula   ⊙ Planetary nebula

MAY

Boötes and Virgo are prominent in both the Northern and Southern Hemispheres in May. Observers south of the equator also have stunning views of Crux and Centaurus, which are now at their highest point above the horizon. In the Northern Hemisphere, the days are lengthening, and once the sky has darkened, Ursa Major stands high and proud in the northern sky.

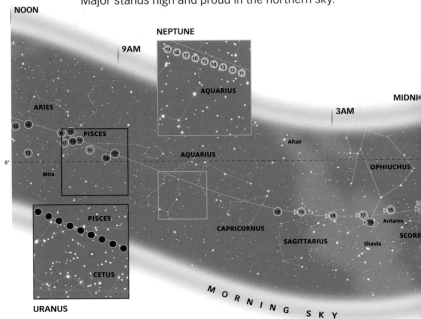

NORTHERN LATITUDES

The bright star Spica and its constellation, Virgo, lie directly to the south in evening skies. Above them is the red giant Arcturus in Boötes. The constellation Ophiuchus lies to the southeast, while the head and body of Leo are still visible in the southwest, sinking below the horizon as the evening progresses.

Antares, a red star that marks the heart of Scorpius, the Scorpion, makes a rare appearance in the northern sky for observers south of 50°N. It lies close to the horizon during the summer months.

Ursa Major is high in the sky to the north. The easily recognized saucepan shape of its asterism the Big Dipper (or Plough) is positioned so that the end of the handle is north, and the pan tips down to the west. Dubhe and Merak, the two brightest stars in Ursa Major, form the side of the pan away from the handle. These are the "pointers," which point the way to Polaris, the Pole Star, which marks the position of the north celestial pole.

The summer stars Vega (in Lyra) and Deneb (in Cygnus) move higher into the northeastern sky as the last of the winter stars, Castor and Pollux (in Gemini), set in the northwest. Lyra is small, but prominent due to its bright star Vega. It is also home to the Ring Nebula (M57), a planetary nebula visible as a misty disk through a small telescope.

METEOR SHOWERS

Early in the month, the Eta Aquarid meteor shower may be visible to observers in lower northern latitudes.

Locating the planets

The chart below shows the positions of the planets in May from 2011 to 2019. The planets are represented by colored dots (see key, right), with the number inside each dot denoting the year. For all planets apart from Mercury, the dot indicates the planet's position on May 15. Mercury is shown only when it is at its greatest elongation.

KEY TO THE PLANETS

- Mercury
- Venus
- Mars
- Jupiter
- Saturn
- Uranus
- Neptune

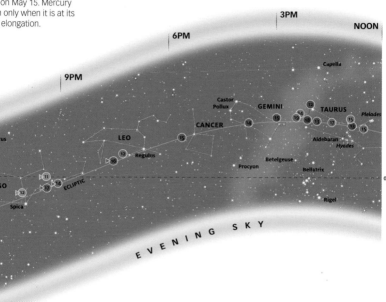

SOUTHERN LATITUDES

Two bright stars, Spica (in Virgo) and Arcturus (in Boötes), are centrally placed when looking south. Arcturus is the lower and brighter one; Spica is almost overhead. To the northwest lies Leo, which will soon make way for the winter constellations.

Southern pointers
The stars Alpha and Beta Centauri (the bright white and blue stars bottom left) form a line pointing towards Crux, the Southern Cross.

Ophiuchus, the Serpent Bearer, is in the eastern sky, with the constellation Serpens, the Serpent, coiled around it. The head, Serpens Caput, lies between Ophiuchus and Bootes; the tail, Serpens Cauda, is close to the eastern horizon.

Centaurus and Crux are high in the sky to the south. The two bright stars of Centaurus—Alpha (α) and Beta (β) Centauri—shine out against the Milky Way and point to the head of Crux. May also offers the first view of the constellation Scorpius in the southeastern sky.

METEOR SHOWERS

The Eta Aquarid meteor shower peaks in the first week of May, when about 35 meteors an hour can be seen radiating from a point in Aquarius.

»

» MAY
NORTHERN LATITUDES

LOOKING NORTH

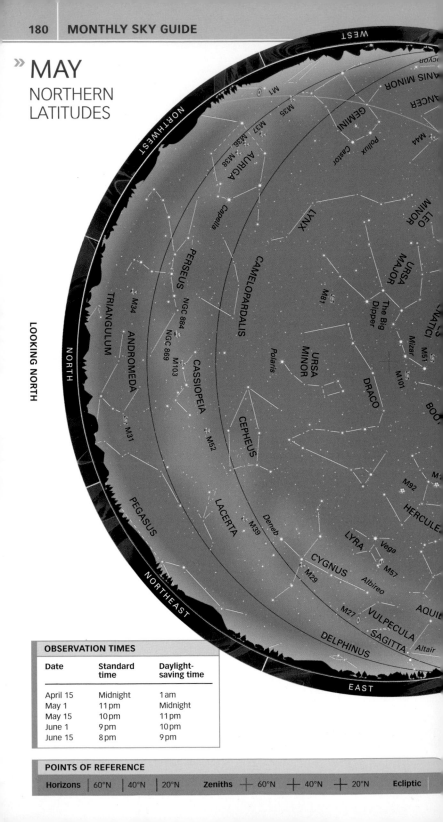

OBSERVATION TIMES		
Date	Standard time	Daylight-saving time
April 15	Midnight	1 am
May 1	11 pm	Midnight
May 15	10 pm	11 pm
June 1	9 pm	10 pm
June 15	8 pm	9 pm

POINTS OF REFERENCE									
Horizons	60°N	40°N	20°N	Zeniths	+ 60°N	+ 40°N	+ 20°N	Ecliptic	

LOOKING SOUTH

STAR MOTION

North — South

STAR MAGNITUDES

-1   0   1   2   3   4   5   ⊙ Variable star

DEEP-SKY OBJECTS

🌀 Galaxy   ✿ Globular cluster   ✿ Open cluster   ☁ Diffuse nebula   ◉ Planetary nebula

MAY
SOUTHERN LATITUDES

LOOKING NORTH

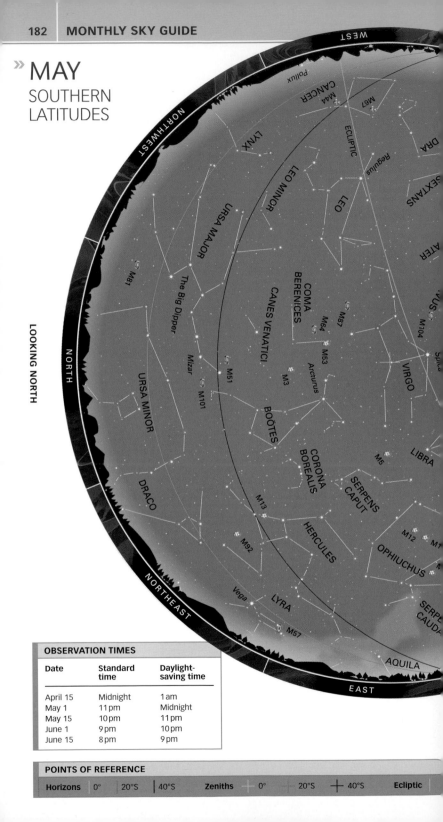

OBSERVATION TIMES		
Date	**Standard time**	**Daylight-saving time**
April 15	Midnight	1 am
May 1	11 pm	Midnight
May 15	10 pm	11 pm
June 1	9 pm	10 pm
June 15	8 pm	9 pm

POINTS OF REFERENCE								
Horizons	0°	20°S	40°S	**Zeniths**	0°	20°S	40°S	**Ecliptic**

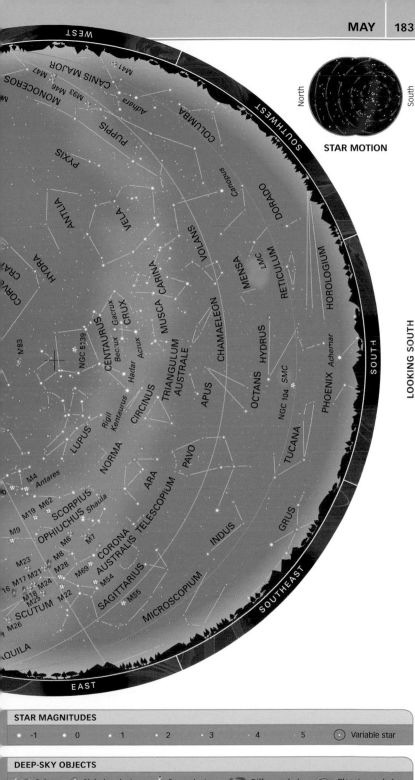

WEST

STAR MOTION

North

South

LOOKING SOUTH

SOUTH

SOUTHWEST

SOUTHEAST

EAST

STAR MAGNITUDES

-1   0   1   2   3   4   5   ⊙ Variable star

DEEP-SKY OBJECTS

Galaxy   Globular cluster   Open cluster   Diffuse nebula   Planetary nebula

JUNE

The mythical hero Hercules moves into the June sky to join Boötes, the Herdsman, and Ophiuchus, the Serpent Bearer. Toward the end of the month, nights are at their shortest in the Northern Hemisphere and at their longest in the Southern Hemisphere. This is because on June 21 or 22, the Sun is at its farthest point north of the celestial equator—the northern summer solstice.

NORTHERN LATITUDES

Looking to the south, Hercules and Boötes sit high in the sky. The bright red star Arcturus makes kite-shaped Boötes easy to find, but Hercules, although large, is not prominent. It is best found by locating Vega a little way to the east. Vega is in the constellation Lyra and is the fifth-brightest star of all, outshone in the June sky only by Arcturus. To the right of Vega is the body of Hercules. Four linked stars that form a distorted square called the Keystone represent his lower torso. His legs point upward, while his head dips toward the horizon. The globular star cluster M13 can just be spotted by the naked eye on a line linking two of the Keystone stars.

As the evening progresses, the front of Scorpius, marked by the red supergiant Antares, rises in the southeast. In the east, meanwhile, the three bright stars of the asterism called the Summer Triangle can be seen. The brightest and highest of these is Vega; the other two are Deneb (in Cygnus) and Altair (in Aquila), which lie closer to the horizon. Looking north, Ursa Minor, the Little Bear, extends above Polaris, the Pole Star, which marks the tip of the bear's tail.

The stars of Ursa Minor have no direct resemblance to a bear—indeed, they take the form of a saucepan in profile. However, their shape echoes that of an asterism in Ursa Major called the Big Dipper (or the Plough)—hence the nickname the Little Dipper. The pan handle is the curved tail of the bear; the pan is the bear's rump. The bear is all but surrounded by the sinuous body of Draco, the Dragon.

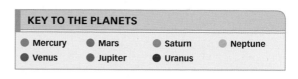

Locating the planets

The chart below shows the positions of the planets in June from 2011 to 2019. The planets are represented by colored dots (see key, right), with the number inside each dot denoting the year. For all planets apart from Mercury, the dot indicates the planet's position on June 15. Mercury is shown only when it is at its greatest elongation.

KEY TO THE PLANETS

- Mercury
- Venus
- Mars
- Jupiter
- Saturn
- Uranus
- Neptune

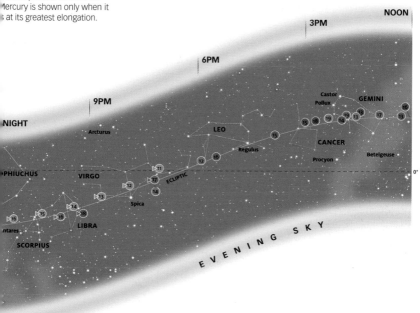

SOUTHERN LATITUDES

In the Southern Hemisphere, Boötes and Hercules lie either side of due north. They can be found by locating the bright stars Arcturus (in Boötes) and Vega (in Lyra), the latter just peeking over the northeastern horizon. On the left, Boötes is upside-down, while to the right, Hercules stands upright, with an arm outstretched above Vega. His lower body is represented by the four stars of the Keystone. The star cluster M13 lies on a line linking two of these stars.

Aquila and its bright star Altair are low in the northeast, while overhead, the distinctive shape of Scorpius, with its curving tail and heart marked by the red supergiant Antares, can be clearly seen.

Scorpius and its neighbor Sagittarius lie in the direction of the center of our galaxy and are crossed by rich star clouds.

From here, the Milky Way flows to the southwest. The globular cluster Omega (ω) Centauri is well placed, as are the dark Coalsack nebula and the Jewel Box (NGC 4755) open cluster, both within Crux.

The Scorpion
The red star Antares marks the heart of Scorpius. The constellation's body curves down through the Milky Way to the bright star clusters M6 and M7.

»

» JUNE
NORTHERN LATITUDES

LOOKING NORTH

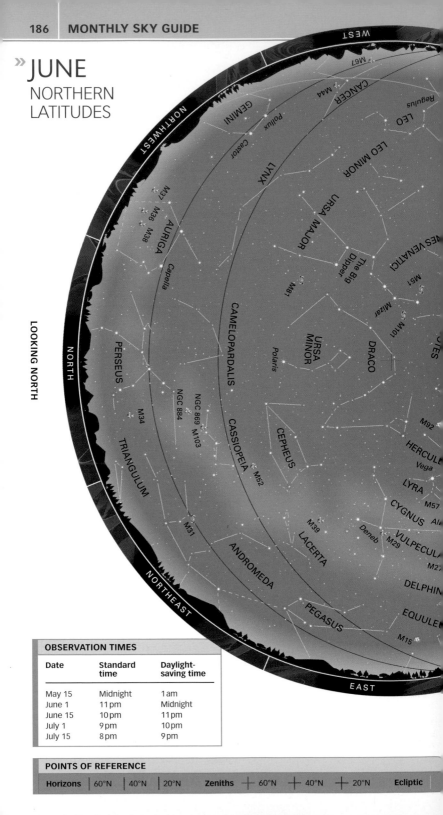

OBSERVATION TIMES		
Date	**Standard time**	**Daylight-saving time**
May 15	Midnight	1 am
June 1	11 pm	Midnight
June 15	10 pm	11 pm
July 1	9 pm	10 pm
July 15	8 pm	9 pm

POINTS OF REFERENCE							
Horizons	60°N	40°N	20°N	**Zeniths** + 60°N + 40°N + 20°N			**Ecliptic**

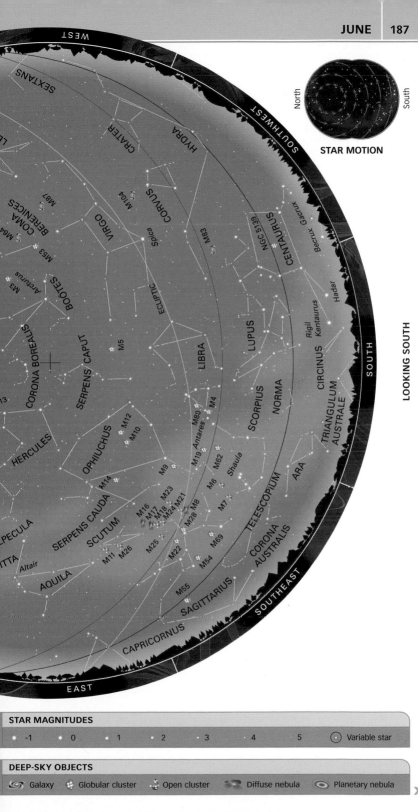

LOOKING SOUTH

STAR MOTION

North   South

STAR MAGNITUDES

● -1   ● 0   ● 1   ● 2   ● 3   · 4   · 5   ⊙ Variable star

DEEP-SKY OBJECTS

🌀 Galaxy   ✿ Globular cluster   ⁂ Open cluster   ☁ Diffuse nebula   ⊙ Planetary nebula

» JUNE
SOUTHERN LATITUDES

LOOKING NORTH

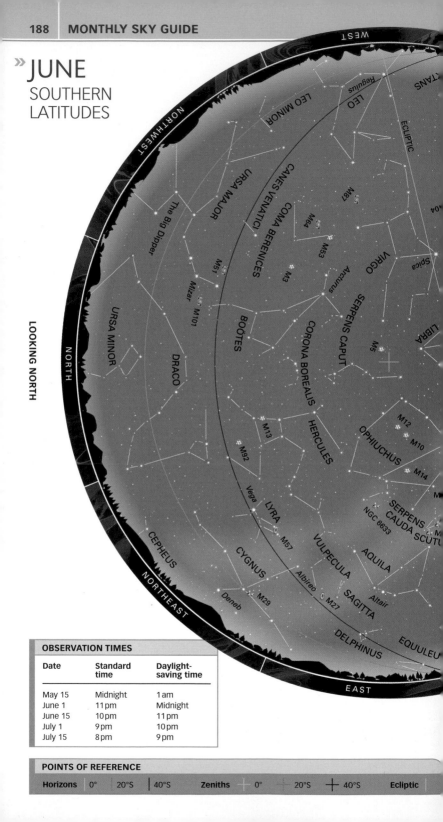

OBSERVATION TIMES

Date	Standard time	Daylight-saving time
May 15	Midnight	1 am
June 1	11 pm	Midnight
June 15	10 pm	11 pm
July 1	9 pm	10 pm
July 15	8 pm	9 pm

POINTS OF REFERENCE

Horizons	0°	20°S	40°S	Zeniths	0°	20°S	40°S	Ecliptic

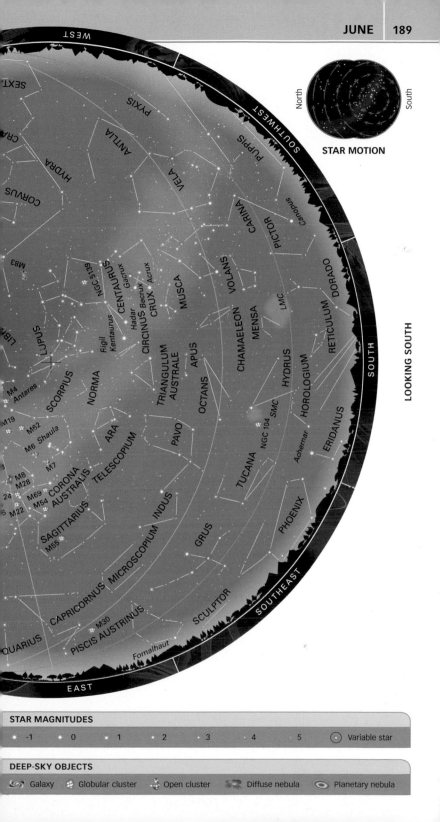

STAR MOTION

LOOKING SOUTH

STAR MAGNITUDES

| ● -1 | ● 0 | ● 1 | • 2 | · 3 | · 4 | · 5 | ⊙ Variable star |

DEEP-SKY OBJECTS

| 🌀 Galaxy | ✷ Globular cluster | ⁛ Open cluster | 🌫 Diffuse nebula | ◉ Planetary nebula |

JULY

Hercules and Ophiuchus remain center stage, while Aquila, the Eagle, has flown into view. Southern observers are treated to the Milky Way, the center of which lies almost directly overhead. Cygnus, the Swan, is high in the sky for northern observers, as is the bright star Vega.

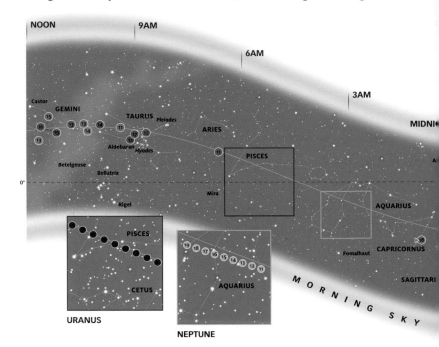

URANUS

NEPTUNE

NORTHERN LATITUDES

Looking north, Ursa Minor reaches up from Polaris, with Draco, the Dragon, coiling around it. Ursa Major lies to the left of Polaris; to the right are the mythical king Cepheus and his queen, Cassiopeia.

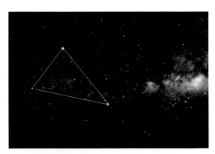

The Summer Triangle
Deneb, Vega, and Altair form the Summer Triangle asterism. Deneb (left) is the dimmest, Vega (top) is the brightest, and Altair (right) is the southernmost.

Arcturus (in Boötes), to the west, and Vega (in Lyra), almost overhead, are the brightest stars in the July sky. Next in brightness is Altair in the southeast, marking the neck of Aquila, the Eagle. The Milky Way flows through Aquila to Cygnus, high in the east. A line of stars forms the swan's body and an intersecting line its wings. This cross shape gives it its other name, the Northern Cross. Cygnus's brightest star, the supergiant Deneb, marks its tail.

Corona Borealis, one of the smallest constellations, is high in the southwest and is flanked by Hercules and Boötes. Ophiuchus is well placed due south, with Hercules above. Observers south of about 45°N will see star-rich Sagittarius and Scorpius at their best above their southern horizons.

Locating the planets

The chart below shows the positions of the planets in July from 2011 to 2019. The planets are represented by colored dots (see key, right), with the number inside each dot denoting the year. For all planets apart from Mercury, the dot indicates the planet's position on July 15. Mercury is shown only when it is at its greatest elongation.

KEY TO THE PLANETS

- Mercury
- Venus
- Mars
- Jupiter
- Saturn
- Uranus
- Neptune

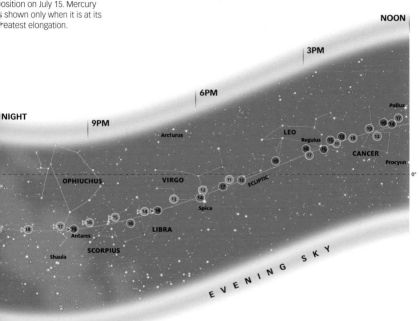

SOUTHERN LATITUDES

For southern observers, Scorpius and Sagittarius are almost overhead. The Scorpion's tail is due south, and the open clusters M6 and M7 are ideally placed for observation. The red supergiant Antares shines out. It is the brightest star in Scorpius and one of the largest stars visible to the naked eye.

Sagittarius lies in the direction of the center of our galaxy. It contains numerous deep-sky objects, including the prominent starfield M24 and the Lagoon Nebula (M8), both of which are visible to the naked eye. M22, the third-brightest globular cluster, is also in Sagittarius; it is easy to find with binoculars. A group of eight stars within the northern reaches of Sagittarius forms the Teapot asterism, but it is difficult to pick out in this star-studded part of the sky. The arc of stars at the forefeet of Sagittarius is easier to see. It is Corona Australis, the Southern Crown, and one of the smallest constellations of all.

Looking north, Ophiuchus, the Serpent Bearer, is high in the sky; below it lies Hercules. Vega in the constellation Lyra is close to the horizon. Adjoining Lyra is the Northern Hemisphere constellation Cygnus. Observers to the north of about 30°S are able to see the complete figure of the flying Swan, with Aquila, the Eagle, above it in the northeastern sky.

METEOR SHOWERS

The Delta Aquarid meteor shower reaches its peak around July 29. Up to 20 rather faint meteors radiate each hour from the southern half of Aquarius.

»

» JULY
NORTHERN LATITUDES

LOOKING NORTH

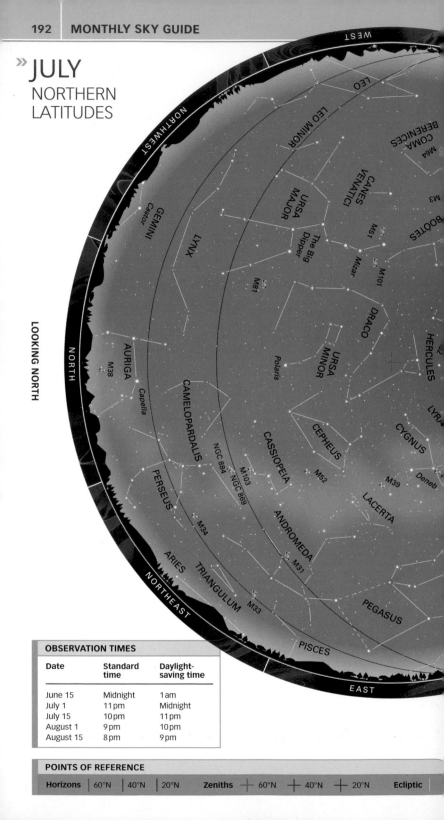

WEST

NORTHWEST

LEO

LEO MINOR

COMA BERENICES

M64

CANES VENATICI

M3

BOÖTES

URSA MAJOR

M51

Mizar

M101

Castor

GEMINI

LYNX

The Big Dipper

M81

DRACO

HERCULES

Polaris

URSA MINOR

LYRA

CYGNUS

NORTH

AURIGA

M38

Capella

CAMELOPARDALIS

CEPHEUS

M52

M39

Deneb

PERSEUS

NGC 884

M103

NGC 869

CASSIOPEIA

LACERTA

M34

ANDROMEDA

M31

ARIES

TRIANGULUM

M33

NORTHEAST

PEGASUS

PISCES

EAST

OBSERVATION TIMES

Date	Standard time	Daylight-saving time
June 15	Midnight	1 am
July 1	11 pm	Midnight
July 15	10 pm	11 pm
August 1	9 pm	10 pm
August 15	8 pm	9 pm

POINTS OF REFERENCE

| Horizons | 60°N | 40°N | 20°N | Zeniths | + 60°N | + 40°N | + 20°N | Ecliptic |

STAR MOTION

North

South

LOOKING SOUTH

WEST

SOUTHWEST

SOUTH

SOUTHEAST

EAST

CORVUS
M104

VIRGO

Spica

HYDRA

M83

CENTAURUS

COMA
BERENICES

SERPENS
CAPUT

Arcturus

BOÖTES

M5

LIBRA

LUPUS

CORONA
BOREALIS

HERCULES

OPHIUCHUS
M12
M10

M80
M19 Antares
M4

SCORPIUS

NORMA

M62

M14

M9

Shaula

ARA

LYRA

M57

SERPENS
CAUDA

M16
M17
M18 M23
M24
M8

M6

M7

M21

Albireo

VULPECULA
M27

SCUTUM
M26
M11
M25

M28

M22

M69
M54

CORONA
AUSTRALIS

TELESCOPIUM

PAVO

SAGITTA
Altair

AQUILA

SAGITTARIUS

M55

DELPHINUS

ECLIPTIC

MICROSCOPIUM

INDUS

M15

EQUULEUS
M2

CAPRICORNUS

PEGASUS

AQUARIUS

M30

PISCIS
AUSTRINUS

» JULY
SOUTHERN LATITUDES

LOOKING NORTH

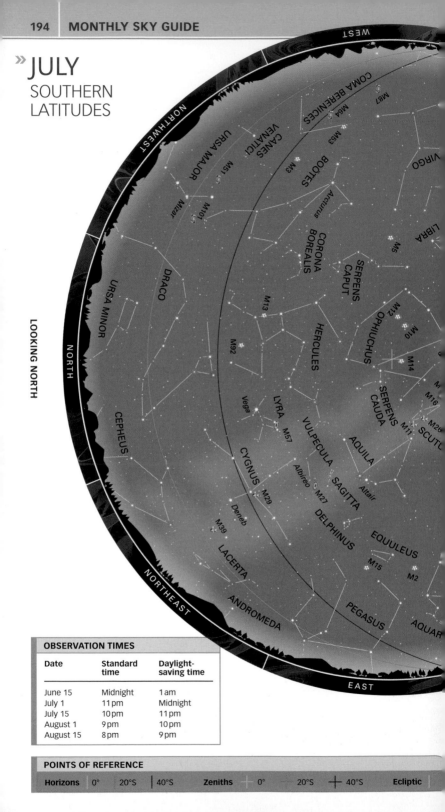

OBSERVATION TIMES

Date	Standard time	Daylight-saving time
June 15	Midnight	1 am
July 1	11 pm	Midnight
July 15	10 pm	11 pm
August 1	9 pm	10 pm
August 15	8 pm	9 pm

POINTS OF REFERENCE

Horizons	0°	20°S	40°S	Zeniths	+ 0°	20°S	+ 40°S	Ecliptic

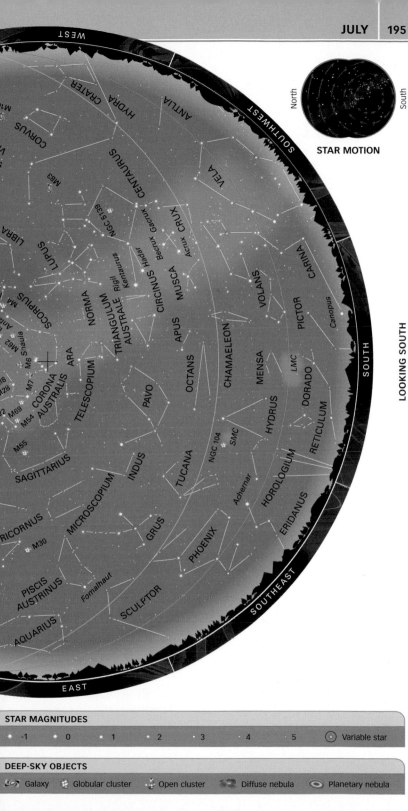

STAR MOTION

LOOKING SOUTH

STAR MAGNITUDES

-1   0   1   2   3   4   5   ⊙ Variable star

DEEP-SKY OBJECTS

🌀 Galaxy   ✦ Globular cluster   ✦ Open cluster   ☁ Diffuse nebula   ⊙ Planetary nebula

AUGUST

This month, northern observers get a glimpse toward the galaxy's center and are treated to the Perseid meteor shower. Sagittarius remains well placed for southern observers, who also have their best chance of observing the Dumbbell Nebula and the Ring Nebula.

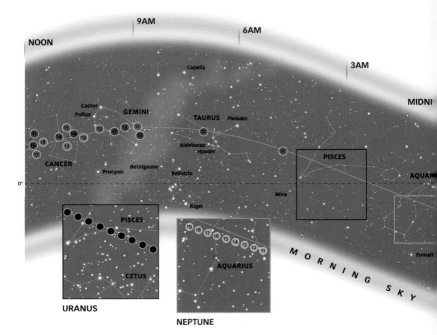

URANUS

NEPTUNE

NORTHERN LATITUDES

For observers in the Northern Hemisphere, the stars of the Summer Triangle lie directly overhead. As the sky darkens, blue-white Vega in Lyra is the first triangle star to appear. To its east is Cygnus, the Swan, which contains the second star, blue-white supergiant Deneb. The third, Altair in Aquila, the Eagle, is to the south.

Sagittarius, the Archer, is a centaur—a mythical beast with the head and torso of a man and the legs and body of a horse. His human half is just above the southern horizon, while his legs are only visible to observers south of about 40°N. Sagittarius marks the center of the galaxy. Here the Milky Way is at its densest and brightest, but a dark horizon is needed to reveal the star-studded path. It may be easier to see this starry path by looking up from

the horizon, even though it is less bright at this point. The path sweeps through Aquila, overhead into Cygnus, and then on to Cassiopeia in the northeastern sky. As it reaches Cygnus, a dark division runs along the length of the Milky Way. This is the Cygnus Rift—an opaque cloud of dust that lies in the foreground relative to the dense Milky Way star clouds, blocking out much of their light.

METEOR SHOWERS

The Perseid meteor shower peaks around August 12, when up to 80 meteors an hour can be seen radiating from a point in Perseus. Although Perseus itself does not clear the eastern horizon before midnight, some pre-midnight meteors may still be seen from its general direction.

Locating the planets

The chart below shows the positions of the planets in August from 2011 to 2019. The planets are represented by colored dots (see key, right), with the number inside each dot denoting the year. For all planets apart from Mercury, the dot indicates the planet's position on August 15. Mercury is shown only when it is at its greatest elongation.

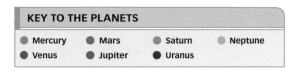

KEY TO THE PLANETS

- Mercury
- Venus
- Mars
- Jupiter
- Saturn
- Uranus
- Neptune

NOON

3PM

NIGHT

9PM

6PM

Altair

Arcturus

LEO

Regulus

OPHIUCHUS

VIRGO

ECLIPTIC

0°

CAPRICORNUS

Spica

LIBRA

Antares

SAGITTARIUS

Shaula

SCORPIUS

EVENING SKY

SOUTHERN LATITUDES

The three bright stars of the Summer Triangle dominate the sky to the north. The constellation of Vulpecula, the Fox, between Aquila and Cygnus, contains the Dumbbell Nebula (M27)—a planetary nebula that can be seen through binoculars. Another planetary nebula, the Ring Nebula (M57) in Lyra, can be seen through a small telescope. To the southeast, the constellation of Tucana, the Toucan, contains the globular cluster 47 Tucanae and the Small Magellanic Cloud.

The starfields of Sagittgarius are still high overhead; from here the galaxy's path flows on through Scorpius to the southwest, then through Lupus to Crux and the southern horizon. Alpha (α) and Beta (β) Centauri are low on the southwestern horizon and mark the front

legs of Centaurus, the Centaur. Alpha Centauri, also known as Rigil Kentaurus, is the third-brightest star in the sky and one of the closest. Beta Centauri, or Hadar, is eleventh-brightest.

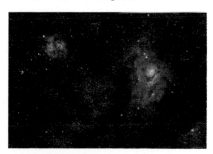

The Lagoon Nebula
The bright, star-forming Lagoon Nebula is one of the many deep-sky objects in Sagittarius. In a dark sky it is visible to the naked eye.

>>

» AUGUST
NORTHERN LATITUDES

LOOKING NORTH

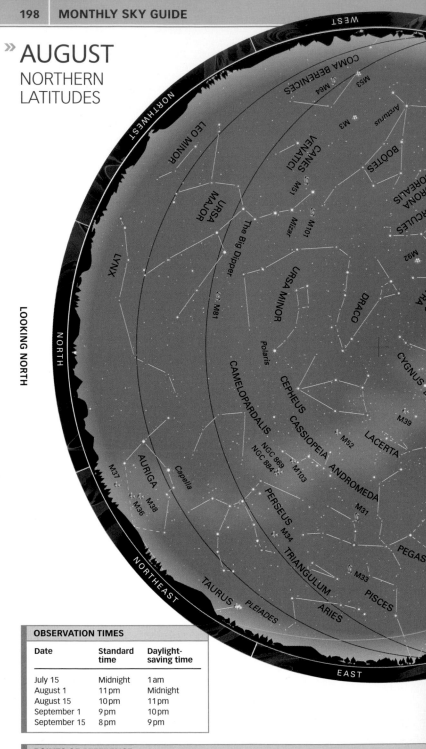

OBSERVATION TIMES

Date	Standard time	Daylight-saving time
July 15	Midnight	1 am
August 1	11 pm	Midnight
August 15	10 pm	11 pm
September 1	9 pm	10 pm
September 15	8 pm	9 pm

POINTS OF REFERENCE

Horizons	60°N	40°N	20°N	Zeniths	+ 60°N	+ 40°N	+ 20°N	Ecliptic

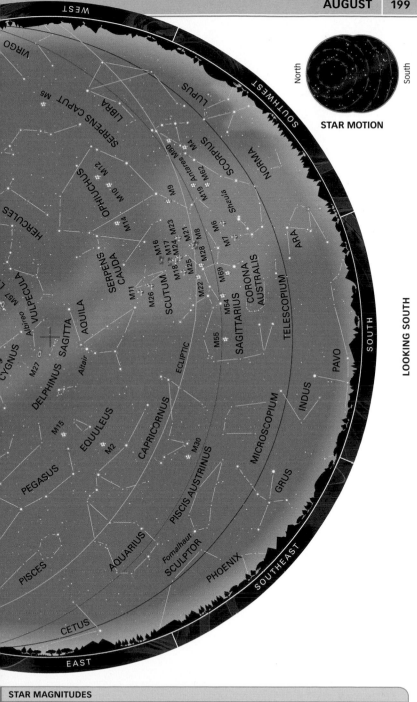

LOOKING SOUTH

STAR MOTION

North

South

STAR MAGNITUDES

-1 0 1 2 3 4 5 ⊙ Variable star

DEEP-SKY OBJECTS

Galaxy Globular cluster Open cluster Diffuse nebula Planetary nebula

»

AUGUST
SOUTHERN LATITUDES

LOOKING NORTH

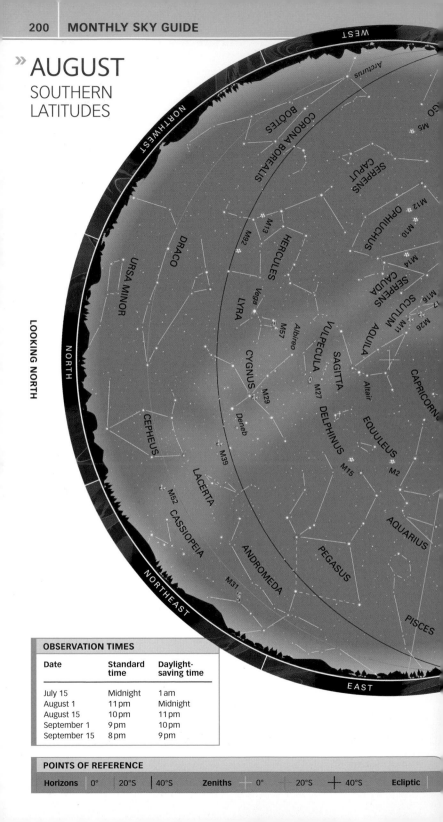

OBSERVATION TIMES

Date	Standard time	Daylight-saving time
July 15	Midnight	1 am
August 1	11 pm	Midnight
August 15	10 pm	11 pm
September 1	9 pm	10 pm
September 15	8 pm	9 pm

POINTS OF REFERENCE

Horizons	0°	20°S	40°S	Zeniths	0°	20°S	40°S	Ecliptic

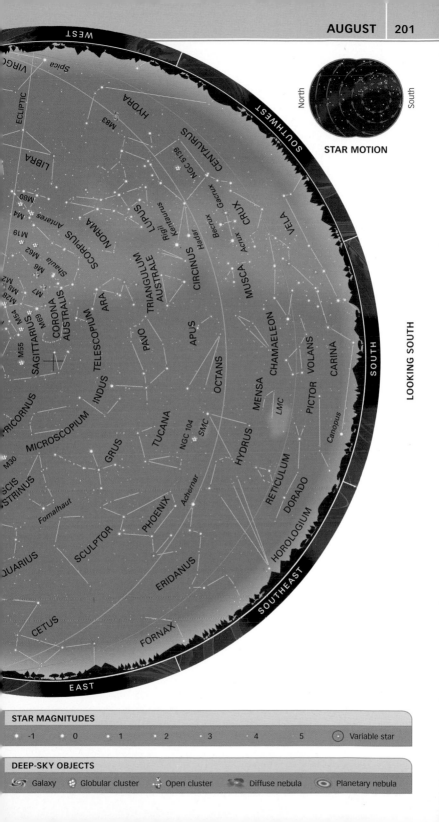

STAR MOTION

North — South

LOOKING SOUTH

WEST

SOUTHWEST

SOUTH

SOUTHEAST

EAST

STAR MAGNITUDES

-1   0   1   2   3   4   5   ⊙ Variable star

DEEP-SKY OBJECTS

🌀 Galaxy   ✿ Globular cluster   ✿ Open cluster   ☁ Diffuse nebula   ⊙ Planetary nebula

SEPTEMBER

The stars of northern autumn and southern spring are now in place, as Capricornus and Aquarius move center stage. On the 23rd of this month, or occasionally on the 22nd, the Sun moves from the northern to the southern half of the celestial sphere. As it crosses the celestial equator, day and night are of equal length.

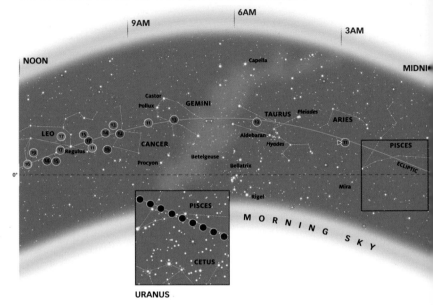

URANUS

NORTHERN LATITUDES

Seen from the Northern Hemisphere, the zodiac constellations Capricornus and Aquarius lie to either side of due south. For those at high northern latitudes, September is the best time to see Capricornus, which is the smallest constellation of the zodiac and one of the faintest. The brightest star in Capricornus, Alpha (α), called Algedi, is a double star. Sharp eyesight or binoculars show a giant of magnitude 3.6, and, six times more distant, a supergiant of magnitude 4.3.

Cygnus, the Swan, remains high overhead, where it is joined by the rising Pegasus, the Winged Horse. The Summer Triangle stars—Deneb (in Cygnus), Vega (in Lyra), and Altair (in Aquila)—remain in view to the west, but Pegasus's presence in the east announces the arrival of fall.

Close to Deneb lies the large, diffuse, but distinctive North America Nebula NGC 7000, which is visible through binoculars in dark skies.

Looking to the north, Ursa Major, the Great Bear, lies beneath Polaris, the Pole Star, while the constellations Cepheus and Cassiopeia are above it. Cepheus is not prominent but is worth seeking out for Delta (δ) Cephei, within the king's head This star is the prototype of the Cepheid variables, a class of yellow supergiant stars that vary in brightness as they change in size. As the star pulsates, its brightness shifts over a five-day cycle.

Only two bright stars can be seen to the north. Capella (in Auriga) lies close to the northeast horizon. A little brighter, and high in the sky to the west, lies Vega.

KEY TO THE PLANETS

- Mercury
- Venus
- Mars
- Jupiter
- Saturn
- Uranus
- Neptune

Locating the planets
The chart below shows the positions of the planets in September from 2011 to 2019. The planets are represented by colored dots (see key, left), with the number inside each dot denoting the year. For all planets apart from Mercury, the dot indicates the planet's position on September 15. Mercury is shown only when it is at its greatest elongation.

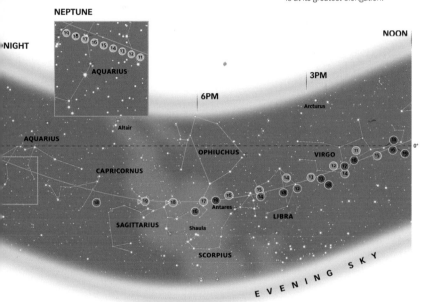

NEPTUNE

NIGHT

AQUARIUS

NOON

3PM

6PM

Arcturus

AQUARIUS

OPHIUCHUS

VIRGO

0°

CAPRICORNUS

SAGITTARIUS

Antares

LIBRA

Altair

Shaula

SCORPIUS

EVENING SKY

SOUTHERN LATITUDES

Capricornus and Aquarius lie high overhead. Neither has especially bright stars, but Aquarius in particular has objects of note. These include M2, a globular cluster that is easily seen through binoculars, and the Helix Nebula (NGC 7293). The latter is believed to be the closest planetary nebula to Earth, and

appears large but diffuse in the sky, covering a third of the apparent width of the Moon. Close by is the bright star Fomalhaut in Piscis Austrinus.

September offers the last chance to see the Summer Triangle in the northwest before Vega and, later, Deneb sink below the horizon. The path of the Milky Way can be seen in the western sky. Sagittarius remains high; below is Scorpius. Lowest of all are Centaurus and Crux, which are sinking below the southwestern horizon.

The blue-white star Achernar is well placed to the southeast. At magnitude 0.45, Achernar marks the end of the mythical river Eridanus. To its right is Tucana, the Toucan, home to the globular cluster 47 Tucanae (NGC 104), and the Small Magellanic Cloud.

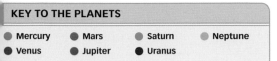

Deep-sky objects in Tucana
The Small Magellanic Cloud and globular cluster 47 Tucanae appear to be the same distance away from us. In fact, the galaxy (bottom), is 210,000 light-years away, while 47 Tucanae (top) is just 13,400 light-years away.

»

» SEPTEMBER
NORTHERN LATITUDES

LOOKING NORTH

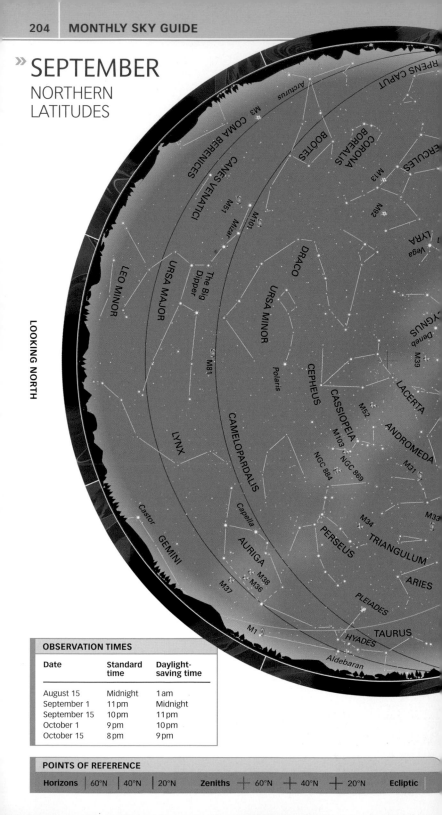

OBSERVATION TIMES

Date	Standard time	Daylight-saving time
August 15	Midnight	1 am
September 1	11 pm	Midnight
September 15	10 pm	11 pm
October 1	9 pm	10 pm
October 15	8 pm	9 pm

POINTS OF REFERENCE

| Horizons | 60°N | 40°N | 20°N | Zeniths | + 60°N | + 40°N | + 20°N | Ecliptic |

STAR MOTION

North

South

LOOKING SOUTH

STAR MAGNITUDES

-1 0 1 2 3 4 5 ⊙ Variable star

DEEP-SKY OBJECTS

🌀 Galaxy ⚜ Globular cluster ⚛ Open cluster ☁ Diffuse nebula ⊙ Planetary nebula

SEPTEMBER
SOUTHERN LATITUDES

LOOKING NORTH

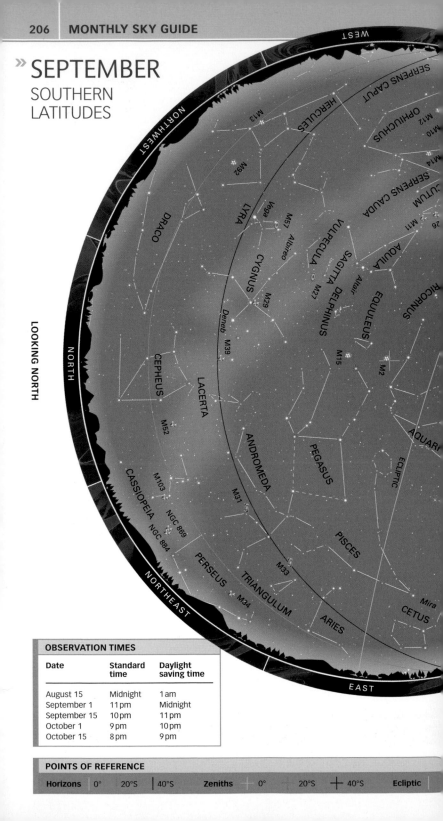

OBSERVATION TIMES

Date	Standard time	Daylight saving time
August 15	Midnight	1 am
September 1	11 pm	Midnight
September 15	10 pm	11 pm
October 1	9 pm	10 pm
October 15	8 pm	9 pm

POINTS OF REFERENCE

| Horizons | 0° | 20°S | 40°S | Zeniths | + 0° | + 20°S | + 40°S | Ecliptic |

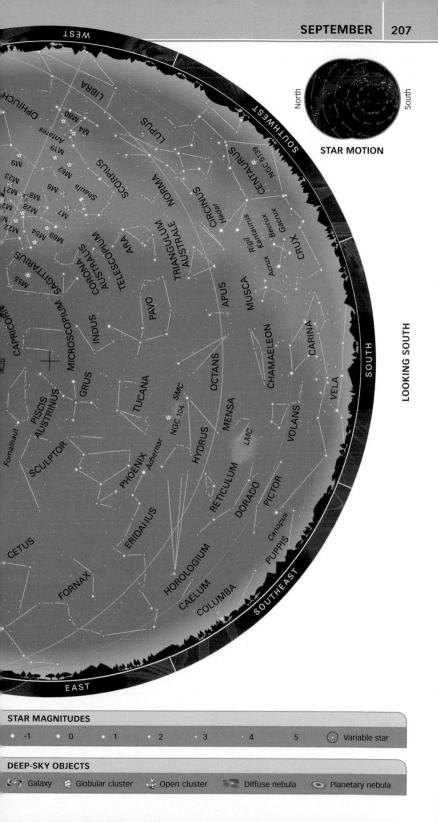

STAR MOTION

North
South

LOOKING SOUTH

WEST

SOUTHWEST

SOUTH

SOUTHEAST

EAST

LIBRA

OPHIUCHUS

M4 M80

Antares

M19

M62

Shaula

M6

M9

M23

M21

M8

M7

M28

M22

M54

M69

M55

SCORPIUS

LUPUS

NORMA

CENTAURUS

NGC 5139

CIRCINUS

Hadar

Rigil Kentaurus

Becrux

Gacrux

Acrux

CRUX

MUSCA

TRIANGULUM AUSTRALE

ARA

TELESCOPIUM

CORONA AUSTRALIS

SAGITTARIUS

APUS

CHAMAELEON

CARINA

PAVO

INDUS

MICROSCOPIUM

CAPRICORN

GRUS

TUCANA

OCTANS

MENSA

VOLANS

VELA

PISCIS AUSTRINUS

Fomalhaut

SMC

NGC 104

Achernar

HYDRUS

LMC

SCULPTOR

PHOENIX

RETICULUM

DORADO

PICTOR

CETUS

ERIDANUS

HOROLOGIUM

CAELUM

COLUMBA

Canopus

PUPPIS

FORNAX

STAR MAGNITUDES

-1   0   1   2   3   4   5   ⊙ Variable star

DEEP-SKY OBJECTS

🌀 Galaxy   ✿ Globular cluster   ⁘ Open cluster   ☁ Diffuse nebula   ◉ Planetary nebula

OCTOBER

The Andromeda Galaxy is now on view to all observers, as is the Great Square of Pegasus, an asterism formed by the brighter stars of the constellations Pegasus and Andromeda. Southern observers can see the Large and Small Magellanic Clouds, while Cassiopeia is well positioned for observers in northern latitudes.

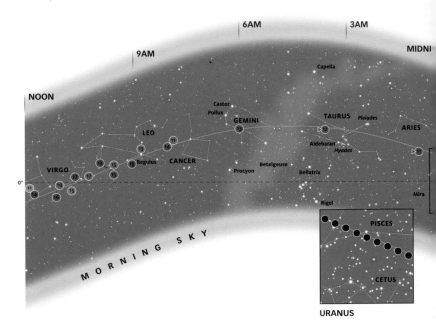

URANUS

NORTHERN LATITUDES

Pegasus and Andromeda lie to either side of due south. The Winged Horse and the Princess are linked by the Great Square of Pegasus. This is formed from three stars in Pegasus and one in Andromeda, and it defines the upper body of the

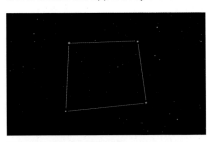

The Great Square of Pegasus
The upper body of Pegasus is formed by an asterism called the Great Square. Here, his head is below the square and his forelegs lie to the right.

mythological Winged Horse. Close to the horse's nose is M15, a globular cluster visible to the naked eye.

Andromeda is home to the Andromeda Galaxy (M31), the largest member of the Local Group galaxy cluster. Its central part can be seen with the naked eye, but a large telescope is needed for its spiral arms.

Looking north, Vega (in Lyra), Polaris, and Capella (in Auriga) trace a line across the sky, while the Milky Way is overhead. Cepheus and Cassiopeia lie above Polaris in their optimum positions for viewing.

METEOR SHOWERS

The Orionid meteor shower peaks around October 20, with some 25 fast meteors a hour. It is best seen after midnight, once Orion has risen in the east.

KEY TO THE PLANETS

- Mercury
- Venus
- Mars
- Jupiter
- Saturn
- Uranus
- Neptune

Locating the planets

The chart below shows the positions of the planets in October from 2011 to 2019. The planets are represented by colored dots (see key, left), with the number inside each dot denoting the year. For all planets apart from Mercury, the dot indicates the planet's position on October 15. Mercury is shown only when it is at its greatest elongation.

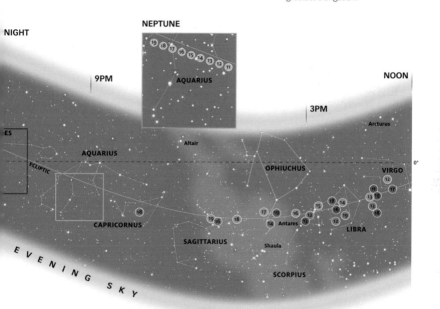

SOUTHERN LATITUDES

Looking north, the Great Square of Pegasus is centrally placed. It is formed by three stars in Pegasus, the Winged Horse (to the left for southern observers), and one star in Andromeda, the Princess (to the right). The horse flies toward the west. His head is defined by a line of stars; below, two lines trace his forelegs.

Andromeda is lower in the sky and to the right of Pegasus. Alpheratz, its brightest star, marks the Princess's head and is the fourth star of the Great Square. The Andromeda Galaxy (M31) is in her left knee. This spiral—similar to but larger than the Milky Way and 2.5 million light-years away—appears as an elongated oval to the naked eye. Its spiral arms and two companion galaxies, M32 and M110, can be seen through a large telescope.

A small telescope will reveal NGC 7662, the Blue Snowball planetary nebula, near Andromeda's right hand.

The bright star Fomalhaut in Piscis Austrinus, the Southern Fish, is almost overhead. Altair, Vega, and Deneb are setting in the northwest, while the summer constellations Taurus and Orion are beginning to appear in the east. Only one bright star, Achernar, is high above the horizon to the south, while four dim constellations depicting birds—Phoenix, Grus, Tucana, and Pavo—are center-stage. Tucana is well placed for observing both the Small Magellanic Cloud and the globular cluster 47 Tucanae, while close to the southeastern horizon is Dorado, home to the Large Magellanic Cloud—the larger and closer of our two companion galaxies. »

» OCTOBER
NORTHERN LATITUDES

LOOKING NORTH

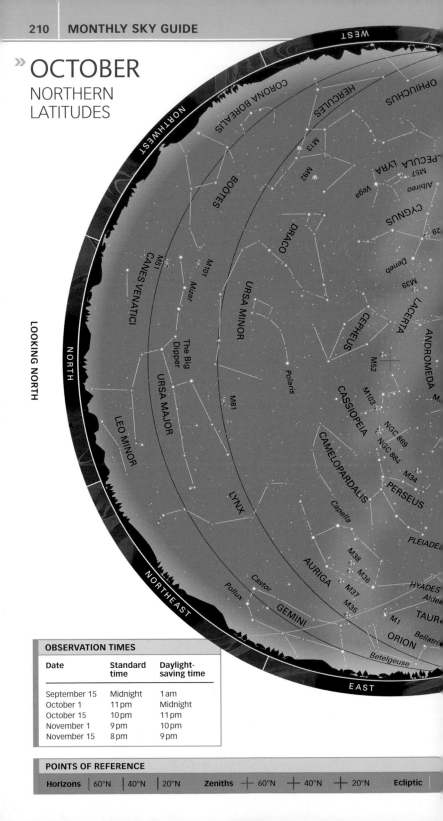

WEST

NORTHWEST

CORONA BOREALIS

HERCULES

OPHIUCHUS

M13

M92

LYRA

M57

Vega

Albireo

CYGNUS

29

BOÖTES

DRACO

Deneb

M39

LACERTA

CANES VENATICI

M51

Mizar

M101

M51

The Big Dipper

URSA MINOR

Polaris

CEPHEUS

M52

ANDROMEDA   M

NORTH

URSA MAJOR

M81

CASSIOPEIA

M103

NGC 869

NGC 884

M34

LEO MINOR

CAMELOPARDALIS

PERSEUS

LYNX

Capella

PLEIADES

AURIGA

M38

M36

HYADES

Aldebaran

M37

Castor

M35

M1

TAURUS

Pollux

GEMINI

Bellatrix

ORION

Betelgeuse

NORTHEAST

EAST

OBSERVATION TIMES

Date	Standard time	Daylight-saving time
September 15	Midnight	1 am
October 1	11 pm	Midnight
October 15	10 pm	11 pm
November 1	9 pm	10 pm
November 15	8 pm	9 pm

POINTS OF REFERENCE

Horizons	60°N	40°N	20°N	Zeniths	+ 60°N	+ 40°N	+ 20°N	Ecliptic

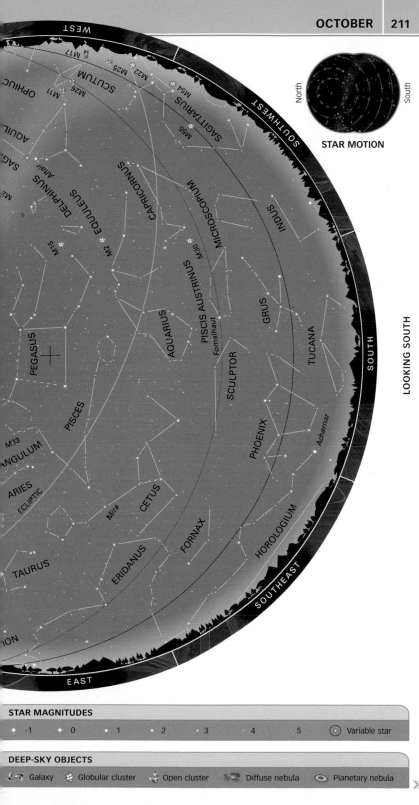

STAR MOTION

North

South

LOOKING SOUTH

STAR MAGNITUDES

-1   0   1   2   3   4   5   ⊙ Variable star

DEEP-SKY OBJECTS

Galaxy   Globular cluster   Open cluster   Diffuse nebula   Planetary nebula

» OCTOBER
SOUTHERN LATITUDES

LOOKING NORTH

OBSERVATION TIMES

Date	Standard time	Daylight-saving time
September 15	Midnight	1 am
October 1	11 pm	Midnight
October 15	10 pm	11 pm
November 1	9 pm	10 pm
November 15	8 pm	9 pm

POINTS OF REFERENCE

Horizons	0°	20°S	40°S	Zeniths	+ 0°	+ 20°S	+ 40°S	Ecliptic

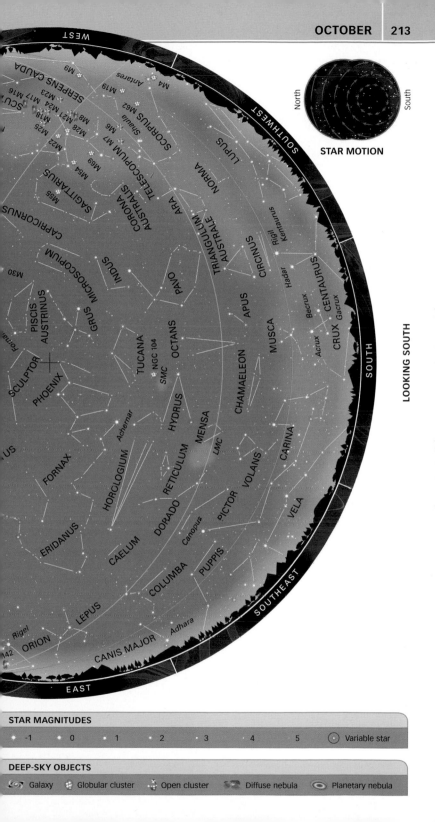

STAR MOTION

North

South

LOOKING SOUTH

STAR MAGNITUDES

-1   0   1   2   3   4   5   Variable star

DEEP-SKY OBJECTS

Galaxy   Globular cluster   Open cluster   Diffuse nebula   Planetary nebula

NOVEMBER

Two celebrated variable stars—Mira in Cetus, the Sea Monster, and Algol in Perseus—are now prominent in the sky for both northern and southern observers. In the northern latitudes, the Milky Way arches overhead, while four birds herald the arrival of the summer stars in the south.

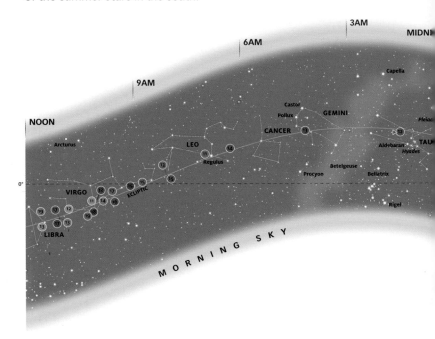

NORTHERN LATITUDES

Perseus and Andromeda are now high overhead, with Andromeda's parents, Cepheus and Cassiopeia, away to the north. Andromeda is well placed for observing the Andromeda Galaxy. Algol in Perseus is an eclipsing binary—a pair of stars in orbit around each other and too close to be seen separately, even through a powerful telescope.

Cetus lies due south; the red giant Mira in the monster's neck is a variable star whose brightness changes in a cycle lasting 11 months. Pisces lies above Cetus; to its right is the Great Square of Pegasus. The Summer Triangle of Altair, Vega, and

Deneb is low in the northwest, while the winter constellations Taurus, Gemini, and Orion are rising in the southeast.

METEOR SHOWERS

The Taurid meteor shower peaks in the first week of November; the Leonid shower peaks around November 17.

Leonid meteors
The Leonid meteor shower radiates from Leo. Like the Taurid shower in Taurus, about 10 meteors per hour are usually visible.

Locating the planets
The chart below shows the positions of the planets in November from 2011 to 2019. The planets are represented by colored dots (see key, left), with the number inside each dot denoting the year. For all planets apart from Mercury, the dot indicates the planet's position on November 15. Mercury is shown only when it is at its greatest elongation.

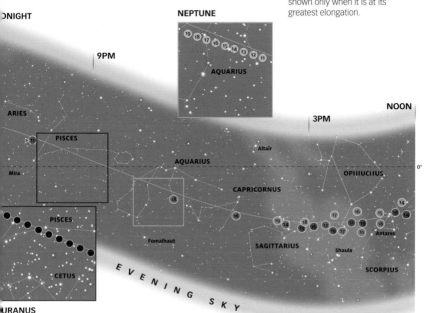

SOUTHERN LATITUDES

The constellation Cetus lies overhead, showcasing the variable Mira—a red giant whose brightness changes between magnitudes 3 and 10 over 11 months.

Pisces and Aries lie either side of due north. Pisces, depicting two fish joined with a cord, is a faint constellation. Its brightest star, alpha (α), marks the knot that ties the fish together, and a telescope reveals it as two stars. Below Pisces and to the northeast are Pegasus and Andromeda, with the Andromeda Galaxy still high enough to be seen.

November offers the chance to see Andromeda and Perseus together. Perseus, who killed the sea monster Cetus and prevented it from devouring Andromeda, is near the horizon in the northeast. The eclipsing binary Beta (β) Persei, or Algol, dips from magnitude 2.1 to 3.4 for 10 hours out of every 69 hours—a change visible to the naked eye.

Achernar in Eridanus is center stage to the south. To its west is Fomalhaut in Piscis Austrinus, the Southern Fish. The fish is the parent of the two in Pisces, and the recipient of water pouring from Aquarius's jug at right. Phoenix, Grus (the crane), Tucana (the toucan), and Pavo (the peacock) lie to the southwest. The Small Magellanic Cloud is in Tucana, and to its left is the Large Magellanic Cloud in Dorado.

Canopus, the white supergiant in Carina, lies to the southeast, with even brighter Sirius (in Canis Major) rising beyond it in the east. The presence of Taurus, Orion, and Canis Major above the horizon is a sign that summer is approaching.

»

» NOVEMBER
NORTHERN LATITUDES

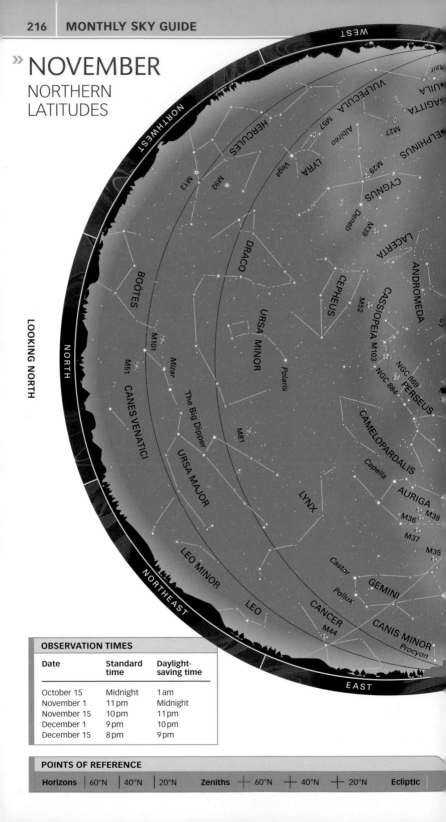

LOOKING NORTH

OBSERVATION TIMES

Date	Standard time	Daylight-saving time
October 15	Midnight	1 am
November 1	11 pm	Midnight
November 15	10 pm	11 pm
December 1	9 pm	10 pm
December 15	8 pm	9 pm

POINTS OF REFERENCE

| Horizons | 60°N | 40°N | 20°N | Zeniths | ✛ 60°N | ✛ 40°N | ✛ 20°N | Ecliptic |

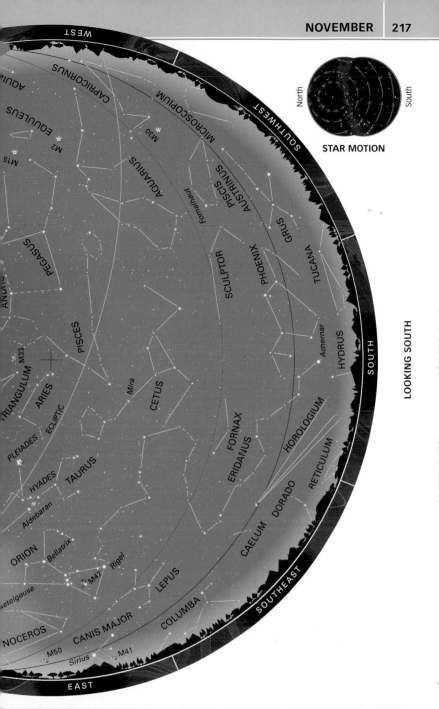

LOOKING SOUTH

STAR MOTION

North

South

STAR MAGNITUDES

-1   0   1   2   3   4   5   ⊙ Variable star

DEEP-SKY OBJECTS

🌌 Galaxy   ⭐ Globular cluster   ✴ Open cluster   🌫 Diffuse nebula   ⊙ Planetary nebula

»

» NOVEMBER
SOUTHERN LATITUDES

LOOKING NORTH

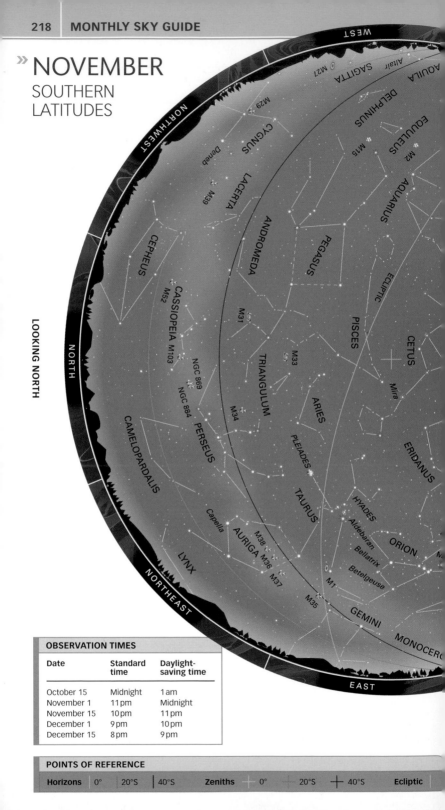

OBSERVATION TIMES

Date	Standard time	Daylight-saving time
October 15	Midnight	1 am
November 1	11 pm	Midnight
November 15	10 pm	11 pm
December 1	9 pm	10 pm
December 15	8 pm	9 pm

POINTS OF REFERENCE

Horizons	0°	20°S	40°S	Zeniths	+ 0°	+ 20°S	+ 40°S	Ecliptic

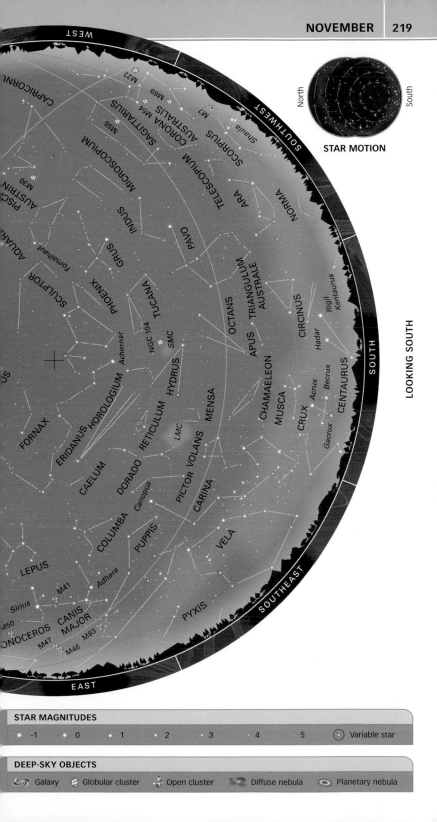

WEST

STAR MOTION

North

South

LOOKING SOUTH

SOUTH

SOUTHWEST

SOUTHEAST

EAST

CAPRICORNUS

M22

M69

M54

M70

Shaula

CORONA AUSTRALIS

SAGITTARIUS

SCORPIUS

M55

MICROSCOPIUM

TELESCOPIUM

ARA

NORMA

M30

PISCIS AUSTRINUS

INDUS

GRUS

PAVO

APUS

TRIANGULUM AUSTRALE

CIRCINUS

Rigil Kentaurus

Fomalhaut

AQUARIUS

PHOENIX

TUCANA

OCTANS

Hadar

SCULPTOR

NGC 104

SMC

CHAMAELEON

MUSCA

Becrux

CENTAURUS

Achernar

HYDRUS

MENSA

CRUX

Acrux

FORNAX

ERIDANUS

HOROLOGIUM

RETICULUM

LMC

Gacrux

CAELUM

DORADO

PICTOR

VOLANS

CARINA

COLUMBA

Canopus

PUPPIS

VELA

LEPUS

M41

Adhara

Sirius

M50

MONOCEROS

CANIS MAJOR

M47

M93

PYXIS

M46

STAR MAGNITUDES

-1   0   1   2   3   4   5   ⊙ Variable star

DEEP-SKY OBJECTS

Galaxy   Globular cluster   Open cluster   Diffuse nebula   Planetary nebula

DECEMBER

Taurus and Orion return to center stage as the year draws to a close. The Sun is at its farthest point south of the celestial equator around December 21–22. In the Northern Hemisphere this is the winter solstice, when the nights are their longest. In the Southern Hemisphere, it is the summer solstice, when the nights are at their shortest.

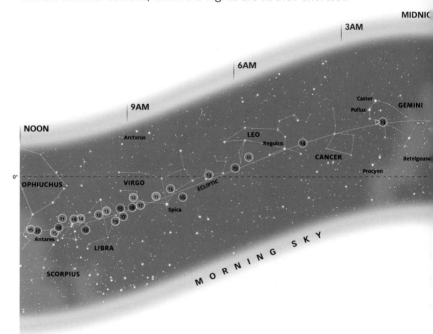

NORTHERN LATITUDES

The stars of winter occupy half the view to the south. Taurus leads the way and is almost due south, while Orion is in the southeast, and Gemini is in the eastern sky. Two particularly bright stars are visible: yellow Capella, in Auriga, is almost

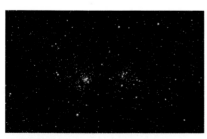

Double Cluster
The Double Cluster is a pair of open star clusters within Perseus: NGC 884 (left) and NGC 869 (right). Each contains hundreds of stars.

overhead, and brilliant-white Sirius, the brightest star of all, is near the southeast horizon. The water constellations, Pisces, the Fish, and Cetus, the Sea Monster, sink toward the western horizon. The stars of Taurus depict a bull's head and shoulders, with the V-shaped Hyades open cluster marking its face and the bright red giant Aldebaran its eye. A second cluster, the Pleiades (M45), marks the bull's back.

To the north, both Ursa Major and Ursa Minor are below Polaris; Andromeda and Perseus remain overhead. A knot of light in Perseus's hand is the Double Cluster.

METEOR SHOWERS

The Geminid meteor shower reaches its peak on December 13, when up to 100 meteors can be seen each hour.

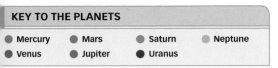

KEY TO THE PLANETS

- Mercury
- Venus
- Mars
- Jupiter
- Saturn
- Uranus
- Neptune

Locating the planets
The chart below shows the positions of the planets in December from 2011 to 2019. The planets are represented by colored dots (see key, left), with the number inside each dot denoting the year. For all planets apart from Mercury, the dot indicates the planet's position on December 15. Mercury is shown only when it is at its greatest elongation.

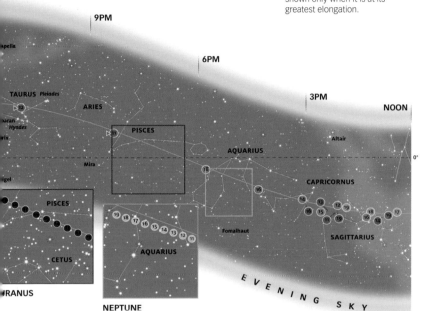

SOUTHERN LATITUDES

Looking north, the spring constellations Aquarius, Pisces, and Pegasus are moving toward the western horizon as the summer stars rise in the east. The inverted figure of Orion, the Hunter, announces that summer is almost here. It is flanked by Taurus to its lower left and Canis Major to the right. Below lies the constellation Gemini, the Twins. Castor and Pollux, the two bright stars marking their heads, are near the northeastern horizon.

Taurus is the front portion of a bull, which, from southern latitudes, is seen facing east. Its shoulders and head are close to the horizon, and its forelegs point upward. The Pleiades open cluster (M45) marks the bull's shoulder. Six of its stars are discernible to the naked eye, but the keen-sighted may see seven, explaining

the cluster's alternative name, the Seven Sisters. Many more stars are visible with optical aid. The bull's face is formed of a second V-shaped cluster, the Hyades, of which more than a dozen stars, as well as the unrelated red giant Aldebaran, are visible to the naked eye.

The meandering path of Eridanus, the River, can be traced from Orion's bright star, Rigel, overhead toward the southern horizon, stopping at Achernar to the right of due south. Brilliant-white Canopus (in Carina) lies to the left of Achernar. Between them, but closer to the horizon, are the Large and Small Magellanic Clouds.

The brightest star of all, Sirius in Canis Major, shines high in the east, forming one corner of a triangle of stars with Betelgeuse in Orion and Procyon in Canis Minor. »

» DECEMBER
NORTHERN LATITUDES

LOOKING NORTH

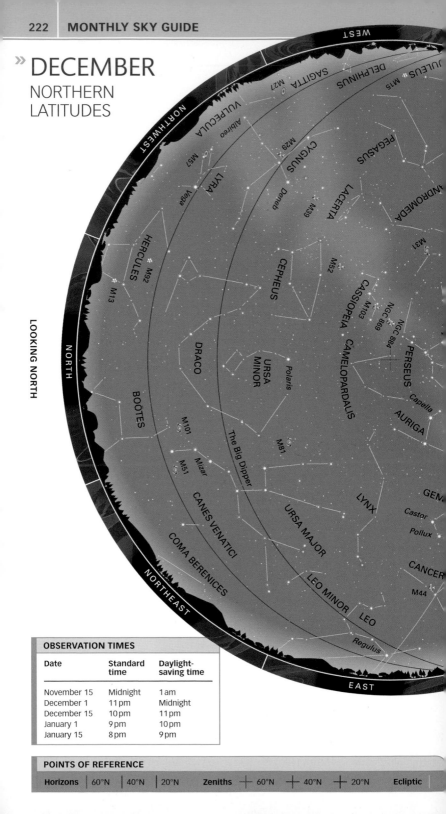

OBSERVATION TIMES

Date	Standard time	Daylight-saving time
November 15	Midnight	1 am
December 1	11 pm	Midnight
December 15	10 pm	11 pm
January 1	9 pm	10 pm
January 15	8 pm	9 pm

POINTS OF REFERENCE

Horizons	60°N	40°N	20°N	Zeniths	+ 60°N	+ 40°N	+ 20°N	Ecliptic

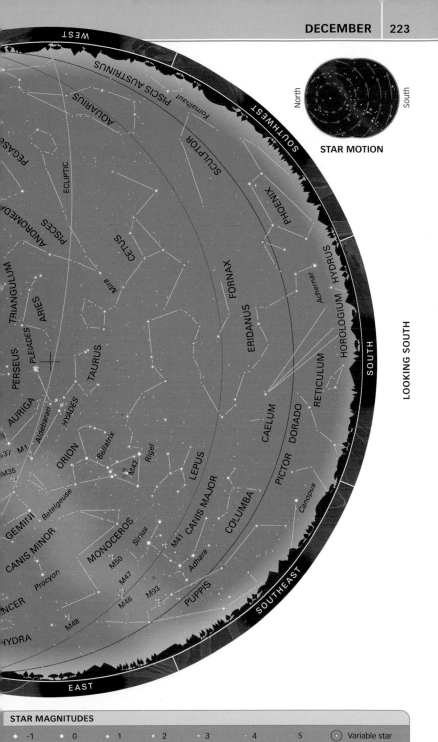

STAR MOTION

North

South

LOOKING SOUTH

WEST

PISCIS AUSTRINUS

AQUARIUS

SOUTHWEST

Fomalhaut

PEGAS...

ECLIPTIC

SCULPTOR

ANDROMED...

PISCES

PHOENIX

TRIANGULUM

ARIES

CETUS

Mira

FORNAX

HYDRUS

Achernar

PERSEUS

PLEIADES

TAURUS

ERIDANUS

HOROLOGIUM

RETICULUM

SOUTH

AURIGA

HYADES

Aldebaran

CAELUM

DORADO

PICTOR

...37 M1

ORION

Bellatrix

M42

Rigel

LEPUS

M35

Betelgeuse

GEMINI

MONOCEROS

Sirius

CANIS MAJOR

COLUMBA

Canopus

CANIS MINOR

Procyon

M50

M47

Adhara

SOUTHEAST

...NCER

M46

M93

PUPPIS

HYDRA

M48

EAST

STAR MAGNITUDES

-1   0   1   2   3   4   5   ⊙ Variable star

DEEP-SKY OBJECTS

Galaxy   Globular cluster   Open cluster   Diffuse nebula   Planetary nebula

»

» DECEMBER
SOUTHERN LATITUDES

LOOKING NORTH

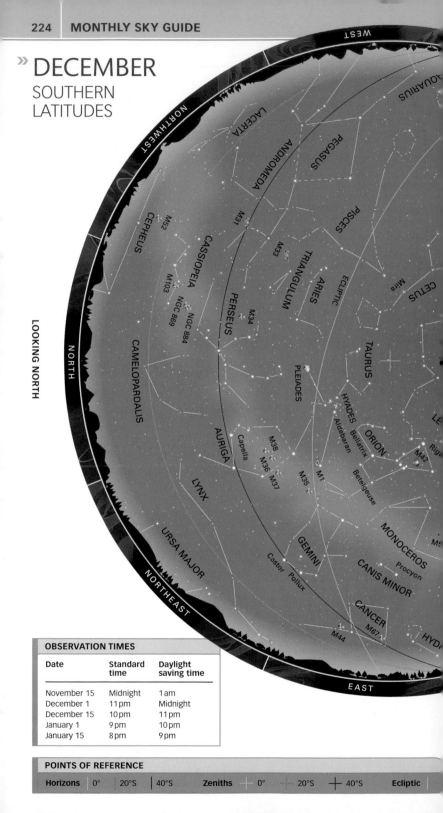

OBSERVATION TIMES

Date	Standard time	Daylight saving time
November 15	Midnight	1 am
December 1	11 pm	Midnight
December 15	10 pm	11 pm
January 1	9 pm	10 pm
January 15	8 pm	9 pm

POINTS OF REFERENCE

Horizons	0°	20°S	40°S	Zeniths	0°	20°S	40°S	Ecliptic

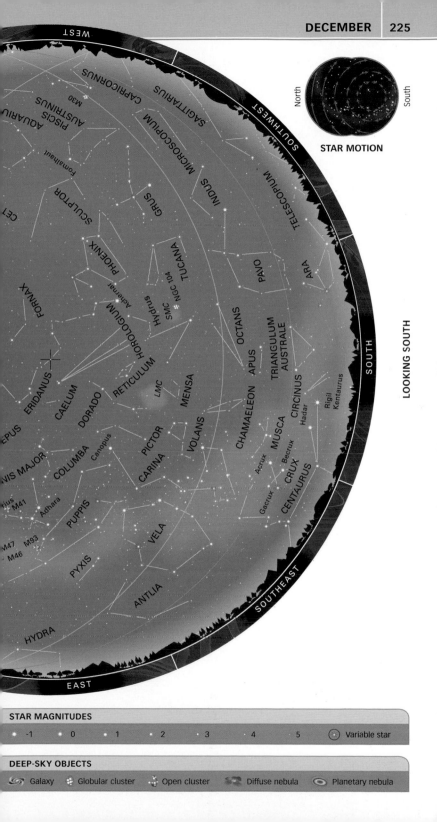

STAR MOTION

North

South

LOOKING SOUTH

STAR MAGNITUDES

-1 0 1 2 3 4 5 ⊙ Variable star

DEEP-SKY OBJECTS

🌀 Galaxy ✦ Globular cluster ⁂ Open cluster ▨ Diffuse nebula ⊙ Planetary nebula

THE
CONSTELLATIONS

The sky in Portugal, with Orion at the top of image, Taurus in the center, and Mars at lower right

HOW TO USE THIS SECTION

This section contains profiles of the 88 internationally recognized constellations. These are arranged by the constellations' position in the sky, starting with those at or near the north celestial pole and ending those at the southern pole.

Constellation profiles

Each entry contains a chart of the constellation. This shows all stars brighter than magnitude 6.5, with labels for stars brighter than magnitude 5. A selection of deep-sky objects is also shown. The text describes the constellations's origin, explains how to find it, and lists the main objects of interest for amateur observers.

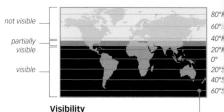

not visible

partially visible

visible

80°N
60°N
40°N
20°N
0°
20°S
40°S
60°S

Visibility

A world map shows the latitudes at which the constellation is visible, partially visible, or not visible at all.

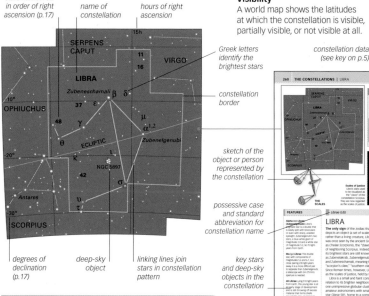

stars numbered in order of right ascension (p.17)

name of constellation

hours of right ascension

Greek letters identify the brightest stars

constellation data (see key on p.5)

constellation border

sketch of the object or person represented by the constellation

possessive case and standard abbreviation for constellation name

degrees of declination (p.17)

deep-sky object

linking lines join stars in constellation pattern

key stars and deep-sky objects in the constellation

Photographs
Some entries include a photograph of the constellation. Lines have been added to these images to join the stars in the constellation pattern. Other photographs show deep-sky objects.

Locator
This shows the position of the constellation on the celestial sphere and its position relative to the Milky Way.

THE GREEK ALPHABET

On most charts, some stars are identified by Greek letters. A constellation's brightest star is usually labeled Alpha, the second brightest Beta, and so on (see p.24).

α *Alpha*	η *Eta*	ν *Nu*	τ *Tau*
β *Beta*	θ *Theta*	ξ *Xi*	υ *Upsilon*
γ *Gamma*	ι *Iota*	ο *Omicron*	φ *Phi*
δ *Delta*	κ *Kappa*	π *Pi*	χ *Chi*
ε *Epsilon*	λ *Lambda*	ρ *Rho*	ψ *Psi*
ζ *Zeta*	μ *Mu*	σ *Sigma*	ω *Omega*

STAR MAGNITUDES

-1.5–0   0–0.9   1.0–1.9   2.0–2.9   3.0–3.9   4.0–4.9   5.0–5.9   6.0–6.9

DEEP-SKY OBJECTS

Galaxy   Globular cluster   Open cluster   Diffuse nebula   Planetary nebula or supernova remnant   Black hole or X-ray binary

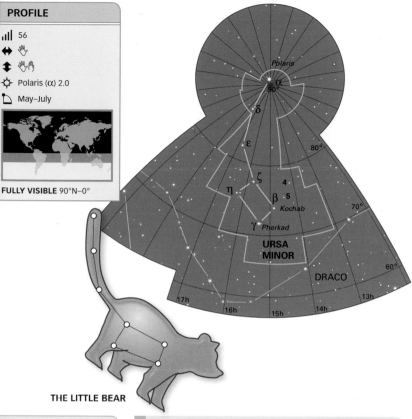

THE LITTLE BEAR

FEATURES

Alpha (α) Ursae Minoris (Polaris) A yellow supergiant star of magnitude 2.0, Polaris lies about 430 light-years away, barely half a degree from the north celestial pole. Until recently, it was classed as a Cepheid variable, but in the past few decades its variations have died away.

Beta (β) Ursae Minoris (Kochab) Ursa Minor's second-brightest star is an orange giant about 100 light-years away from Earth.

Ursae Minoris (UMi)

URSA MINOR

This constellation is a constant feature of the northern sky, never rising or setting but instead spinning around the north celestial pole once every 24 hours. With a shape that mimics the brighter and larger Big Dipper (or Plough) pattern of Ursa Major (pp.238–39), Ursa Minor is also known as the Little Dipper. It

NORTHERN HEMISPHERE

has been recognized since 600 BCE, when it was introduced by the Greek astronomer Thales of Miletus. It is named after Ida, a nymph who nursed the infant Zeus, although it is not clear why she is depicted as a little bear.

Ursa Minor is best known for containing the star Polaris or the Pole Star. Polaris occupies a point in the sky very close to the north celestial pole, in the "tail" of the constellation. The axis of Earth's rotation almost points to it, so that its position marks the direction of "true north" as seen from anywhere in the Northern Hemisphere.

PROFILE

.ıll 8

↔ 🖑🖑🖐🖐

↕ 🖑🖐🖐

☼ Thuban (α) 3.7

🗓 April–August

FULLY VISIBLE 90°N–4°S

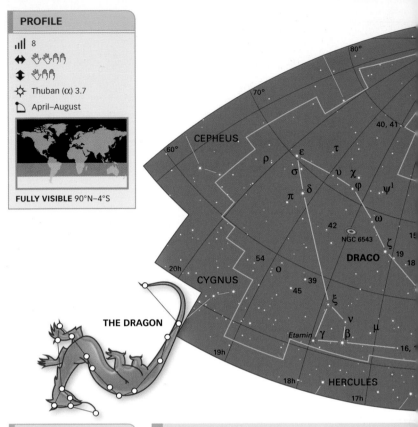

CEPHEUS

DRACO

THE DRAGON

CYGNUS

HERCULES

Etamin

FEATURES

Alpha (α) Draconis (Thuban) A blue-white giant about 300 light-years away, Thuban is Draco's brightest star, at an unimpressive magnitude 3.7. Due to precession (p.17), Thuban was the northern pole star 5,000 years ago.

Nu (ν) Draconis Marking the "head" of the dragon, Nu (ν) is an attractive double star for viewing with binoculars. It consists of twin white stars with a combined magnitude of 4.9.

16 and 17 Draconis This pair of stars, of magnitudes 5.1 and 5.5, is easily divided with binoculars, but a small telescope will show that the brighter star is itself a double.

NGC 6543 (the Cat's Eye Nebula) This is one of the brightest planetary nebulae.

Draconis (Dra)

DRACO

A large constellation, Draco wraps itself around Ursa Minor and the north celestial pole. According to Greek mythology, it represents a dragon that guarded an orchard of golden apples belonging to the Hesperides, the daughters of the Titan Atlas. Draco was killed by Hercules, and in the night sky the constellation Hercules stands on Draco's head. Draco also represents the dragon that guarded the golden fleece and was killed by Jason and the Argonauts.

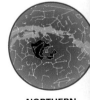

NORTHERN HEMISPHERE

The constellation is the eighth largest in the sky, but despite its size it has no stars brighter than magnitude 2. It is also lacking in nebulae and star clusters, but it does at least contain some faint galaxies and one interesting planetary nebula. Draco is home to one of the lesser annual meteor showers, the Draconids, which peak around October 9 every year.

URSA MINOR

λ

6   κ   4

URSA
MAJOR

10h

11h

12h

α   10
Thuban

13h

BOÖTES

14h

15h

The Cat's Eye Nebula
This planetary nebula is best seen
with a telescope. Large instruments
reveal a complex structure of gas
bubbles. This view was captured
by the Hubble Space Telescope.

Dragon and bear
The long body of Draco curls around
the stars of Ursa Minor, the Little
Bear. The head of the dragon can
easily be identified at the bottom.

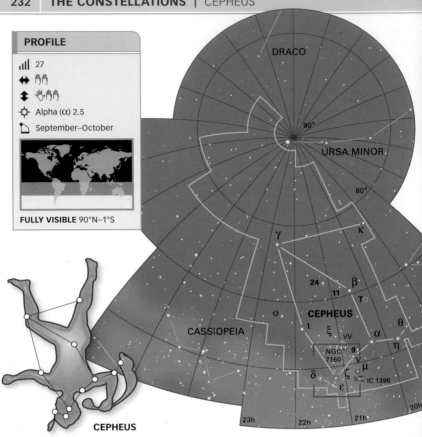

CEPHEUS

FEATURES

Beta (β) Cephei A blue giant with a faint companion, this is Cepheus's second-brightest star. Its brightness varies in a 4.6-hour cycle but only by 0.1 magnitude.

Delta (δ) Cephei This is a prototype star for a class of variables called Cepheids. An aging yellow supergiant, it is passing through a phase of its life where it expands and contracts repeatedly. It changes brightness between magnitudes 3.5 and 4.4 in a little under five days and nine hours.

Mu (μ) Cephei Called the Garnet Star because of its blood-red color, Mu (μ) Cephei is a red supergiant. Like Delta (δ) Cephei, it is a variable star, but it is less predictable, varying between magnitudes 3.4 and 5.1 in a period of about two years.

Cephei (Cep)

CEPHEUS

The constellation of Cepheus
lies in the far-northern sky between
the prominent Cassiopeia and
Draco. Its main stars form a distorted
tower or steeple shape, but this
ancient Greek star pattern in fact
represents the mythical King
Cepheus of Ethiopia, who was
the husband of the vain queen

**NORTHERN
HEMISPHERE**

Cassiopeia and father of the princess Andromeda.
    Cepheus is not a prominent constellation—it contains
no stars brighter than magnitude 2.5. However, the
northern reaches of the Milky Way pass through this
part of the sky, so Cepheus is not without interest. In
particular, it contains several interesting variable stars,
the best known of which are the highly reliable Delta (δ)
Cephei and the less predictable Mu (μ) Cephei (see chart
opposite), or The Garnet Star, a red supergiant surrounded
by IC 1396, a large region of star-forming nebulosity.

DELTA (δ) AND MU (μ) CEPHEI

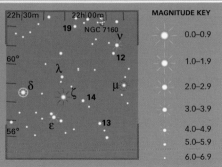

MAGNITUDE KEY	
●	0.0–0.9
●	1.0–1.9
●	2.0–2.9
●	3.0–3.9
•	4.0–4.9
•	5.0–5.9
·	6.0–6.9

HENRIETTA LEAVITT

The remarkable properties of Cepheid variable stars, named after Delta (δ) Cephei, were discovered in the early 20th century by Henrietta Leavitt. She identified Cepheids in the Small Magellanic Cloud (SMC) and reasoned that they were all at more or less the same distance, so their magnitudes would reflect their true luminosities. This revealed the link between luminosity and variation period, which can be used to calculate distance. Her discovery is integral to our knowledge of the size of the Universe.

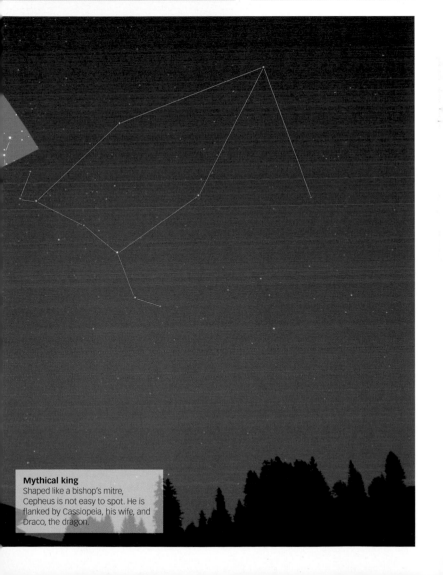

Mythical king
Shaped like a bishop's mitre, Cepheus is not easy to spot. He is flanked by Cassiopeia, his wife, and Draco, the dragon.

CEPHEUS

CASSIOPEIA

50
48

70°
ι
ω
ψ

60°
ε
NGC 637
SN 1572
4

CAMELO-
PARDALIS
IC 1805
NGC 559
M52
δ
γ
κ
τ
Cas A
1

IC 1848
NGC 663
M103
χ
υ
12
β
ρ
NGC 7789

LACERTA

50°
φ
NGC
457
η
α
Shedir
σ

PERSEUS
θ
ζ
λ
ν
ξ

ANDROMEDA
23h

0h

o
π

CASSIOPEIA

PROFILE

📶 25

↔ 🖐🖐

↕ 🖐🖐

☼ Shedir (α), Gamma (γ) 2.2

▢ October–December

FEATURES

Gamma (γ) Cassiopeiae
This hot, rapidly rotating star occasionally throws off rings of gas from its equator, causing unpredictable changes in its brightness. It currently lies at magnitude 2.2, making it the equal-brightest star in the constellation.

Rho (ρ) Cassiopeiae This highly luminous, yellow-white supergiant fluctuates between 4th and 6th magnitudes every 10 or 11 months. It is more than 10,000 light-years away, which is exceptionally distant for a naked-eye star.

M52 One of several open clusters in Cassiopeia that can be seen with small instruments, M52 is visible through binoculars as a somewhat elongated patch of light. Its individual stars can be seen with a small telescope.

Cassiopeiae (Cas)

CASSIOPEIA

A distinctive constellation of the northern sky, Cassiopeia lies on the opposite side of the Pole Star from Ursa Major, against a rich part of the Milky Way between Perseus and Cepheus. The large W-shape formed by its five main stars is easily recognizable. Due to its high northern declination, the

NORTHERN HEMISPHERE

constellation is circumpolar from many parts of the Northern Hemisphere, which means it is always visible somewhere in the sky and never sets below the horizon.

According to Greek mythology, Cassiopeia was the queen of the ancient kingdom of Ethiopia, wife to King Cepheus, and mother to Andromeda. She angered the Nereids, daughters of the sea-god Poseidon, which resulted in the sacrifice of Andromeda. Cassiopeia contains a variety of clusters and nebulae that can be observed by the amateur astronomer.

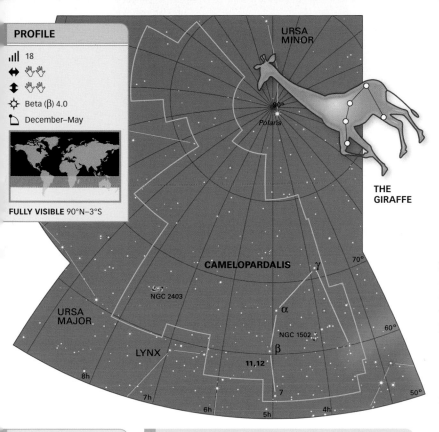

FEATURES

Alpha (α) Camelopardalis
Despite its designation, Alpha (α) is the constellation's second brightest star. It is a blue supergiant, but because it lies about 3,000 light-years away, it shines at only magnitude 4.3.

Beta (β) Camelopardalis
Outshining Alpha (α) with a magnitude of 4.0, Beta (β) is a yellow supergiant about 1,000 light-years away, with a faint, magnitude-8.6 companion star.

NGC 1502 This small open star cluster, with about 45 members, is visible through binoculars and is some 3,100 light-years from Earth.

NGC 2403 This spiral galaxy lies at a distance of 12 million light-years. A small telescope will show it as an 8th-magnitude, elliptical smudge.

Camelopardalis (Cam)

CAMELOPARDALIS

Introduced in the year 1613
by the Dutch astronomer and theologian Petrus Plancius, this faint constellation of the northern sky represents a giraffe. Its Latin name is a hybrid word meaning "leopard camel." Its brightest stars are Alpha (α) and Beta (β) Camelopardalis. The constellation's long neck can be

NORTHERN HEMISPHERE

visualized as stretching around the north celestial pole toward Ursa Minor and Draco. It is most easily found by locating one of its neighbors, Ursa Major.

Although Camelopardalis covers a large area of the sky, it lies at some distance from the Milky Way, so it contains no stars brighter than magnitude 4 and no bright clusters or nebulae. One of its most attractive features, covering an area five times the diameter of the full Moon, is a trail of stars known as Kemble's Cascade, tumbling across the sky toward the star cluster NGC 1502.

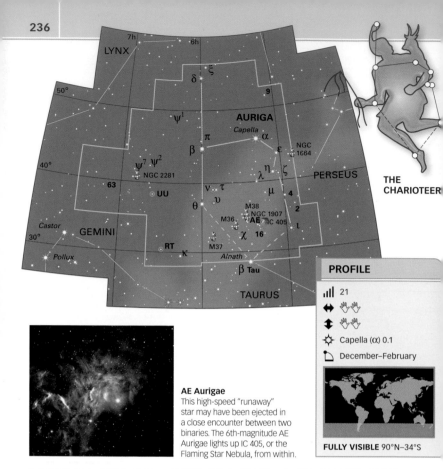

LYNX

7h   6h

50°

δ   ξ

9

ψ¹

AURIGA

Capella

π

α

β

ε

NGC 1664

η ζ

PERSEUS

40°

ψ⁷ ψ²

λ

NGC 2281

63

θ

ν τ

μ 4

UU

υ

M38

2

M36   NGC 1907   IC 405

AE

ι

Castor

30°

GEMINI

χ 16

Pollux

RT

κ

M37

Alnath

β **Tau**

TAURUS

THE
CHARIOTEER

AE Aurigae
This high-speed "runaway"
star may have been ejected in
a close encounter between two
binaries. The 6th-magnitude AE
Aurigae lights up IC 405, or the
Flaming Star Nebula, from within.

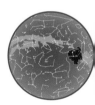
FEATURES

Alpha (α) Aurigae (Capella)
The sixth-brightest star in
the sky, Capella shines at
magnitude 0.1 and is just 42
light-years from Earth. It is a
binary system, composed of
twin yellow giants that orbit
each other in 104 days, far
too close for a telescope to
separate them.

**Epsilon (ε) Aurigae
(Almaaz)** The northernmost
"Kid" is a curious eclipsing
binary (p.127), whose eclipses
last for about one year out
of 27. This intensely luminous
supergiant is orbited by a dark
partner that is both huge and
semiopaque—perhaps a
star orbited by a disk of
planet-forming material.

Zeta (ζ) Aurigae This
southwestern member of a
trio of stars forming "the Kids"
is also an eclipsing binary.

Aurigae (Aur)

AURIGA

This constellation lies in the
Milky Way between Gemini and
Perseus, to the north of Orion. It
is easily identified in the northern
sky by the yellowish presence of
Alpha (α) Aurigae (Capella) the
most northerly 1st-magnitude
star. It is usually said to represent
Erichthonius, an ancient king of

**NORTHERN
HEMISPHERE**

Athens and a skilled charioteer. Interestingly, Auriga also
has another set of associations. It is said to represent a
goat that suckled the infant god Zeus, while Epsilon (ε),
Eta (η), and Zeta (ζ) Aurigae are known as "the Kids."

Auriga's southernmost star, once called Gamma (γ)
Aurigae, is shared with Taurus and is now officially known
as Beta (β) Tauri. It marks the northern tip of the celestial
bull's horns. The Milky Way passes diagonally across the
constellation, making it rich in interesting stars and
clusters, including the open clusters M36, M37, and M38.

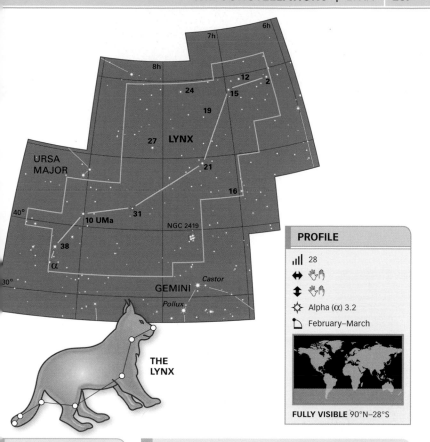

PROFILE

📶 28

↔ 🖐🖐

↕ 🖐🖐

☼ Alpha (α) 3.2

🗓 February–March

FULLY VISIBLE 90°N–28°S

THE LYNX

FEATURES

Alpha (α) Lyncis Lying around 150 light-years from Earth, this is a red giant of magnitude-3.2.

12 Lyncis This white star appears at a faint magnitude 4.9 to the naked eye, but a small telescope will reveal a blue-white companion of magnitude 7.3. Larger instruments show that the brighter star is itself a binary star of magnitudes 5 and 6, which orbit each other every 700 years. This triple-star system sits 140 light-years from Earth.

NGC 2419 A faint globular cluster of magnitude 10, NGC 2419 is only visible through a telescope of moderate size. It lies 210,000 light-years away from Earth, far more distant than the Milky Way's other globular clusters.

Lyncis (Lyn)

LYNX

A relatively late addition to the classical constellations, this faint northern group of stars is surprisingly large. Lynx covers an area greater than the constellation Gemini. It was introduced by the Polish astronomer Johannes Hevelius in the 1680s in order to fill the gap between Ursa Major and

NORTHERN HEMISPHERE

Auriga. Hevelius is reputed to have named it Lynx because only the lynx-eyed would be able to see it—he himself was renowned for his sharp eyesight. However, the animal he drew on his star chart bore little resemblance to a real lynx. Likewise, this chain of faint stars bears no resemblance to the European wildcat.

In clear-sky conditions, naked-eye observers will see little more than its brightest star, Alpha (α) Lyncis, but Lynx does contain interesting double and multiple stars that will reward telescope users.

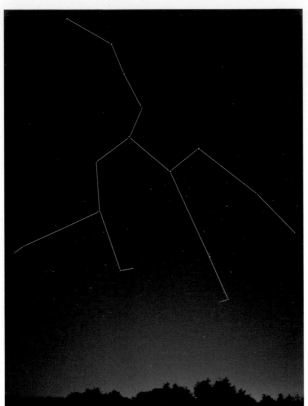

Familiar shape
The saucepan shape formed by seven stars of the Big Dipper is one of the most easily recognized shapes in the sky, but it makes up only part of the constellation pattern of Ursa Major.

BOÖTES

FEATURES

Alpha (α) Ursae Majoris (Dubhe) A yellow giant, Dubhe lies just over 100 light-years away and shines at magnitude 1.8. A line from Beta (β) Ursae Majoris (Merak) through Dubhe points toward the pole star, Polaris, in Ursa Minor.

Zeta (ζ) Ursae Majoris (Mizar) This is a famous double star. Its neighbor Alcor just happens to lie in the same direction, but a small telescope will show that Mizar is also a true binary with a much closer companion.

M81 (Bode's Galaxy) This bright spiral galaxy is just 10 million light-years away, but still requires a small telescope. The elongated shape of M82, the Cigar Galaxy, can be spotted nearby. It is found one diameter of the Moon away from M81.

Ursae Majoris (UMa)

URSA MAJOR

One of the best-known

constellations of the northern sky, Ursa Major's seven brightest stars form the familiar pattern of the Big Dipper, or Plow, a useful signpost to other stars. However, the constellation's fainter stars extend far beyond this central pattern— Ursa Major is, in fact, the third-largest constellation in the entire sky. It was regarded by many ancient civilizations as a bear, which in Greek myth was associated with the beautiful Callisto, who was transformed into a bear by the goddess Artemis.

NORTHERN HEMISPHERE

Most of the bright stars of the Big Dipper are members of the Ursa Major Moving Cluster—an open star cluster so close to Earth (about 70 light-years away) that its stars appear widely scattered in the sky. Beyond them lie several nearby galaxies that make rewarding targets for amateur astronomers.

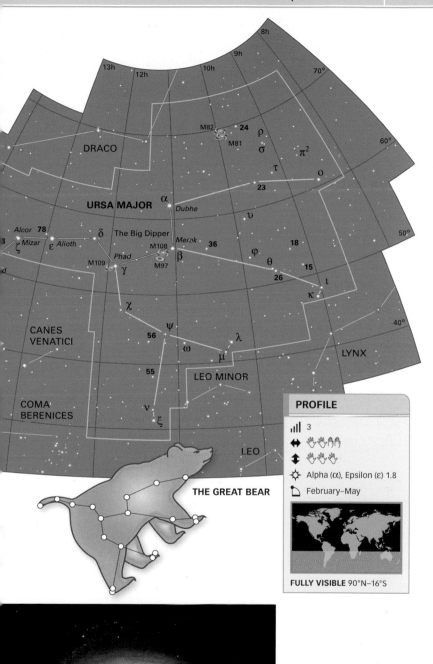

DRACO

M82 **24** ρ

M81 σ

τ π² o

23

URSA MAJOR α υ
Dubhe

Alcor **78** δ The Big Dipper Merak **36** 18

3 ζ Mizar ε Alioth M108 φ θ

M109 Phad β **15**

γ M97 **26** ι

χ κ

CANES ψ λ
VENATICI **56** ω μ LYNX

55 LEO MINOR

COMA
BERENICES ν ζ

LEO

THE GREAT BEAR

PROFILE

▮▮▮ 3

↔ 🖐🖐🖐

↕ 🖐🖐

☼ Alpha (α), Epsilon (ε) 1.8

▢ February–May

FULLY VISIBLE 90°N–16°S

Bode's Galaxy
This large, tightly wound spiral
galaxy, as seen by the Hubble Space
Telescope, has an unusually bright
core—an "active galactic nucleus"
whose activity is triggered as
material is dragged into a black hole.

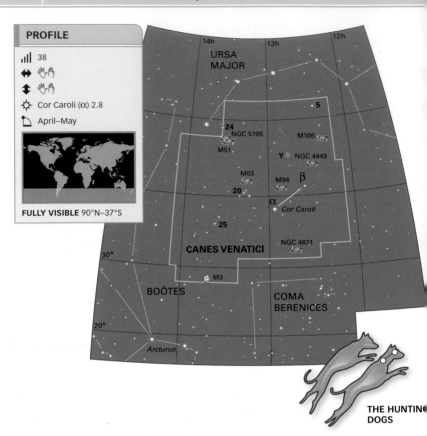

PROFILE

▮▮▮ 38

↔ 🖐🖐

↕ 🖐🖐

☼ Cor Caroli (α) 2.8

🗓 April–May

FULLY VISIBLE 90°N–37°S

URSA MAJOR

14h 13h 12h

5

24
NGC 5195
M51
M106
Y ⊙ *NGC 4449*
M63
M94 β
20
α
Cor Caroli
25
NGC 4631
CANES VENATICI
M3
BOÖTES
30°
COMA BERENICES
20°
Arcturus

THE HUNTING DOGS

FEATURES

Alpha (α) Canum Venaticorum (Cor Caroli) Binoculars will show that this star, whose name is usually shortened to simply Cor Caroli, is a wide binary system having two white stars with magnitudes 2.8 and 5.6. It lies 82 light-years away from Earth.

M3 One of the northern sky's best globular clusters, M3 appears as a fuzzy star in binoculars and a hazy ball of light through small telescopes.

M51 (the Whirlpool Galaxy) This spectacular spiral galaxy is bright and relatively close— some 31 million light-years away. It lies face to Earth, so binoculars or, ideally, a small telescope will show the bright core, while medium-sized instruments will reveal traces of the spiral arms that give the galaxy its name.

Canum Venaticorum (CVn)

CANES VENATICI

Lying in the northern sky between Boötes and Ursa Major, Canes Venatici's position beneath the "handle" of the Big Dipper (pp.238–39) makes it relatively easy to identify. It depicts a pair of hunting dogs used by the herdsman Boötes to chase the Great and Little Bears, represented by the constellations

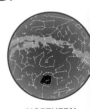

NORTHERN HEMISPHERE

Ursa Major and Ursa Minor, around the north celestial pole. It was created by the Polish astronomer Johannes Hevelius at the end of the 17th century.

A small constellation, Canes Venatici has no memorable pattern of stars to identify it, but it does contain the relatively bright star Cor Caroli Regis Martyris "the Heart of King Charles the Martyr," named in honor of King Charles I of England, who was executed at the end of the English Civil War of the 1640s. This constellation is also home to the bright Whirlpool Galaxy, M51.

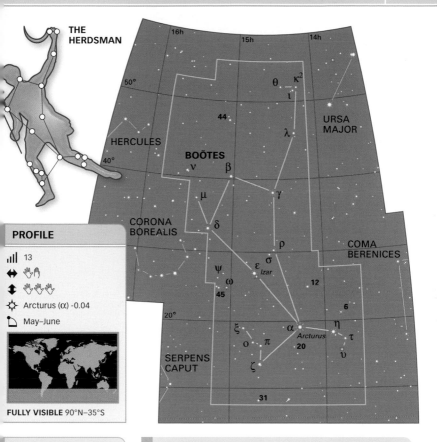

THE
HERDSMAN

16h 15h 14h

50°

θ κ²
ι

44

λ

URSA
MAJOR

HERCULES

BOÖTES

ν β

40°

μ

γ

CORONA
BOREALIS

δ

ρ

COMA
BERENICES

ψ

ε
Izar
ω

σ

12

45

6

20°

ξ

α
Arcturus

η
τ

ο
π

20

υ

SERPENS
CAPUT

ζ

31

PROFILE

📶 13

✋ 🖐🖐

↕ 🖐🖐🖐

☼ Arcturus (α) -0.04

🗓 May–June

FULLY VISIBLE 90°N–35°S

FEATURES

Alpha (α) Boötis (Arcturus)
At magnitude 0.04, Arcturus
is one of the closest and
brightest stars in the sky. It is
an orange giant nearing the
end of its life and is located
just 36 light-years away.

Epsilon (ε) Boötis (Izar)
Also called Pulcherrima, this is
a beautiful double star. A small
telescope will split it to reveal
an orange giant of magnitude
2.7 accompanied by a blue star
of magnitude 5.1. The pair lies
about 203 light-years away.

Tau (τ) Boötis
This uninspiring star of
magnitude 4.5 is notable as a
host to one of the first planets
discovered beyond our solar
system. Tau (τ) is a yellow star
quite similar to the Sun and 51
light-years from Earth. A giant
planet, three times the size of
Jupiter, orbits it every 3.3 days.

Boötis (Boo)

BOÖTES

A large and prominent kite-
shaped constellation, Boötes
extends from the constellation
Draco and the "handle" of the Big
Dipper (pp.238–39) in Ursa Major
toward the constellation Virgo in
the south. According to Greek
mythology, Boötes was the
herdsman who drove the bear Ursa

**NORTHERN
HEMISPHERE**

Major. He was the son of Zeus and Callisto, Zeus's lover,
who is represented by Ursa Major.

Boötes contains Arcturus, the brightest star north of
the celestial equator and the fourth-brightest star in the
entire sky. It lacks bright clusters, nebulae, and galaxies
but has some targets visible through moderately large
telescopes. Faint stars in the northern part of Boötes
once formed the now-defunct constellation Quadrans
Muralis, which gave its name to the Quadrantid meteor
shower that radiates from this area every January.

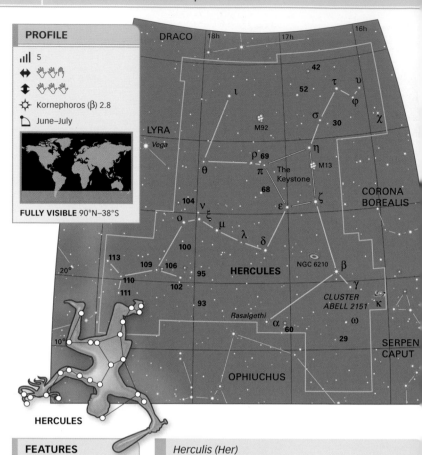

HERCULES

FEATURES

Alpha (α) Herculis (Rasalgethi) A double-star system, Rasalgethi means "the kneeler's head" in Arabic. It contains two stars that orbit one another. One is a huge red giant, so large that it has become unstable and varies in brightness between magnitudes 2.8 and 4.0. The other star is a smaller giant, shining steadily at magnitude 5.3. The pair is located 380 light-years away from Earth.

M13 This bright globular cluster is the finest in the northern sky. It contains a knot of 300,000 closely packed stars, about 25,000 light-years away from Earth. M13 can be glimpsed with the naked eye. Viewed through binoculars, it appears as a fuzzy ball. A small telescope will reveal some of the more loosely packed stars around its edges.

Herculis (Her)

HERCULES

A large but relatively faint constellation, Hercules sits in the northern sky between the bright stars Arcturus in Boötes and Vega in Lyra. It is the fifth-largest constellation in the sky and represents the Greek demigod Hercules. This strong man is depicted clothed in a lion's pelt,

NORTHERN HEMISPHERE

brandishing a club in one hand and the severed head of the watchdog Cerberus in the other, while kneeling with one foot on the head of Draco, the celestial dragon.

The constellation's most distinctive feature is the Keystone, a quadrilateral asterism composed of the stars Epsilon (ε), Zeta (ζ), Eta (η), and Pi (π) Herculis. Fainter stars extending from the Keystone form the hero's arms. Some of the constellation's other features include the distant and faint Hercules galaxy cluster and the spectacular globular cluster M13.

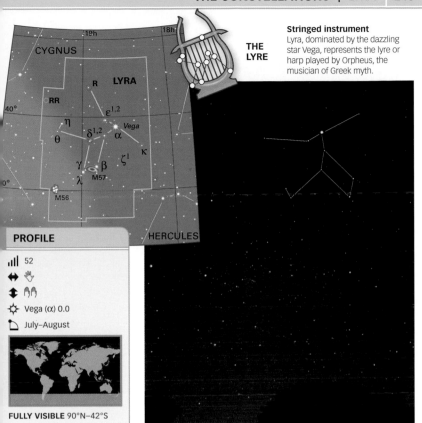

Stringed instrument
Lyra, dominated by the dazzling star Vega, represents the lyre or harp played by Orpheus, the musician of Greek myth.

THE LYRE

PROFILE

▁▂▃ 52

↔ 🖐

↕ 🖐🖐

☼ Vega (α) 0.0

📖 July–August

FULLY VISIBLE 90°N–42°S

FEATURES

Alpha (α) Lyrae (Vega)
Just 25 light-years away from Earth, this white star shines at magnitude 0.0, indicating that it is some 50 times as luminous as the Sun. It is surrounded by a disk of dusty debris that may be left over from the formation of a planetary system.

Epsilon (ε) Lyrae This famous multiple star splits into a double when viewed through binoculars, but a small telescope will show that each of these components is also double, making Epsilon (ε) a "double-double" system.

M57 (the Ring Nebula)
The most famous planetary nebula in the sky is a delicate shell of gas cast off by a dying star 2,000 light-years away. M57 shines at magnitude 8.8, and is best seen through a small telescope.

Lyrae (Lyr)

LYRA

The compact constellation of Lyra lies on the edge of the Milky Way, close to the distinctive cross shape of Cygnus. Although it is one of the smaller constellations, Lyra is easily spotted in the northern sky, thanks to the presence of the fifth-brightest star in the sky, the brilliant white Vega. This constellation has been

NORTHERN HEMISPHERE

known since ancient times and represents the lyre, a stringed musical instrument. According to Greek mythology, the lyre was invented by the Olympian god Hermes and later used by the hero Orpheus to charm the gods of the underworld in his efforts to win back his lost wife Eurydice.

Vega forms one corner of the Summer Triangle of the northern sky—the other two being Deneb in Cygnus and Altair in Aquila. The Lyrid meteors radiate from a point nearby around April 21–22 every year.

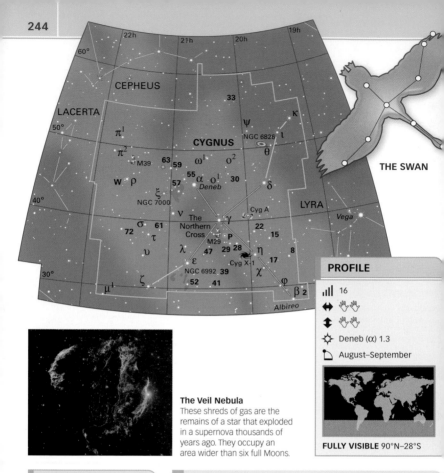

CEPHEUS

LACERTA

CYGNUS

THE SWAN

LYRA

The Veil Nebula
These shreds of gas are the
remains of a star that exploded
in a supernova thousands of
years ago. They occupy an
area wider than six full Moons.

PROFILE

.ıll 16

☆ Deneb (α) 1.3

August–September

FULLY VISIBLE 90°N–28°S

FEATURES

Alpha (α) Cygni (Deneb)
Around 1,500 light-years from
Earth, the brightest star in
Cygnus is nearly 160,000 times
brighter than the Sun. Although
this supergiant is one of the
sky's most luminous stars at
magnitude 1.3, it is outshone
by nearby Vega.

Beta (β) Cygni (Albireo) This
is a beautiful double star of
contrasting colors that marks
the bill of the swan. A pair
of binoculars will split it into
yellow and blue stars of
magnitudes 3.1 and 4.7.

NGC 6992 (the Veil Nebula)
Consisting of several shredded
gaseous remnants arranged in
a ragged loop, this nebula is
1,500 light-years from Earth. It
is the 6,000-year-old remnant
of a supernova explosion and
is part of a large, complex
nebula called the Cygnus Loop.

Cygni (Cyg)

CYGNUS

One of the most prominent
constellations of the northern sky,
Cygnus lies in a rich area of the
Milky Way. It represents a swan
flying down the Milky Way, recalling
the Greek God Zeus' disguise while
seducing Leda, the wife of the King
of Sparta. The shape formed by the
stars at the center of the pattern is

**NORTHERN
HEMISPHERE**

also known as the Northern Cross. Although Cygnus
lacks globular clusters and bright galaxies, it does contain
many nebulae and open clusters and is an impressive
sight when seen through a pair of binoculars or a
low-powered telescope.

The constellation is home to Cygnus X-1, a strong
X-ray source thought to mark the site of a black hole
orbiting another star. Perhaps its most distinctive
feature is the Great Rift—a relatively close, dark cloud
of dust that blocks light from the Milky Way.

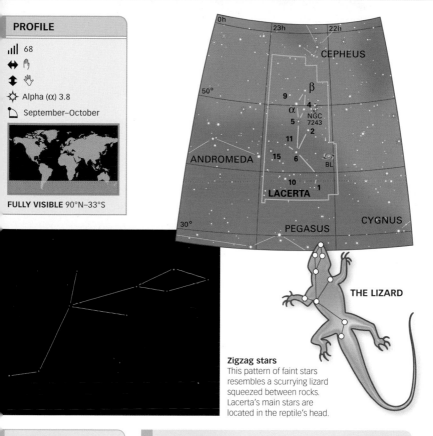

THE LIZARD

Zigzag stars
This pattern of faint stars resembles a scurrying lizard squeezed between rocks. Lacerta's main stars are located in the reptile's head.

FEATURES

Alpha (α) Lacertae This blue-white star shines at magnitude 3.8 and lies 102 light-years away from Earth, indicating that it is roughly 27 times as luminous as the Sun.

NGC 7243 A loose group of blue-white stars, thought to lie about 2,800 light-years away from Earth, NGC 7243 is so scattered that some astronomers suspect it is not a true open cluster at all.

BL Lacertae This strange and rapidly varying starlike object is in fact a blazar—a distant elliptical galaxy with a flickering source of radiation at its surface that varies in brightness between magnitudes 12 and 16. Jets of gas shoot from the center of BL Lacertae directly toward Earth, making it appear deceptively starlike.

Lacertae (Lac)

LACERTA

A small, obscure constellation, Lacerta straddles the northern Milky Way between Cassiopeia and Cygnus. It was introduced by the Polish astronomer Johannes Hevelius in 1687 and represents a scurrying lizard. Its size means that it contains few significant deep-sky objects, but its position among the dense Milky

NORTHERN HEMISPHERE

Way star clouds makes it the site of occasional nova explosions, in which a star system rapidly undergoes a huge increase in brightness.

Lacerta is also famed for the object BL Lacertae or BL Lac. Once thought to be a peculiar 14th-magnitude variable star, this is in fact the prototype for a class of objects called blazars. These are distant galaxies with supermassive black holes at their cores that are gulping down material from their surroundings and spitting it out in jets that are aligned directly toward Earth.

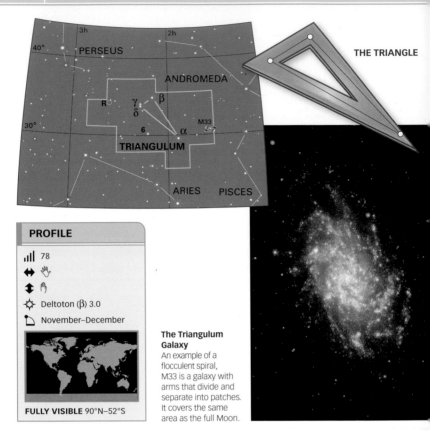

THE TRIANGLE

PROFILE

📶 78

↔ 🖐

↕ 🖐

☼ Deltoton (β) 3.0

📄 November–December

FULLY VISIBLE 90°N–52°S

The Triangulum Galaxy
An example of a flocculent spiral, M33 is a galaxy with arms that divide and separate into patches. It covers the same area as the full Moon.

FEATURES

Beta (β) Trianguli (Deltoton)
Triangulum's brightest star shines at magnitude 3.0 and lies about 135 light-years away.

Alpha (α) Trianguli (Rasalmothallah) With a magnitude of 3.4, this white star lies some 65 light-years away. Although named Alpha (α), it is not Triangulum's brightest star.

6 Trianguli This yellow star has a magnitude of 5.2 and a companion of magnitude 7, which can be seen through a small telescope.

M33 (the Triangulum Galaxy) The constellation's finest sight is this spiral galaxy, one of the closest of its kind at a distance of 2.7 million light-years. In spite of its proximity, M33 is hard to spot because its light is thinly spread.

Trianguli (Tri)

TRIANGULUM

A small northern constellation, Triangulum fills the gap between Perseus, Andromeda, and Aries. It consists of little more than an elongated triangle of three insignificant stars but is relatively easy to spot because of its compact size. Despite its lack of obvious features, Triangulum has an ancient

NORTHERN HEMISPHERE

origin. Greek astronomers originally saw it as a version of their letter "delta." They later visualized it as the Nile river delta or the island of Sicily.

It does not contain any clusters or nebulae, but it doe include one showpiece object—the nearby galaxy M33, known simply as the Triangulum Galaxy. At about a quarter of the diameter of the Milky Way Galaxy and the Andromeda Galaxy, M33 is the third-largest member of our Local Group of galaxies (p.146). It may be trapped in orbit around its much larger neighbor in Andromeda.

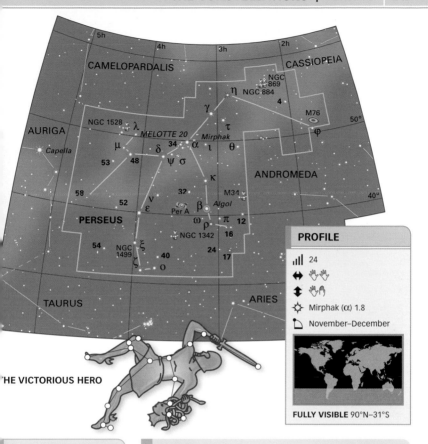

THE VICTORIOUS HERO

FEATURES

Persei (Per)

Alpha (α) Persei (Mirphak)
Binoculars reveal that
this magnitude-1.8 yellow
supergiant of lies at the heart
of a cluster of fainter blue
stars. It sits 590 light-years
from Earth.

Beta (β) Persei (Algol) This
famous variable star was the
first eclipsing binary to be
identified. It is a triple star with
two components that pass in
front of each other every 2.87
days, causing its overall
brightness to dip from
magnitude 2.1 to 3.4 for
about 10 hours.

**NGC 869, NGC 884 (the
Double Cluster)** About 7,000
light-years away from Earth,
this twin open cluster is a
spectacular sight in binoculars
and can be seen with the
naked eye as a bright knot in
the Milky Way.

PERSEUS

A prominent northern
constellation, Perseus lies in
the Milky Way between the
constellations Cassiopeia and
Auriga. It is named after the Greek
mythological hero Perseus, who
rescued Andromeda from the sea
monster Cetus and killed the
serpent-haired gorgon Medusa.

**NORTHERN
HEMISPHERE**

Perseus is depicted with his left hand holding the
gorgon's head, marked by the famous variable star Algol.
His right hand brandishes a sword, represented by the
twin star clusters NGC 869 and NGC 884. Dense star
clouds and a multitude of young blue and white stars
around Alpha (α) Persei make Perseus a rewarding target
for binoculars or a small telescope. Alpha (α) Persei lies
at the heart of Perseus OB1 association, also called
Melotte 20, a cluster of stars scattered over an area
several times the width of the full Moon.

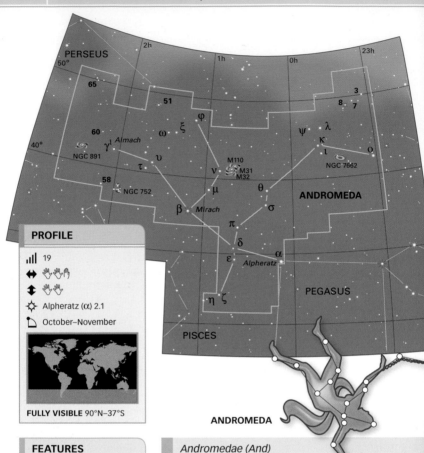

PROFILE

📶 19

↔ ✋✋🖐

↕ ✋✋

☼ Alpheratz (α) 2.1

🗓 October–November

FULLY VISIBLE 90°N–37°S

FEATURES

**Alpha (α) Andromedae
(Alpheratz)** Sometimes also
referred to as Delta (δ) Pegasi,
Alpheratz is a blue-white star
97 light-years away.

**Gamma (γ) Andromedae
(Almach)** Through small
telescopes, Almach appears
as a contrasting double,
with yellow and blue stars
of magnitudes 2.3 and 4.8.
Larger telescopes will also
show the blue star's fainter,
6th-magnitude companion.

**M31 (the Andromeda
Galaxy)** At 2.5 million
light-years from Earth, this is
the most distant object visible
to the naked eye, appearing
like a fuzzy, 4th-magnitude
star. Binoculars or a small
telescope reveal an elliptical
disk—the bright central area
of a huge spiral galaxy larger
than the Milky Way.

ANDROMEDA

Andromedae (And)

ANDROMEDA

Easy to find because of its link
to the Square of Pegasus, this
celebrated constellation of the
northern sky depicts Andromeda,
the daughter of the mythical Queen
Cassiopeia and King Cepheus of
Ethiopia. According to Greek myth,
Cassiopeia had boasted that she
was more beautiful than the

**NORTHERN
HEMISPHERE**

daughters of the sea-god Poseidon. This angered
Poseidon, who sent Cetus, a sea monster, to attack
Ethiopia. To appease Poseidon, Cassiopeia and Cepheus
had Andromeda chained to a rock by the sea as a
sacrifice, but she was rescued by the hero Perseus.

Andromeda consists of several strings of relatively faint
stars, meeting at Alpheratz, its brightest star on the corner
of the Square of Pegasus. It contains little of interest to the
amateur astronomer and owes its fame almost entirely to
the Andromeda Galaxy (M31) and its satellites.

PERSEUS AND ANDROMEDA

Princess Andromeda lies at the heart of a story that encompasses many constellations of the northern sky. According to Greek myth, she was chained to a rock and offered as a sacrifice to the sea monster Cetus, in atonement for the boastfulness of her mother, Cassiopeia, the wife of King Cepheus. The hero Perseus, flying home after slaying the Gorgon Medusa, came to Andromeda's rescue, swooping down in his winged sandals and using Medusa's head to turn the monster to stone.

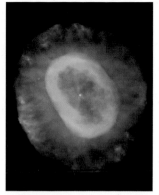

NGC 7662
This planetary nebula is roughly 1,800 light-years away and about one-third of a light-year across. It appears starlike, with only slight nebulosity, through a small telescope, but larger instruments reveal its bluish disk.

The Andromeda Galaxy
The nucleus of M31 is visible to the naked eye, but larger telescopes or long-exposure photographs are needed to capture the stunning detail of the galaxy's spiral arms and silhouetted dust lanes.

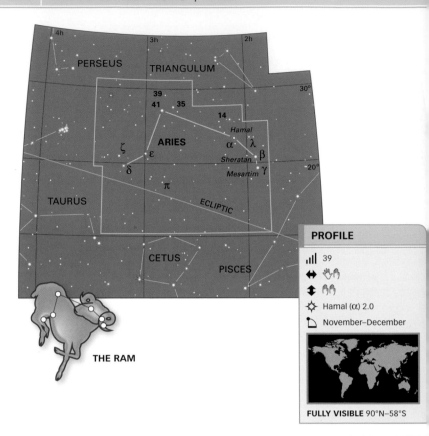

THE RAM

PROFILE

▮▮▮ 39

↔ ✋🤏

↕ 🤏🤏

☼ Hamal (α) 2.0

◹ November–December

FULLY VISIBLE 90°N–58°S

FEATURES

Alpha (α) Arietis (Hamal) A yellow giant star about 66 light-years from Earth, Hamal shines at magnitude 2.0. Its popular name is derived from the Arabic for "lamb." Its diameter is 15 times greater than that of the Sun.

Gamma (γ) Arietis (Mesartim) One of the first stars discovered to be a double, Mesartim was found by the English scientist Robert Hooke in 1664. Small telescopes will separate it to reveal twin white components, each of magnitude 4.8. The stars orbit each other some 200 light-years away.

Lambda (λ) Arietis This is another double star. Binoculars will reveal that the white primary star of magnitude 4.8 has a yellow companion of magnitude 7.3.

Arietis (Ari)

ARIES

Found between Pisces and Taurus, Aries is an inconspicuous zodiac constellation. It represents the golden ram of Greek mythology, whose fleece Jason and the Argonauts sought. Even before Greek times, ancient astronomers visualized it as a crouching ram. Its most recognizable features

NORTHERN HEMISPHERE

are three stars near the border with Pisces: Alpha (α), Beta (β), and Gamma (γ) Arietis, of 2nd, 3rd, and 4th magnitudes.

Aries contains little of interest to amateur observers, but is historically significant because more than 2,000 years ago, the vernal equinox—the point at which the ecliptic crosses the celestial equator—lay on the border of Aries and Pisces. The effects of precession have since moved the vernal equinox through Pisces toward Aquarius, but it is still known as the First Point of Aries.

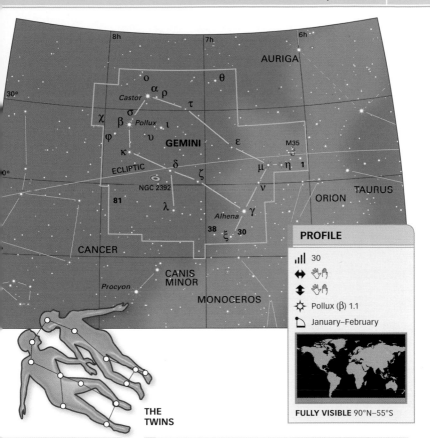

**THE
TWINS**

PROFILE

.ıll 30

↔ 🖐

↕ 🖐

☼ Pollux (β) 1.1

⌐ January–February

FULLY VISIBLE 90°N–55°S

FEATURES

**Beta (β) Geminorum
(Pollux)** A single yellow star,
Pollux is about 34 light-years
away from Earth. At magnitude
1.1, it is brighter than Alpha
(α) Geminorum.

**Alpha (α) Geminorum
(Castor)** This is a fascinating
multiple-star system with
overall magnitude 1.6. A small
telescope will split it into two
white stars, while a larger one
reveals a faint red companion.
Each of these stars is itself a
double, although none can
be separated visually, giving
Castor a total of six stars.

M35 This open cluster can
be spotted with the naked
eye and is a good target for
binoculars, through which
it appears as an elongated,
elliptical patch of light with
the same diameter as the
full Moon.

Geminorum (Gem)

GEMINI

This prominent zodiacal
constellation of the northern
sky, Gemini lies between the
constellations Taurus and Cancer.
It represents the mythical twins
Castor and Pollux, who were the
sons of Queen Leda of Sparta and
the brothers of Helen of Troy. They
were among the Argonauts, who

**NORTHERN
HEMISPHERE**

went in search of the golden fleece of Aries, the ram.
Gemini is easily identifiable because of its two brightest
stars, named after the twins.

Even though it is labeled Beta (β) Geminorum, Pollux is
brighter than Castor or Alpha (α) Geminorum. The feet of
the twins lie bathed in the Milky Way, and the constellation
contains several interesting deep-sky objects, including
the open cluster M35 and the faint but beautiful Eskimo
Nebula, NGC 2392. Around mid-December each year,
the Geminid meteors radiate from a point near Castor.

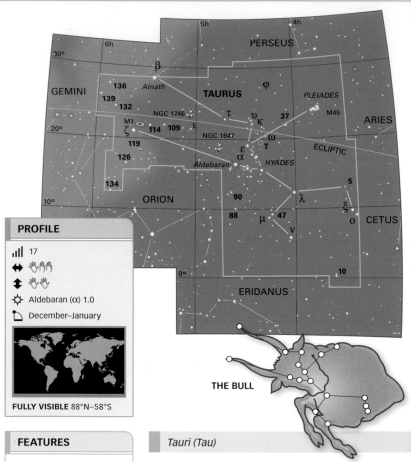

THE BULL

FEATURES

Alpha (α) Tauri (Aldebaran)
A red giant lying some
65 light-years from Earth,
Aldebaran shines at around
magnitude 1.0. Its brightness
varies because this elderly
star has become unstable.

The Hyades This V-shaped
star cluster lies well beyond
Aldebaran, some 160 light-
years away. Binoculars will
reveal stunning starscapes.

M1 (the Crab Nebula) M1
is the shredded remnant
of a star that exploded
as a supernova in 1054.

M45 (the Pleiades)
Naked-eye observers usually
see six of the so-called "Seven
Sisters," but binoculars or a
telescope show many more
hot blue stars. The cluster is
just 50 million years old and
lies 400 light-years away.

Tauri (Tau)

TAURUS

One of the oldest constellations,
Taurus has been recognized since
Babylonian times. It is a zodiacal
constellation lying between Aries
and Gemini and represents a
bull—the disguise used by Zeus
when he abducted Princess Europa
of Phoenicia. Just north of the
celestial equator, Taurus charges

**NORTHERN
HEMISPHERE**

toward the hunter Orion in the skies, the upper of its two
outstretched horns linking it to neighboring Auriga.

Taurus's brightest star is Aldebaran, and it contains
two prominent open clusters: the Pleiades, or M45, and
the more scattered Hyades, which has about 200 stars
and marks the V-shaped face of the bull. The constellation
contains excellent targets for binoculars and small
telescopes, including M1, the brightest supernova
remnant in the sky. In November, the Taurid meteors
seem to radiate from a point south of the Pleiades.

THE LOST PLEIAD

Also known as the Seven Sisters, the Pleiades represents the seven daughters of the Titan Atlas in Greek mythology. But only six stars are easily visible to the naked eye, and two myths explain the "missing" Pleiad. One says that the fainter sister is Merope, the only one of the seven Pleiades to marry a mortal. Alternatively, it may be Electra, hiding her face from the fate of Troy, the city founded by her brother, Orestes. However, despite these legends the faintest named member is Asterope.

The Pleiades
Although this bright star cluster is popularly known as the Seven Sisters, it contains nine named stars—the seven sisters and their parents, Atlas and Pleione.

Hyades and Pleiades
The Hyades (lower left) is the larger of these two dazzling star clusters; the Pleiades (upper right) is a tighter bunch that appears hazy at first glance.

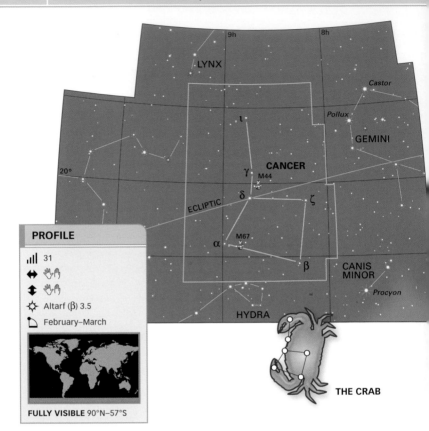

THE CRAB

PROFILE

.ıll 31

↔ 🖐🖐

↕ 🖐🖐

☼ Altarf (β) 3.5

▢ February–March

FULLY VISIBLE 90°N–57°S

FEATURES

Beta (β) Cancri (Altarf)
Cancer's brightest star, Altarf, is an orange giant, 290 light-years from Earth and noticeably brighter than Alpha (α) at magnitude 3.5.

Alpha (α) Cancri (Acubens)
With a name that means "the claw," this star is actually fainter than nearby Beta. It is a white star of magnitude 4.2, some 175 light-years from Earth.

M44 (Praesepe) This is a group of 50 young stars spread across an area of the sky three times the size of the full Moon. Although their combined light makes them easily visible to the naked eye, binoculars are needed to reveal individual stars within the group.

Cancri (Cnc)

CANCER

The faintest of the twelve zodiacal constellations, Cancer has an indistinct pattern but is still easy to find because of its location between the brighter stars of Leo and Gemini. According to Greek mythology, Cancer represents a crab that attacked Hercules during his fight with Hydra. The crab was crushed underfoot during the struggle.

NORTHERN HEMISPHERE

Since the Milky Way does not run through Cancer, this region of the sky is relatively barren, with no nebulosity, globular clusters, or planetary nebulae. However, it does contain two open clusters: the relatively loose M44 and the more concentrated M67. M44 is also known as the Beehive Cluster, or Praesepe (the Manger). Its stars lie about 570 light-years from Earth. In Greek myth, it was regarded as a manger of hay from which two donkeys, represented by nearby Gamma (γ) and Delta (δ) Cancri, were feeding.

PROFILE

FULLY VISIBLE 90°N–48°S

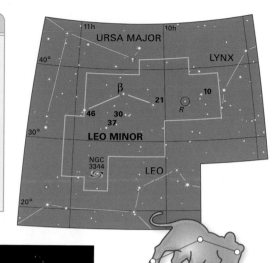

THE LITTLE LION

NGC 3344
This attractive face-on spiral has tightly wound inner spiral arms that are hard to distinguish. Despite its overall magnitude of 10.5, its spread-out nature makes it a difficult target for small telescopes.

FEATURES

46 Leonis Minoris The constellation's brightest star is an orange giant of magnitude 3.8. It lies about 80 light-years away from Earth and is nearing the end of its life. Due to an error, this star is not labeled Alpha (α).

Beta (β) Leonis Minoris The second-brightest star in Leo Minor is a yellow giant shining at magnitude 4.2. Its distance of 190 light-years reveals that in reality it is considerably more luminous than 46 Leonis Minoris.

R Leonis Minoris Lying to the west of 21 Leonis Minoris, this is a pulsating red giant with a period of 372 days. At its peak magnitude of 6.3, it is easily spotted through binoculars, but at its dimmest it fades beyond the reach of small telescopes.

Leonis Minoris (LMi)

LEO MINOR

This faint constellation, squeezed between Leo and Ursa Major was invented by the Polish astronomer Johannes Hevelius around 1680. He claimed that its pattern of stars resembled Leo, but the resemblance is far from obvious—the brightest stars actually form a rough triangle—and

NORTHERN HEMISPHERE

it seems that Hevelius was simply keen to fill a gap in the sky for his great star atlas, *Uranographia*. Nevertheless, the constellation is usually seen as representing a lion cub.

Lying in a relatively sparse region of the sky, Leo Minor contains few objects of interest to amateur astronomers. Its brightest star, 46 Leonis Minoris, should have been assigned the Greek letter α, but was missed by the English astronomer Francis Bailey while he was assigning letters to the constellation's stars in 1845. However, its second-brightest star has been assigned the Greek letter β.

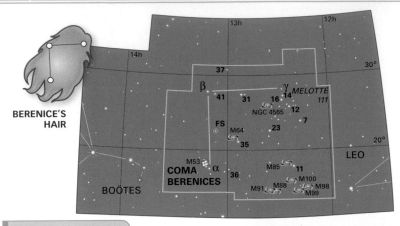

BERENICE'S HAIR

BOÖTES

COMA BERENICES

LEO

PROFILE

⏹ 42

↔ 𝄞𝄞

↕ 𝄞𝄞

☼ Beta (β) 4.2

◗ April–May

FULLY VISIBLE 90°N–56°S

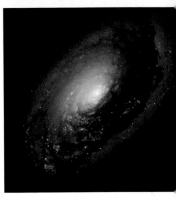

The Black Eye Galaxy
This galaxy appears as an elliptical patch of light through a small telescope; the central dust cloud is best seen through a large telescope with an aperture of 6 in (150 mm) or more.

FEATURES

M53 The brighter of two globular clusters in Coma Berenices, M53 is some 56,000 light-years away. Although it is visible through binoculars, it is best seen with a small telescope.

M64 (the Black Eye) Coma's brightest galaxy is nicknamed the Black Eye after the dark dust cloud at its core. Tilted at an angle to Earth, it lies around 19 million light-years away.

Melotte 111 (the Coma Star Cluster) This open star cluster is Coma's main feature. It consists of a scattered group of faint stars stretching southward in a fan shape from Gamma (γ) Comae Berenices. It is one of the closest open star clusters to Earth, with more than 20 of its stars visible to the naked eye.

Comae Berenices (Com)

COMA BERENICES

The stars in this region of the sky originally marked the tail of neighboring Leo, the lion. In the mid-16th century, the Dutch cartographer Gerardus Mercator transformed them into the new constellation Coma Berenices (often shortened to simply Coma). Lying between Leo and Boötes, this

NORTHERN HEMISPHERE

constellation represents the flowing locks of Queen Berenice of Egypt, which she cut off as a tribute to the gods after the safe return of her husband, King Ptolemy III, from a battle during the 3rd century BCE.

Coma has relatively faint stars, but it contains deep-sky objects, including the nearby star cluster Melotte 111 and more distant galaxies that are visible wit binoculars and small telescopes. Some of these galaxies are overspill from the relatively nearby Virgo Cluster, whil others are part of the more distant Coma Cluster.

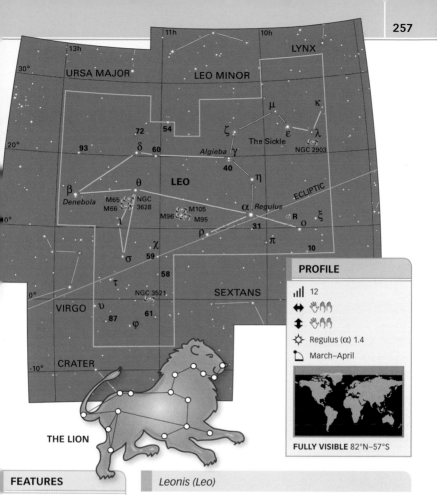

THE LION

PROFILE

.ıll 12
↔ 🖐🖐🖐
↕ 🖐🖐🖐
☼ Regulus (α) 1.4
🗓 March–April

FULLY VISIBLE 82°N–57°S

FEATURES

Leonis (Leo)

Alpha (α) Leonis (Regulus)
Meaning "little king" in Latin, Regulus is a bright, blue-white star, located almost 80 light-years away from Earth. It lies at the foot of the Sickle and shines at magnitude 1.4. It has a companion star of magnitude 7.8 that can be seen through binoculars.

Gamma (γ) Leonis (Algieba)
This attractive double star consists of two yellow giants some 170 light-years away. Easily split in small telescopes, the brighter star is magnitude 2.0 and the fainter 3.2. The stars orbit each other every 600 years or so.

R Leonis This red giant lies some 3,000 light-years away. It varies in brightness over 312 days, usually staying below naked-eye visibility but peaking at magnitude 4.

LEO

This large zodiacal constellation lies just north of the celestial equator. It represents the Nemaean Lion slain by the Greek hero Hercules, and its outline stars do bear a marked resemblance to a crouching lion. A group of six stars forming the lion's head and chest is also known as the Sickle because of the hook-shaped pattern they form. The Leonid meteors radiate from this part of the constellation every November.

NORTHERN HEMISPHERE

Leo lacks nebulae or clusters because it lies away from the Milky Way and the plane of our galaxy, but it does contain a few bright galaxies. It also contains the dwarf star Wolf 359, which is the fourth-nearest star to our own star, the Sun, at a distance of only 7.8 light-years. However, this is too faint a target for amateur instruments.

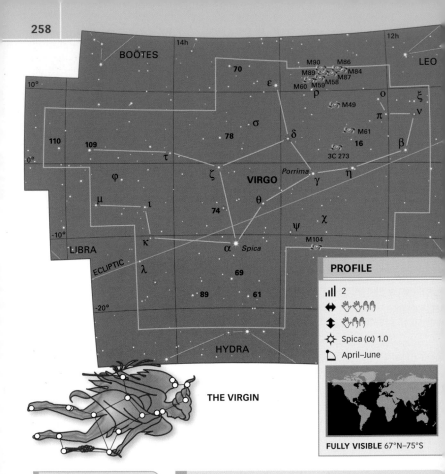

BOÖTES

LEO

70

M90 M86
M89 M84
M60 M59 M87
M58

ρ

M49

10°

14h

12h

ε

σ

78

δ

o

ξ

ν

π

M61

16

β

110

109

τ

0°

3C 273

φ

ζ

VIRGO

Porrima

γ

η

μ

ι

θ

74

ψ

χ

-10°

κ

α Spica

M104

LIBRA

λ

ECLIPTIC

69

89

61

-20°

HYDRA

THE VIRGIN

PROFILE

📶 2

↔ 🖐🖐🤟🤟

↕ 🖐🤟🤟

☼ Spica (α) 1.0

🗓 April–June

FULLY VISIBLE 67°N–75°S

FEATURES

Alpha (α) Virginis (Spica)
A bright star with an average
magnitude of 1.0, Spica lies
about 260 light-years away
from Earth. It is actually an
indivisible binary, in which
a smaller companion affects
the brighter star's shape and
causes its brightness to vary
as it rotates.

M87 This giant elliptical galaxy
lies at the heart of the Virgo
Cluster of galaxies. It shines
at magnitude 8.1 and lies
50 million light-years away.

**M104 (the Sombrero
Galaxy)** A bright galaxy some
28 million light-years away,
M104 is much closer than the
Virgo Cluster. It is an edge-on
spiral with a dark lane of dust
across its central bulge, best
seen through a large telescope.

Virginis (Vir)

VIRGO

The second-largest constellation
after Hydra, Virgo straddles the
celestial equator and is the only
female sign of the zodiac. It lies
southeast of the more easily
identifiable Leo and is associated
with many virgin goddesses,
including Dike, the Greek goddess
of justice, with nearby Libra
representing the scales of justice.

**NORTHERN
HEMISPHERE**

The constellation is also associated with Demeter, the
goddess of harvest. Virgo's brightest star, Spica, marks
an ear of wheat held in her hand.

Virgo lacks bright star clusters and nebulae, but make
up for this with the galaxies it contains. When looking in
this direction, we are looking out of our own galaxy
toward the Virgo Cluster, a cluster of more than 2,000
galaxies some 55 million light-years away. Many of these
are visible through amateur telescopes.

THE MAIDEN GODDESS

Virgo is often identified as Dike or Iustitia, the Greek goddess of justice, who abandoned Earth and flew up to heaven when human behavior deteriorated. In this interpretation, neighboring Libra represents her scales of justice. However, from ancient Mesopotamian times in the early first millennium BCE, the constellation was also associated with fertility and the harvest, and ancient Greek astronomers also interpreted the figure as the harvest goddess Demeter or Ceres, clutching an ear of wheat marked by Virgo's brightest star, Spica.

The Sombrero Galaxy
This galaxy, named for its resemblance to a Mexican hat, is a spiral with an unusually large central hub, seen almost edge on. A ring of silhoutted dust forms a dark rim around the galaxy's disk. This image was captured by the Hubble Space Telescope.

Markarian's Chain
In one part of the Virgo Cluster, several galaxies form a smooth chain. This is named after the Armenian astrophysicist B. E. Markarian, who first proved that the galaxies are gravitationally linked rather than a chance alignment.

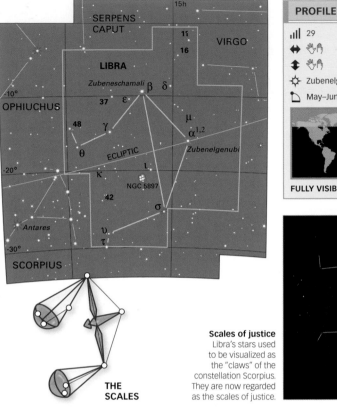

SERPENS CAPUT

15h

11

16

VIRGO

LIBRA

Zubeneschamali β δ

-10°

OPHIUCHUS

37 ε

48 γ

μ

α^{1,2}

θ

ECLIPTIC

Zubenelgenubi

-20°

ι

κ

NGC 5897

42

σ

Antares

υ

-30°

τ

SCORPIUS

**THE
SCALES**

PROFILE

.ıll 29

↔ ✋

↕ ✋

☼ Zubenelgenubi 2.8

◻ May–June

FULLY VISIBLE 60°N–90°S

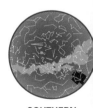

Scales of justice
Libra's stars used
to be visualized as
the "claws" of the
constellation Scorpius.
They are now regarded
as the scales of justice.

FEATURES

**Alpha (α) Librae
(Zubenelgenubi)** Libra's
brightest star is a double that
is easily split with binoculars
or even with sharp, unaided
eyesight. Zubenelgenubi's two
stars, a blue-white giant of
magnitude 2.8 and a white star
of magnitude 5.2, lie 70 light-
years from Earth.

Mu (μ) Librae This double
star, with components of
magnitudes 5.6 and 6.7, is a
close pairing 235 light-years
away. It is a more difficult pair
to separate than Zubenelgenubi;
a telescope with 3in (75mm)
aperture is needed.

48 Librae Lying 510 light-years
from Earth, this young star is at
an early stage of development
and is still throwing off excess
material that forms shells
around the star, causing it to
vary slightly in brightness.

Librae (Lib)

LIBRA

The only sign of the zodiac that
depicts an object (a set of scales)
rather than a living creature, Libra
was once seen by the ancient Greeks
as *Chelae Scorpionis*, the "claws"
of neighboring Scorpius. Indeed,
its brightest stars are still known
as Zubenelakrab, Zubenelgenubi,
and Zubeneschamali, meaning the

**SOUTHERN
HEMISPHERE**

"scorpion's claw," "southern claw," and "northern claw."
Since Roman times, however, Libra has been interpreted
as the scales of justice, held by the figure of nearby Virgo.

Libra is a small and faint constellation, best located in
relation to its brighter neighbors in the sky. Apart from
one unimpressive globular cluster, it has little to offer
amateur astronomers with small telescopes. The faint
star Gliese 581, home to a complex system of planets,
lies to the north of Beta (β), but is beyond the reach of
most amateur astronomers.

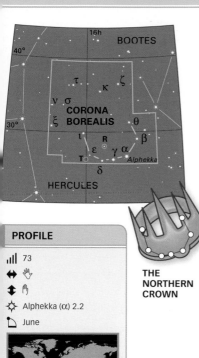

THE NORTHERN CROWN

Celestial crown
The most prominent stars of this constellation form an arc shape that resembles a magnificent crown in the night sky.

PROFILE

📶 73

↔ ✋

↕ ✋

☼ Alphekka (α) 2.2

📖 June

FULLY VISIBLE 90°N–50°S

FEATURES

Alpha (α) Coronae Borealis (Alphekka) This challenging eclipsing binary star of average magnitude 2.2 varies in brightness by just 0.1 magnitude as its components pass in front of one another in a 17.4-day cycle.

R Coronae Borealis This intriguing variable is normally just visible to the naked eye, enclosed by the curve of the crown and shining at magnitude 5.8. Every few years, this yellow supergiant, lying 6,000 light-years away, suddenly dips in brightness, going beyond the range of most amateur telescopes.

T Coronae Borealis Also known as the Blaze Star, this is one of the brightest and most reliable recurrent novae, brightening from magnitude 11 to around 2 every few decades.

Coronae Borealis (CrB)

CORONA BOREALIS

A small but distinctive

constellation, Corona Borealis is situated between Boötes and Hercules in the northern sky. It consists of a chain of seven faint stars arranged in a horseshoe shape, representing the crown worn by the mythical Princess Ariadne of Crete during her

NORTHERN HEMISPHERE

wedding to the god Dionysius. This ancient constellation was recognized by the Greek–Egyptian astronomer Ptolemy in his original list of 48 star patterns around the 2nd century CE.

Making up for a lack of deep-sky objects for amateur observers, Corona Borealis contains a number of attractive double stars and interesting variables. It also contains a huge cluster of more than 400 galaxies called Abell 2065, but at a distance of 1.5 billion light-years none is brighter than magnitude 16.

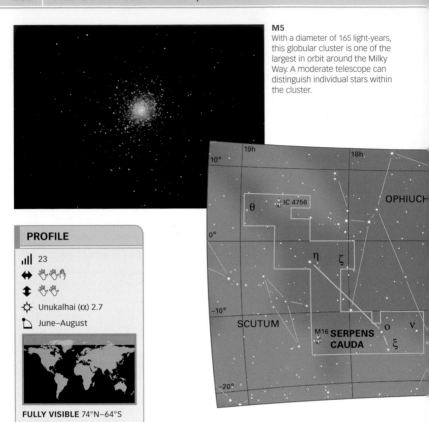

M5
With a diameter of 165 light-years, this globular cluster is one of the largest in orbit around the Milky Way. A moderate telescope can distinguish individual stars within the cluster.

PROFILE

- ⬛ 23
- ↔ 🖐🖐🖐
- ↕ 🖐🖐
- ☼ Unukalhai (α) 2.7
- 🗓 June–August

FULLY VISIBLE 74°N–64°S

FEATURES

Alpha (α) Serpentis (Unukalhai) Situated in Serpens Caput, the constellation's brightest star lies 70 light-years away. It is an orange giant of magnitude 2.7.

M5 About 13 billion years old, this attractive globular cluster is one of the oldest in the night sky. It lies 24,500 light-years away from Earth and is just visible to the naked eye on dark nights, hovering around magnitude 5.6.

M16 This open cluster of about 60 stars lies 8,000 light-years away at the heart of the large, faint Eagle Nebula—a huge cloud of gas and dust from which the stars have recently been born. It appears as a hazy patch of light about the size of the full Moon.

Serpentis (Ser)

SERPENS

The only constellation that is made up of two parts, Serpens, the snake, lies on either side of the constellation Ophiuchus, the serpent bearer. It is split into Serpens Cauda, the snake's tail, and Serpens Caput, the head. In Greek mythology, snakes were considered a symbol of rebirth because they shed their skins. Ophiuchus, meanwhile, represents the great healer Asclepius, who was reputedly able to revive the dead.

NORTHERN HEMISPHERE

The Milky Way passes through Serpens Cauda, while Serpens Caput offers clear views into intergalactic space which gives the two halves of the constellation distinctly different characters. The tail contains bright star clusters including M16 at the heart of the Eagle Nebula. The head's brightest deep-sky objects include M5, a globular cluster and a number of distant galaxies.

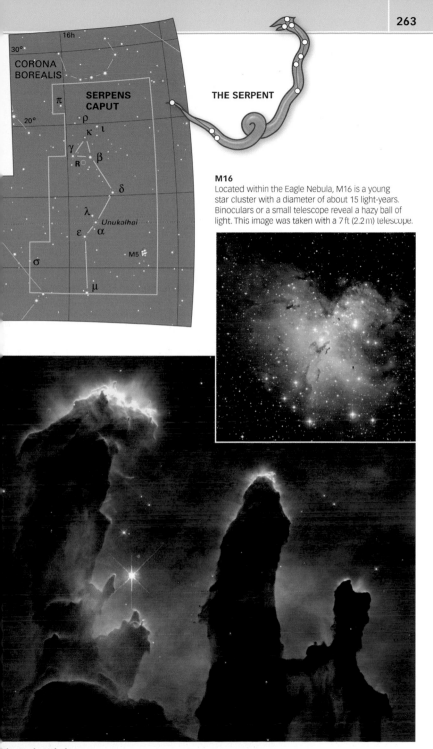

CORONA
BOREALIS

SERPENS
CAPUT

THE SERPENT

16h

30°

20°

π

ρ

κ ι

γ

R

β

δ

λ

Unukalhai

ε α

σ

M5

μ

M16

Located within the Eagle Nebula, M16 is a young
star cluster with a diameter of about 15 light-years.
Binoculars or a small telescope reveal a hazy ball of
light. This image was taken with a 7 ft (2.2 m) telescope.

The Eagle Nebula
This Hubble Space Telescope image shows columns of gas and dust within the
Eagle Nebula known as the Pillars of Creation. Stars form in these dark clouds,
then emerge into open space as their radiation blows away surrounding material.

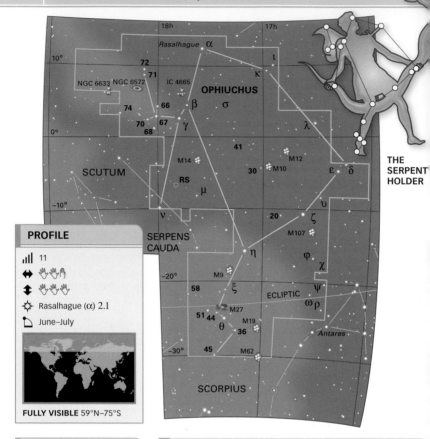

THE
SERPENT
HOLDER

PROFILE

.ıll 11

↔ 🤚🤚✋

↕ 🤚🤚🤚

☼ Rasalhague (α) 2.1

📅 June–July

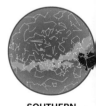

FULLY VISIBLE 59°N–75°S

FEATURES

Alpha (α) Ophiuchi (Rasalhague) Ophiuchus's brightest star is this white giant of magnitude 2.1, about 50 light-years away from Earth.

Rho (ρ) Ophiuchi This fine multiple star is still embedded in the faint gas from which it was formed. Binoculars show two wide companions close to the magnitude-5.0 primary star, while a small telescope reveals that the primary has a closer neighbor of magnitude 5.9.

Barnard's Star The most celebrated star in Ophiuchus and the fastest-moving in the sky, Barnard's Star is found near Beta (β) Ophiuchi. It moves so fast that it crosses a Moon's width of the sky every 200 years and, at just 6 light-years away from Earth, is the second-closest star to the Sun.

Ophiuchi (Oph)

OPHIUCHUS

A large, ancient constellation straddling the celestial equator, Ophiuchus depicts a man holding a snake and is often associated with Asclepius, the Greek god of healing. The snake itself is represented by the neighboring constellation Serpens. The stars of Ophiuchus are faint and its pattern indistinct,

SOUTHERN HEMISPHERE

so it is best located by looking to the south of Hercules, or to the north of Scorpius, located in the south.

The path of the ecliptic currently passes through this region of the sky, and the Sun can be found here in the first half of December. Despite this, Ophiuchus is not regarded as a true member of the zodiac. Although its stars are faint, the constellation contains a number of interesting objects and is especially rich in globular clusters. It also contains large regions of nebulosity, making it a popular target for astrophotographers.

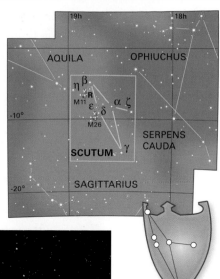

THE SHIELD

The Wild Duck Cluster
From a dark location, M11, lying some 5,600 light-years from Earth, can be seen with the naked eye as a bright, misty patch of the Milky Way.

FEATURES

Delta (δ) Scuti This pulsating giant of magnitude 4.7 is 260 light-years away from Earth. It is the prototype for a class of rapidly changing variable stars, although it only varies by about 0.1 magnitude over each 4.6-hour cycle.

R Scuti This is a variable star with a slower period of change than Delta (δ) Scuti, and its changes are easier to follow. It is a yellow supergiant that varies from magnitude 4.5 at its peak, down to magnitude 8.8, the limit of binocular visibility. This cycle lasts 144 days.

M11 (the Wild Duck Cluster) This rich open cluster in the Scutum Star Cloud is located just south of Beta (β) Scuti. It is visible to the naked eye and rewarding through binoculars. When seen with a telescope, the stars form a fan shape.

Scuti (Sct)

SCUTUM

Small, kite-shaped, and lying to the south of the celestial equator, the constellation Scutum was invented in the 17th century by the Polish astronomer Johannes Hevelius. He originally named it Scutum Sobiescianum, meaning Sobieski's Shield, in honor of his patron, King John Sobieski of

SOUTHERN HEMISPHERE

Poland. Now known simply as Scutum, the constellation is best located by searching between neighboring Aquila and Sagittarius.

Crossed by a bright region of the Milky Way, Scutum contains rich star clouds, ideal for sweeping with binoculars. Indeed, the Scutum Star Cloud is the brightest part of the Milky Way outside Sagittarius. Highlights for amateur astronomers include the beautiful M11, or the Wild Duck Cluster, and R Scuti, an impressive long-period variable star.

PROFILE

.ıll 86

↔ ✋

↕ 🖐

✦ Gamma (γ) 3.5

🗓 August

FULLY VISIBLE 90°N–69°S

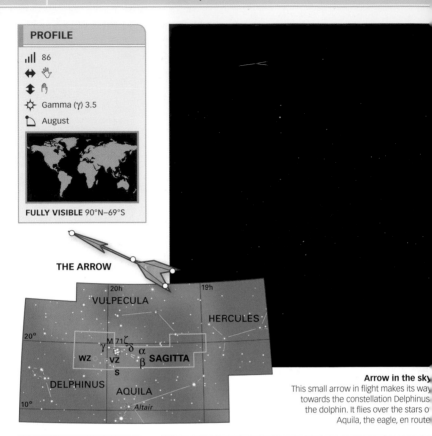

THE ARROW

VULPECULA

HERCULES

20h   19h

20°

γ M 71 ζ
δ α
WZ   VZ   β SAGITTA
S

DELPHINUS AQUILA

10°   Altair

Arrow in the sky
This small arrow in flight makes its way
towards the constellation Delphinus,
the dolphin. It flies over the stars of
Aquila, the eagle, en route

FEATURES

Gamma (γ) Sagittae The
brightest star in Sagitta is
an orange giant of magnitude
3.5. It lies at a distance of
175 light-years from Earth
at the tip of the arrow.

**Alpha (α) (Sham) and
Beta (β) Sagittae** These
twin yellow stars, both of
magnitude 4.4, are genuine
neighbors in space around
470 light-years away.

S Sagittae This yellow
supergiant lies 4,300
light-years from Earth and is
a pulsating variable. It varies
in brightness between
magnitudes 5.5 and 6.2
every 8.38 days.

M71 This star cluster is usually
classed as a globular but has a
loose structure that suggests it
may be an open cluster. It lies
about 13,000 light-years away.

Sagittae (Sge)

SAGITTA

The third-smallest constellation
in the sky, Sagitta lies in the Milky
Way between Vulpecula and Aquila
in the northern sky. It is best
located by looking to one side
of the wqually compact but more
distinctive constellation Delphinus,
the celestialdolphin. Despite the
presence of some Milky Way star

**NORTHERN
HEMISPHERE**

clouds and the dark stream of the Great Rift (a dust cloud
that appears to divide the Milky Way), this constellation
has few objects of interest to amateur astronomers.

In spite of its small size and relative obscurity,
Sagitta was known and recognized as an arrow by
the ancient Greeks. However, it has nothing to do
with the better-known celestial archer Sagittarius.
Instead, Greek authorities believed it represented
an arrow shot by Apollo, Hercules, or Eros toward
Cygnus and Aquila.

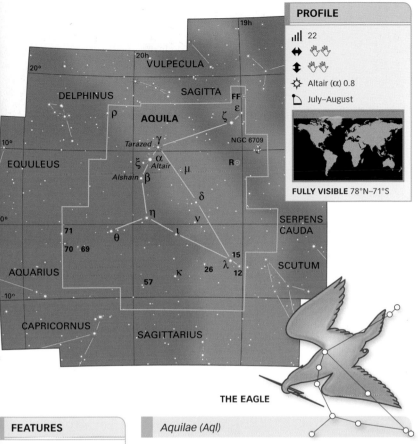

THE EAGLE

PROFILE

ıll 22

⟷ 🖐🖐

⇕ 🖐🖐

☼ Altair (α) 0.8

▱ July–August

FULLY VISIBLE 78°N–71°S

FEATURES

Alpha (α) Aquilae (Altair)
Lying just 17 light-years away, this is the 12th-brightest star in the sky. Shining at magnitude 0.8, it is one of the closest bright stars to Earth. It marks the neck of the eagle and is flanked by Beta (β) and Gamma (γ) Aquilae.

Beta (β) Aquilae (Alshain)
This star and Gamma (γ) (Tarazed) are the near-twins that flank Altair. Alshain is actually the fainter of the two, shining at magnitude 3.7 compared to Tarazed's 2.7. Alshain is just 49 light-years away, while the giant Tarazed has a distinctly orange color and is five times farther away.

NGC 6709 Some 3,000 light-years away, this open cluster appears through binoculars as a bright knot in the Milky Way star clouds.

Aquilae (Aql)

AQUILA

This constellation lies on the celestial equator in a rich area of the Milky Way near Cygnus, Scutum, and Sagittarius. It also shares borders with no fewer than nine other constellations. It is easily located thanks to its central bright star, Altair, and its near-twin companions, Alshain and Tarazed.

NORTHERN HEMISPHERE

Altair forms one corner of the northern Summer Triangle (p.190), a pattern of stars completed by Vega (in Lyra) and Deneb (in Cygnus). Aquila has few deep-sky objects of note.

The constellation has been identified as an eagle in flight for at least 3,000 years. It may represent either the eagle that carried the thunderbolts of the Greek God Zeus or the God himself, taking the form of a bird in order to carry off the youth Ganymede, identified as neighboring Aquarius.

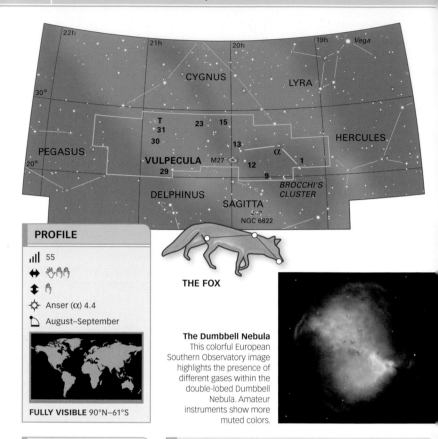

THE FOX

The Dumbbell Nebula
This colorful European Southern Observatory image highlights the presence of different gases within the double-lobed Dumbbell Nebula. Amateur instruments show more muted colors.

PROFILE

- ▂▄▆ 55
- ↔ 🖐🤏🤏
- ↕ 🤏
- ☼ Anser (α) 4.4
- ▱ August–September

FULLY VISIBLE 90°N–61°S

FEATURES

Alpha (α) Vulpeculae (Anser) With a modest magnitude of 4.4, Anser is Vulpecula's brightest star. It is a red giant about 250 light-years away from Earth.

Brocchi's Cluster (the Coathanger) This small group of 10 stars lies on Vulpecula's southern border. It is also known as the Coathanger because of its shape. Its members hover at the limit of naked-eye visibility and are an attractive sight in binoculars.

M27 (the Dumbbell Nebula) This is the brightest planetary nebula in the sky and the easiest to spot. Appearing as a rounded patch, it is one-quarter of the diameter of the full Moon and about 1,000 light-years away. A small telescope will reveal the hourglass shape from which it gets its name.

Vulpeculae (Vul)

VULPECULA

A small, faint northern

constellation, Vulpecula lies in the Milky Way south of Cygnus. It was introduced in the late 17th century by the Polish astronomer Johannes Hevelius under the name of *Vulpecula cum Anser* (the fox with the goose), but its name has since been simplified to Vulpecula.

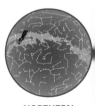

NORTHERN HEMISPHERE

It consists of a handful of faint stars with no obvious pattern and is best located by looking to one side of the adjacent brighter constellation Pegasus.

Despite its relative obscurity, Vulpecula contains several impressive objects for binocular and small telescope users, including M27, or the Dumbbell Nebula, and two open star clusters—Brocchi's Cluster and NGC 6822. It is also home to the radio source PSR 1919+21, the first pulsar to be discovered. Once considered inexplicable, this is now known to be a rapidly spinning neutron star.

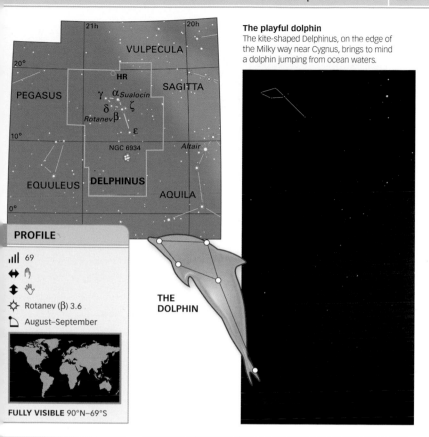

The playful dolphin
The kite-shaped Delphinus, on the edge of the Milky way near Cygnus, brings to mind a dolphin jumping from ocean waters.

THE DOLPHIN

PROFILE

.ıll 69

↔ 🖑

↕ 🖑

☆ Rotanev (β) 3.6

🗓 August–September

FULLY VISIBLE 90°N–69°S

FEATURES

Beta (β) Delphini (Rotanev)
Slightly brighter than Alpha (α), Rotanev is a pure white star of magnitude 3.6 that lies 72 light-years away from Earth.

Alpha (α) Delphini (Sualocin) A hot, blue-white star, Sualocin is located 190 light-years away from Earth and shines at magnitude 3.8.

Gamma (γ) Delphini This attractive double star, about 125 light-years away from Earth, consists of two yellow-white stars of magnitudes 4.3 and 5.1. They are easily separated with a small telescope.

Delphini (Del)

DELPHINUS

This small but distinctive constellation is situated between Aquila and Pegasus. According to Greek myth, Delphinus represents a dolphin that saved the poet and musician Arion from drowning after he leapt into the sea to escape robbers onboard a ship. Alternatively, the constellation is

NORTHERN HEMISPHERE

said to depict one of the dolphins sent by Poseidon, the god of the sea, to bring the sea nymph Amphitrite to him to marry. It is one of the original constellations listed by the Greek–Egyptian astronomer Ptolemy.

Its brightest stars are Alpha (α) and Beta (β)—first given the names Sualocin and Rotanev in a star catalog compiled in 1814. When read backward these names reveal the words Nicolaus Venator, the Latinized name of Niccolo Cacciatore, an Italian assistant astronomer, who cheekily named the stars after himself.

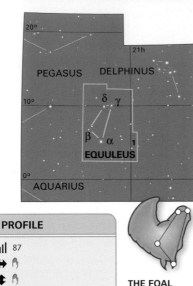

THE FOAL

PROFILE

📶 87

↔ ♓

↕ ♓

☼ Kitalpha (α) 3.9

🗓 September

FULLY VISIBLE 90°N–77°S

The foal's head
Equuleus consists of a small area of faint stars that can easily be overlooked. It is located between the constellations Pegasus and Delphinus and is visualized as a foal's head.

FEATURES

Alpha (α) Equulei (Kitalpha)
This yellow giant of magnitude 3.9 lies 190 light-years away from the Sun and is 75 times more luminous than it.

Epsilon (ε) Equulei This triple star combines a chance alignment with a genuine binary system. A small telescope will reveal that the primary star, of magnitude 5.4, has a companion of magnitude 7.4 that just happens to lie in the same direction. The fainter star is actually the closer of the two, lying 125 light-years away compared to 200 light-years for the primary. Larger telescopes reveal that the primary is actually a double star.

Equulei (Equ)

EQUULEUS

This is the second-smallest constellation in the sky, and its stars are relatively faint. It represents the head of a young horse and since ancient times has been seen as a companion to the nearby, larger horse-shaped constellation Pegasus. Equuleus was catalogued along with 48 other constellations by the Greek–Egyptian astronomer Ptolemy in the 2nd century CE. One Greek myth associated it with the swift-footed foal Celeris, the offspring or brother of the better-known constellation Pegasus.

NORTHERN HEMISPHERE

Equuleus is best located by looking in the wedge of sky between Epsilon (ε) Pegasi, in the southwest corner of Pegasus, and the diamond-shaped constellation Delphinus. Equuleus includes double and multiple stars, and its brightest star is the yellow giant Kitalpha, or Alpha (α) Equulei.

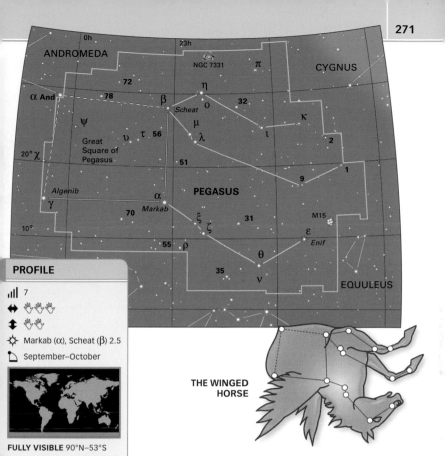

ANDROMEDA
72
78
α And
ψ
Great
Square of
Pegasus
20° χ
Algenib
γ
70   Markab
55 ρ
35

0h
23h
NGC 7331
β
Scheat
μ
υ τ 56
51
α
ξ ζ
θ
ν

η
π
32
o
λ
9
31

CYGNUS
κ
ι
2
1
M15
ε
Enif

PEGASUS

EQUULEUS

10°

THE WINGED HORSE

FEATURES

Alpha (α) Pegasi (Markab)
Around 140 light years away, this blue-white star is normally the brightest star in the constellation and shines at magnitude 2.5.

Beta (β) Pegasi (Scheat)
This red giant is different in color to the other stars in the Great Square of Pegasus. It lies 200 light-years from Earth and varies unpredictably—usually shining at around magnitude 2.7, but sometimes outshining Markab and occasionally becoming fainter than Gamma (γ) Pegasi.

M15 A bright globular cluster of magnitude 6.2, M15 is easily spotted through binoculars. More than 30,000 light-years away, it is one of the densest star clusters. It contains nine pulsars, the remains of ancient supernova explosions.

Pegasi (Peg)

PEGASUS

One of the largest constellations,
Pegasus covers an empty area of sky to the north of the zodiacal constellations Aquarius and Pisces. It is very easy to find, because its four bright stars (one of which is shared with the neighboring constellation Andromeda) form the Great Square of Pegasus.

NORTHERN HEMISPHERE

Famous as the winged horse in Greek mythology, Pegasus was born from the blood of the gorgon Medusa, after she was killed by the Greek hero Perseus. Although only the forequarters of the horse are indicated by stars, Pegasus is still the seventh-largest constellation in the sky. It lies well away from the Milky Way, but interesting objects within its boundaries include the fine globular cluster M15 and the yellow main sequence star 51 Pegasi—the first star discovered to have an extrasolar planet.

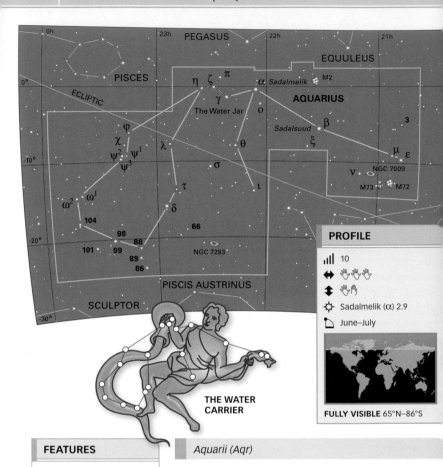

THE WATER CARRIER

FEATURES

M2 The brighter of Aquarius's two globular clusters, M2 shines at magnitude 6.5 and lies 37,000 light-years away from Earth.

NGC 7009 (the Saturn Nebula) Another planetary nebula, this is 3,000 light-years away and shines at magnitude 8.0. It appears to be of a size similar to the disk of Saturn, and when seen through a small telescope it looks like a greenish disk.

NGC 7293 (the Helix Nebula) This is our nearest planetary nebula, some 300 light-years away. At roughly the size of the full Moon, its light is spread across a large area, so it is hard to identify except in clear and dark skies. It is best seen through binoculars, which have a wide field of view.

Aquarii (Aqr)

AQUARIUS

One of the oldest zodiacal constellations, Aquarius lies between Capricornus and Pisces, near the celestial equator. Its pattern is indistinct, but its brightest star, Alpha (α) or Sadalmelik, and the Y-shaped pattern formed by a group of four stars known as the Water Jar are useful aids for locating

NORTHERN HEMISPHERE

it. In early May each year, the Eta (η) Aquarid meteor shower radiates from this area.

Aquarius has been seen as a figure pouring water from a jug since the second millennium BCE. According to Greek mythology, this figure depicts Ganymede, a young shepherd kidnapped by Zeus in the form of an eagle, represented by nearby Aquila, and carried away to Mount Olympus to serve as cupbearer to the gods. This constellation contains some great deep-sky objects, including the closest planetary nebula to Earth.

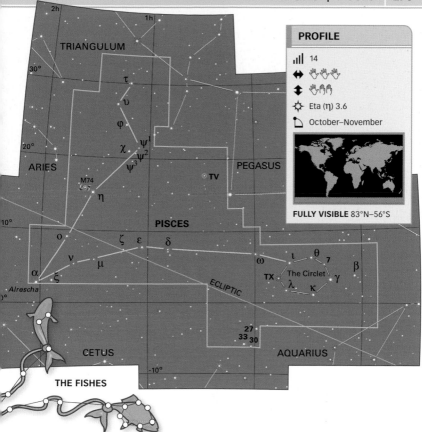

TRIANGULUM

ARIES

M74

η

PISCES

Alrescha

CETUS

THE FISHES

PEGASUS

TV

The Circlet

ECLIPTIC

AQUARIUS

PROFILE

📶 14

↔ ✋✋✋

↕ ✋✋

☼ Eta (η) 3.6

▱ October–November

FULLY VISIBLE 83°N–56°S

FEATURES

Eta (η) Piscium The brightest star in Pisces, Eta (η) is a yellow supergiant of magnitude 3.6, brighter than either of Alpha's components, and more than twice their distance at 300 light-years.

Alpha (α) Piscium (Alrescha) The second-brightest star of the constellation marks the point where the tails of Pisces' fish join. Just 140 light-years away from Earth, this white double star's components shine at magnitudes of 4.2 and 5.2, giving a combined magnitude of 3.8. Its stars are too close to separate with small telescopes.

M74 This beautiful spiral galaxy, 25 million light-years away, is face-on to Earth. Its light is so spread out that it is quite a challenging target for small telescope users to spot.

Piscium (Psc)

PISCES

This zodiacal constellation lies between the constellations Aquarius and Aries and depicts two fish whose tails are tied with a cord. According to a Greek myth, the fish represent the goddess Aphrodite and her son Eros, who transformed into fish and plunged into the Euphrates to escape the fearsome monster Typhon. This constellation's most distinctive feature is the asterism known as the Circlet, a ring of seven stars lying south of the Great Square of Pegasus (p.271). This ring marks the body of one of the fish.

NORTHERN HEMISPHERE

Although Pisces is a large constellation, it does not contain many objects of interest to the amateur astronomer. However, it does contain the vernal equinox—the point where the Sun crosses the celestial equator in March each year. This is used on star maps as the origin of the celestial coordinate system.

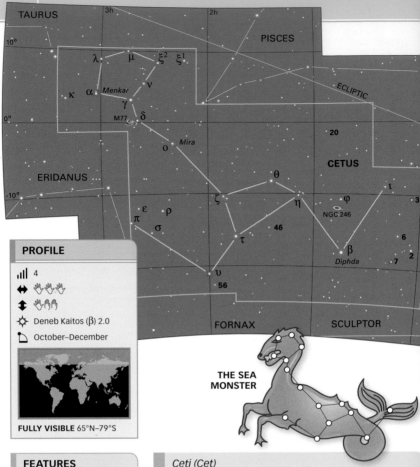

PROFILE

Ill 4

↔ 🖐️🖐️🖐️

↕ 🖐️🖐️

☼ Deneb Kaitos (β) 2.0

🗋 October–December

FULLY VISIBLE 65°N–79°S

FEATURES

Omicron (o) Ceti (Mira) With a name derived from the Latin for "wonderful," Mira is among the most prominent variable stars in the sky. It varies between magnitudes 10 and 2 over a cycle of 332 days. This unstable red giant undergoes a long but regular cycle of pulsations. Hence, depending on how much it pulsates it can be either a naked-eye star or one that is visible only with a telescope.

Tau (τ) Ceti One of the closest stars to Earth, Tau is also the most Sun-like of stars due to its temperature and brightness. It lies just 11.9 light-years away, surrounded by a swarm of asteroids and comets. Technically, it is a yellow subdwarf. If it has planets in orbit around it, they could be prime candidates for extraterrestrial life.

Ceti (Cet)

CETUS

One of the original 48 Greek constellations, Cetus straddles the celestial equator. It is large but relatively faint, and is best found by looking southwest of the more distinctive Taurus. Although its name suggests that it depicts a whale, Cetus is represented on old star charts as an unlikely looking, almost comical, hybrid sea monster.

SOUTHERN HEMISPHERE

The constellation is not easy to identify, because it can vary significantly in appearance depending on the brightness of its most famous star, Mira or Omicron (o) Ceti. In 1596, the Dutch astronomer David Fabricius noticed what he thought was a new star in Cetus. It was soon realized that it was not a new star, but that Mira had increased in brightness in a long but regular cycle of pulsations, making it the first variable star to be discovered.

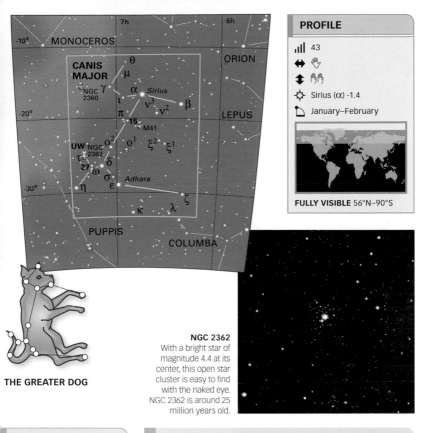

PROFILE

⏸ 43

↔ ✋

↕ 🖐🖐

☼ Sirius (α) -1.4

📖 January–February

FULLY VISIBLE 56°N–90°S

NGC 2362
With a bright star of magnitude 4.4 at its center, this open star cluster is easy to find with the naked eye. NGC 2362 is around 25 million years old.

THE GREATER DOG

FEATURES

Alpha (α) Canis Majoris (Sirius) The famous "Dog Star" is 23 times more luminous than the Sun. It shines at a brilliant magnitude -1.4 because it lies just 8.6 light-years away. Sirius is a binary system: the primary star is orbited by a faint, white dwarf called Sirius B.

Beta (β) Canis Majoris (Mirzam) At magnitude 2.0, Mirzam is far outshone by Sirius. In reality, this blue giant, which lies 500 light-years from Earth, is a far more luminous star than Sirius.

M41 This open cluster of stars, lying around 2,300 light-years away, is visible to the naked eye as a hazy patch the size of the full Moon. Binoculars distinguish its brightest stars, while telescopes show chains of stars radiating from its center.

Canis Majoris (CMa)

CANIS MAJOR

Following obediently behind Orion, the hunter, on his journey across the sky, Canis Major is host to Sirius, the brightest star in the entire sky and one of our closest neighboring stars. This ancient constellation is depicted as the larger of Orion's two hunting dogs, the other being Canis Minor.

SOUTHERN HEMISPHERE

According to another interpretation, Canis Major is Laelaps, a mythical fleet-footed dog, trapped in endless pursuit of its prey, the Teumessian fox.

Sirius, along with Procyon in Canis Minor and Betelgeuse in Orion, forms a triangle around the celestial equator. The rising of Sirius alongside the Sun was used by the ancient Egyptians to predict the timing of the Nile's annual flooding. Since the constellation lies across the Milky Way, it contains several star clusters and other features of the deep sky.

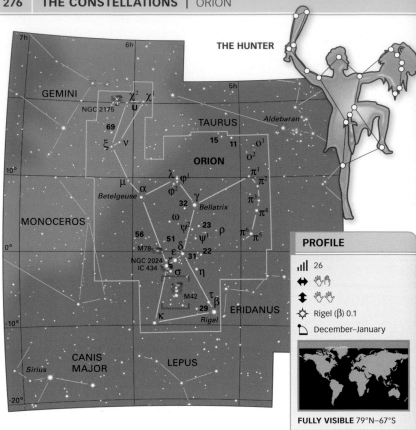

THE HUNTER

FEATURES

Beta (β) Orionis (Rigel)
Apart from rare occasions when Betelgeuse is at its maximum magnitude 0.0, Rigel is Orion's brightest star. It is a brilliant blue-white supergiant of magnitude 0.1, about 770 light-years away.

Alpha (α) Orionis (Betelgeuse) Some 430 light-years from Earth, this is one of the brightest red giants in the sky. Its magnitude varies unpredictably between magnitudes 0.0 and 1.3.

M42 (the Orion Nebula)
Marking the "sword" to the south of Orion's belt, M42 is a huge star-forming region about 1,500 light-years away. Visible to the naked eye, it is a beautiful sight through binoculars or a small telescope, with the Trapezium cluster of newborn stars lying within it.

Orionis (Ori)

ORION

One of the most glorious constellations in the sky, Orion represents a giant hunter or warrior facing a charging bull, Taurus, and followed by his dogs, Canis Major and Canis Minor. An ancient constellation, the Syrians knew it as Al Jabbar, the Giant, and the ancient Egyptians as Sahu, the soul of Osiris.

NORTHERN HEMISPHERE

Orion's position on the celestial equator ensures that it can be seen from all over the world. Its most distinctive feature is Orion's belt, formed by a line of three 2nd-magnitude stars almost exactly on the celestial equator. Orion also contains two of the brightest and best known stars—Rigel and Betelgeuse. A group of stars and nebulosity marks the sword hanging from Orion's belt and contains the great star-forming region of the Orion Nebula. Every October, the Orionid meteors seem to radiate from a point near Orion's border with Gemini.

THE ORION NEBULA REGION

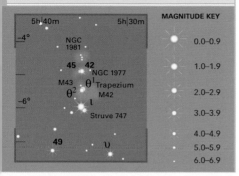

ORION THE HUNTER

In Greek mythology, Orion was the son of the sea god Poseidon, famed for his strength and handsome looks. The Greek poet Homer described him as a great hunter brandishing a club of bronze, who was often seen as the companion of Artemis, goddess of the hunt. In one legend, the boastful Orion claimed he could overcome any animal on Earth, so the Earth goddess brought forth a scorpion to sting the hunter to death. Orion was thus placed opposite the constellation of Scorpius in the sky.

The Orion Nebula
This image was captured by the Hubble Space Telescope, but even through binoculars this nebula reveals its spectacular flowerlike structure. It is one of several nebulae around Orion's belt and sword.

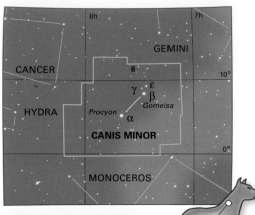

THE LITTLE DOG

Lone star
Unlike the distinctive constellation of the greater dog (Canis Major), Canis Minor consists of little more than its brightest star, Procyon.

FEATURES

Alpha (α) Canis Minoris (Procyon) This white star is the eighth-brightest in the night sky. Its name means "before the dog" in Greek, because from Mediterranean latitudes it rises shortly before the brilliant Dog Star, Sirius. Somewhat less luminous than Sirius and slightly farther away at 11.4 light-years, Procyon shines at magnitude 0.34. It is seven times more luminous than the Sun.

Beta (β) Canis Minoris (Gomeisa) Lying about 150 light-years away, this blue-white star of magnitude 2.9 is far more distant and much more radiant than Procyon. Its name is derived from the Arabic for "the bleary-eyed woman," which refers to the weeping sister of Sirius, whom he left behind to flee for his life.

Canis Minoris (CMi)

CANIS MINOR

One of the ancient constellations listed by the Greek–Egyptian astronomer Ptolemy in the 2nd century CE, Canis Minor lies virtually on the celestial equator. It is usually identified as the smaller of Orion's two hunting dogs, although it has also been identified with the hounds of Diana, the goddess of the hunt. It is easily located by its brightest star—Procyon,

SOUTHERN HEMISPHERE

or Alpha (α) Canis Minoris. Procyon forms a large triangle with two other 1st-magnitude stars: Betelgeuse in Orion and Sirius in Canis Major.

Coincidentally, Procyon and Sirius are both roughly the same distance from Earth, so the difference in their brightness is representative of a true difference in their luminosity. Despite its prominence, Canis Minor is one of the smaller constellations and contains little of particular note to small telescope users.

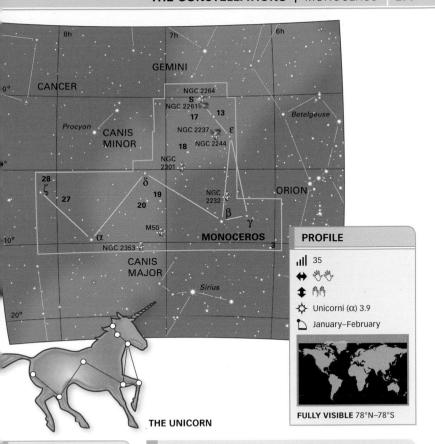

THE UNICORN

FEATURES

Alpha (α) Monocerotis (Unicorni) This bright, orange giant star lies about 175 light-years from Earth and shines at magnitude 3.9.

Beta (β) Monocerotis (Eite) This beautiful triple star is separated in a small telescope to reveal a chain of three blue-white stars of magnitude 5.

M50 This is one of many open clusters populating a rich band of the Milky Way as it passes through Monoceros. Small telescopes will reveal its individual stars.

NGC 2244 This star cluster lies at the heart of the Rosette Nebula—a glowing gas cloud and outlying part of a huge star-forming complex centered on Orion. It is a diffuse nebula, but can be seen with good binoculars on a dark night.

Monocerotis (Mon)

MONOCEROS

The W-shape of the constellation is easy to overlook because of the presence of its brighter neighbours, but Monoceros can be easily located with reference to the constellations Orion and Canis Major. It sits on the celestial equator in the middle of a triangle formed by the brilliant first-magnitude stars

SOUTHERN HEMISPHERE

Betelgeuse in Orion, Procyon in Canis Minor, and Sirius in Canis Major.

The origins of the constellation, depicting the mythical unicorn, are unknown, but it was first written about in the 17th century. Its introduction has been attributed to the Dutch astronomer Petrus Plancius in 1613 and to the German scientist Jakob Bartsch in 1624, but some have claimed that it has a more ancient origin. Monoceros is rich in star clusters and nebulae because the Milky Way runs through it, but it lacks bright stars.

PROFILE

📶 1

↔ 🖐🖐🖐🖐🖐

↕ 🖐🖐🖐🖐

☼ Alphard (α) 2.0

🗓 February–June

FULLY VISIBLE 54°N–83°S

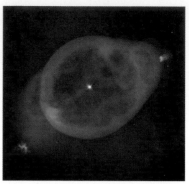

NGC 3242
Also known as the "Ghost of Jupiter" on account of its planetlike appearance, this planetary nebula is best seen with a small telescope and appears distinctly bluish to observers.

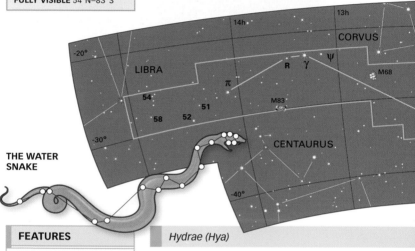

THE WATER SNAKE

Hydrae (Hya)

FEATURES

Alpha (α) Hydrae (Alphard)
This star's name means "the solitary one," reflecting its location in an otherwise blank area of sky. An orange giant of magnitude 2.0, it lies around 175 light-years from Earth.

Epsilon (ε) Hydrae This binary star has contrasting yellow and blue components of magnitudes 3.4 and 6.7 that can be separated with a moderate-sized telescope.

M48 An open cluster, M48 lies close to Hydra's border with the richer starfields of Monoceros. It contains about 80 stars and is just visible to the naked eye in dark skies.

M83 This face-on spiral galaxy, 15 million light-years away, has a bright central nucleus that can be spotted with ease through a small telescope.

HYDRA

The largest constellation in the night sky is a hard-to-trace chain of stars of mostly average brightness. Hydra stretches for more than a quarter of the way around the sky. Its head lies south of Cancer and just north of the celestial equator, while its tail is in the Southern Hemisphere between Libra and Centaurus. Its brightest star, Alphard, marks its heart.

SOUTHERN HEMISPHERE

The constellation's name translates as the water snake. In legend it is associated with Corvus, the crow, and Crater, the cup, both of which lie along its back. Hydra also shares its name with the multiheaded monster fought by the mythical hero Hercules.

Despite its size, there is little to mark out Hydra other than the group of six moderate stars that form its head. It is largely devoid of deep-sky objects, apart from some distant galaxies that are mostly difficult to spot.

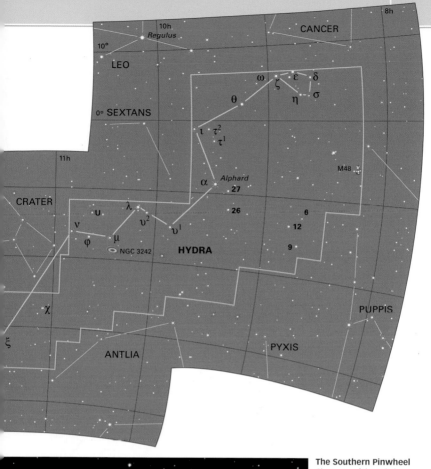

The Southern Pinwheel
Close-ups of the spiral galaxy M83, such as this one captured with a very large telescope, reveal sharply defined pinkish regions of star formation in its spiral arms. M83 has roughly half the diameter of our Milky Way.

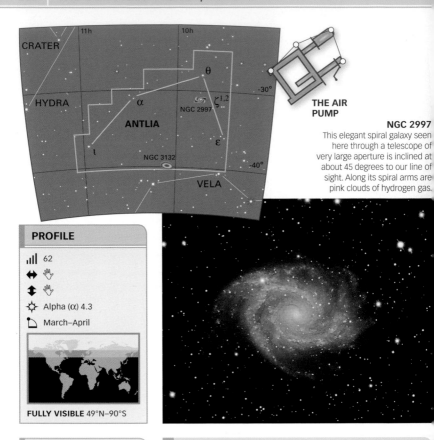

CRATER

HYDRA

ANTLIA

θ

α

ζ¹,²

NGC 2997

ε

ι

NGC 3132

VELA

THE AIR PUMP

NGC 2997
This elegant spiral galaxy seen here through a telescope of very large aperture is inclined at about 45 degrees to our line of sight. Along its spiral arms are pink clouds of hydrogen gas.

PROFILE

- ▪▪▪▪ 62
- ↔ 🖐
- ↕ 🖐
- ☼ Alpha (α) 4.3
- 🗓 March–April

FULLY VISIBLE 49°N–90°S

FEATURES

Alpha (α) Antliae This orange giant is 500 times more luminous than the Sun, but at a distance of 365 light-years from Earth, it shines at a weak magnitude of 4.3.

Theta (θ) Antliae At magnitude 4.8, Theta (θ) is the constellation's second-brightest star. It is actually a double, consisting of white and yellow stars of magnitudes 5.6 and 5.7. Lying 385 light-years away, they are too close for small telescopes to separate them.

NGC 3132 (the Eight-burst Nebula) Sometimes referred to as the Southern Ring Nebula, this planetary nebula straddles the boundary of Antlia and Vela about 2,000 light-years away from Earth. It shines at magnitude 8 and is a good target for small telescopes.

Antliae (Ant)

ANTLIA

This is a faint constellation to the south of Hydra, best found by looking to the northeast of the Milky Way as it passes through Puppis. The French astronomer Nicolas de Lacaille introduced this constellation for his 1756 map of the southern skies. Like most of his constellations, it honors a

SOUTHERN HEMISPHERE

scientific invention—in this case, the air pump used by the French scientist Denis Papin and the British physicist Robert Boyle for their experiments on gases.

Antlia contains no named stars, bright clusters, or nebulae, and is of little interest to those with binoculars or a small telescope. However, it contains a relatively close cluster of galaxies, known as the Antlia Cluster. This cluster has around 230 member galaxies, and its brighter members, around 32 million light-years away, can be observed with larger amateur telescopes.

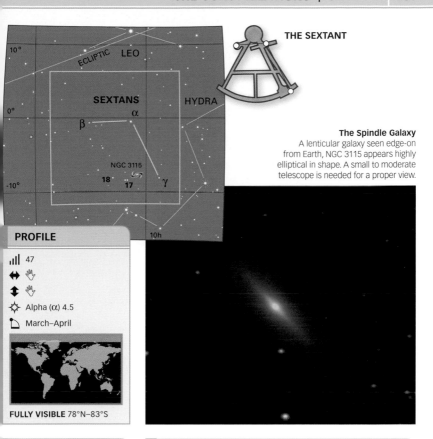

THE SEXTANT

The Spindle Galaxy
A lenticular galaxy seen edge-on from Earth, NGC 3115 appears highly elliptical in shape. A small to moderate telescope is needed for a proper view.

PROFILE

.ıll 47

↔ ✋

↕ ✋

☼ Alpha (α) 4.5

🗓 March–April

FULLY VISIBLE 78°N–83°S

FEATURES

Alpha (α) Sextantis This blue-white giant star is some 340 light-years from Earth. Due to this distance, Alpha (α) shines at a relatively weak magnitude 4.5.

Beta (β) Sextantis Another blue-white giant, Beta (β) is more luminous than Alpha (α), but only reaches magnitude 5.1 in our skies because it lies 520 light-years away.

NGC 3115 (the Spindle Galaxy) One of the closest large galaxies, NGC 3115 lies some 14 million light-years away. Its huge, bulging disk of stars appears elliptical or spindle-shaped because it is viewed edge-on from Earth, giving the Spindle Galaxy its popular name. The combined light of its stars reaches magnitude 8.5, making it just visible in binoculars.

Sextantis (Sex)

SEXTANS

Small and indistinct, the constellation Sextans lies exactly on the celestial equator. However, its pattern is easy to find since it lies to the south of the bright star Regulus in the constellation Leo. This constellation was introduced in 1687 by the Polish astronomer Johannes Hevelius. It represents

SOUTHERN HEMISPHERE

the sextant, a scientific instrument used by navigators, along with an accurate clock, to determine their positions. The sextant continued to be used for charting locations onboard ships right up to the era of electronic navigation.

Sextans lies away from the Milky Way and has no bright or named stars, clusters, or nebulosity. However, it contains several galaxies, including one that lies just beyond our own Local Group (p.146). Most of these galaxies require large telescopes for a detailed view.

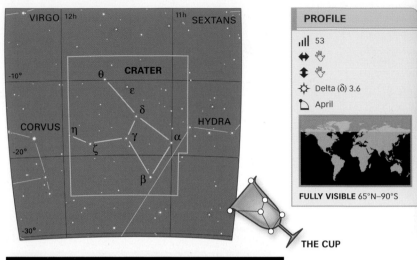

THE CUP

PROFILE

📶 53

↔ ✋

↕ ✋

☼ Delta (δ) 3.6

🗓 April

FULLY VISIBLE 65°N–90°S

NGC 3981
This elegant barred spiral galaxy, just visible through a small telescope, lies around 82 million light-years away and has about two-thirds the diameter of the Milky Way.

FEATURES

Delta (δ) Crateris Crater's brightest star has ended up with the designation Delta (δ) through historical accident. An orange giant of magnitude 3.6, it lies 62 light-years from Earth.

Gamma (γ) Crateris This white star, 75 light-years away and of magnitude 4.1, has a faint binary companion that can be seen through a small telescope.

Alpha (α) Crateris Significantly fainter than Gamma (γ) Crateris at magnitude 4.1, Alpha (α) is a yellow giant about 175 light-years away from Earth.

Crateris (Crt)

CRATER

A faint constellation located between Virgo and Hydra, Crater is relatively easy to locate thanks to its distinctive bow-tie shape. It represents the drinking cup of the Greek god Apollo. Legend links it to the crow of Corvus—the constellation that lies directly to its east—and to the water snake of Hydra below. All three constellations are among the 48 ancient star patterns listed by the Greek–Egyptian astronomer Ptolemy in the 2nd century CE. Later astronomers attempted to introduce two other constellations to the area—Felis, the cat, and Noctua, the owl. However, these have since been dropped in favor of Ptolemy's earlier patterns.

SOUTHERN HEMISPHERE

Although larger than Corvus, Crater lacks bright stars, clusters, and nebulae, but it does have some galaxies that can be observed with large amateur telescopes.

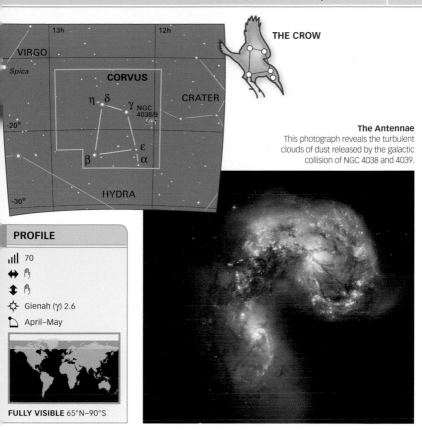

THE CROW

The Antennae
This photograph reveals the turbulent clouds of dust released by the galactic collision of NGC 4038 and 4039.

PROFILE

📶 70

↔ ✋

↕ ✋

☼ Gienah (γ) 2.6

📅 April–May

FULLY VISIBLE 65°N–90°S

FEATURES

Gamma (γ) Corvi (Gienah)
Corvus's brightest star, Gamma (γ) is a blue-white star of magnitude 2.6, lying at a distance of 220 light-years. It shares its common name, Gienah, with Epsilon (ε) Cygni.

Delta (δ) Corvi This double star is a good target for small telescopes. The bright blue-white primary is orbited by a deeper blue or even purple star of magnitude 9.2.

Alpha (α) Corvi (Alchiba)
Despite its Greek letter designation, Alpha (α) is outshone by Gamma (γ), Beta (β), and Delta (δ) Corvi. It is a white star 52 light-years away, shining at magnitude 4.0.

NGC 4038 and 4039 (Antennae Galaxies) These colliding galaxies are barely visible with small telescopes.

Corvi (Crv)

CORVUS

This constellation lies to the southwest of the bright star Spica, in Virgo, and its four brightest stars define its roughly rectangular shape. In Greek mythology, Corvus was a crow sent by Apollo to fill his cup with water from a spring (the cup is represented by the neighboring constellation Crater).

SOUTHERN HEMISPHERE

However, Corvus stopped to eat figs and returned late, clutching a water snake (Hydra) in its claws. It blamed the snake for blocking the spring. Angered at the lie, Apollo flung the cup, the crow, and the snake into the sky, where they lie preserved.

Corvus has no bright clusters or nebulae of interest to amateur astronomers. Although there are galaxies, they are too faint for most amateur telescopes. It does, however, contain NGC 4038 and NGC 4039, a pair of colliding galaxies that can be seen with large instruments.

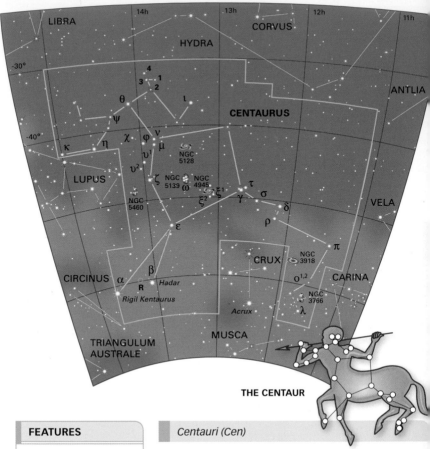

THE CENTAUR

FEATURES

Alpha (α) Centauri (Rigil Kentaurus) At 4.3 light-years from Earth, this is the closest star system to the Sun. Binoculars reveal that it is a multiple, consisting of yellow and orange stars of magnitudes 0 and 1.3 respectively. A third member, Proxima Centauri of magnitude 11, can be spotted with a good telescope.

Omega (ω) Centauri (NGC 5139) Despite its stellar name, Omega (ω) is a globular cluster. It is a tight ball of several million stars, lying 17,000 light-years away and shining at magnitude 3.7. A small telescope will reveal its brightest individual members.

NGC 5128 (Centaurus A) This bright galaxy is an active elliptical, about 15 million light-years away, that gives out strong radio signals.

Centauri (Cen)

CENTAURUS

This prominent constellation of the southern skies is located between Lupus and Scorpius on one side, and Carina and Vela on the other. Centaurus represents the mythological centaur Chiron, a creature with the torso of a man and the legs of a horse. The constellation of Crux sits neatly

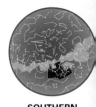

SOUTHERN HEMISPHERE

between the centaur's forelegs and hindlegs. Chiron is one of two centaurs in the sky (the other is Sagittarius, the archer). He was the wisest of the centaurs and tutor to many Greek heroes.

A rich stream of the Milky Way passes through Centaurus, which also contains the sky's brightest globular cluster, Omega (ω) Centauri, and an unusual galaxy NGC 5128, also known as the radio source Centaurus A. It also contains Alpha (α) Centauri—the nearest star system to the Sun and the third brightest star in the sky.

PROFILE

📶 9

↔ ✋ ✋ ✋

↕ ✋ ✋

☼ Rigil Kentaurus (α) 0.0

📄 April–June

FULLY VISIBLE 25°N–90°S

NGC 5139
Seen here from the observatory in La Silla, Chile, this is by far the largest globular cluster orbiting our galaxy. Astronomers believe that it may be the surviving core of an ancient dwarf galaxy otherwise consumed by the Milky Way.

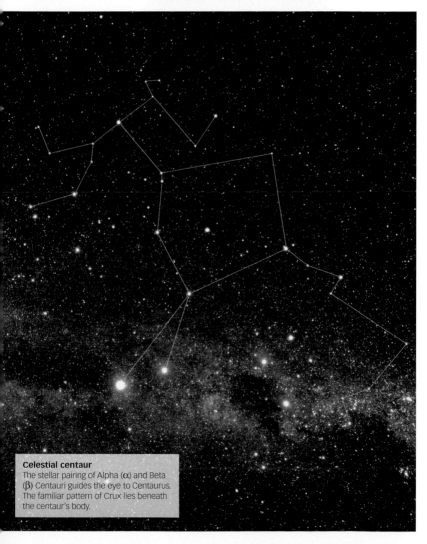

Celestial centaur
The stellar pairing of Alpha (α) and Beta (β) Centauri guides the eye to Centaurus. The familiar pattern of Crux lies beneath the centaur's body.

PROFILE

📶 46

↔ 🖐

↕ 🖐

☼ Alpha (α) 2.3

📖 May–June

FULLY VISIBLE 34°N–90°S

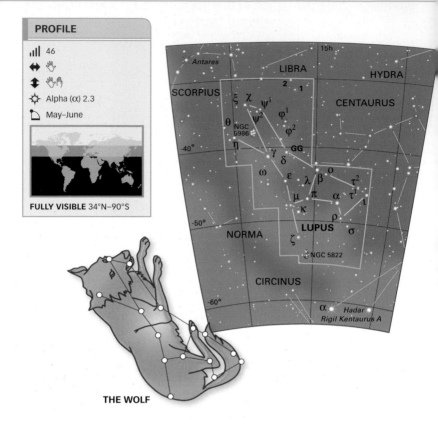

THE WOLF

FEATURES

Alpha (α) and Beta (β) Lupi Lupus's two brightest stars are almost identical and are close neighbors in space. Both are blue giants about 650 light-years away, but Alpha is slightly closer, giving it a magnitude of 2.3 compared to Beta's 2.7.

Mu (μ) Lupi A multiple star, this is one of the easiest to split. Small telescopes will easily show that the primary, a blue-white star of magnitude 4.3, has a companion of magnitude 7. Larger telescopes will reveal that the primary is itself a double, composed of twin stars of magnitude 5.1.

NGC 5822 This large and rich open cluster contains more than 100 stars. Lying about 2,600 light-years away, it has an overall magnitude of 7.0 and can be seen with binoculars.

Lupi (Lup)

LUPUS

A constellation of the southern sky, Lupus lies on the edge of the Milky Way between Centaurus and Scorpius. Despite being relatively bright, it is cursed with a complex jumble of stars that can make it rather hard to identify. It was one of the original 48 constellations familiar to the ancient Greeks. They

SOUTHERN HEMISPHERE

visualized it as a wild animal of unspecified nature, impaled by neighboring Centaurus, the centaur, on the end of a long pole called a thyrsus. In consequence, Centaurus and Lupus were often regarded as a combined constellation. The identification of Lupus as a wolf only seems to have become common during Renaissance times.

The constellation's location in the Milky Way means that it contains numerous double stars and interesting objects for amateur observers.

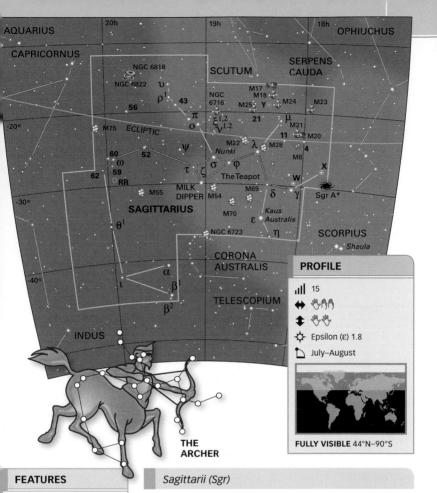

AQUARIUS
20h
19h
18h
OPHIUCHUS
CAPRICORNUS
NGC 6818
NGC 6822
SCUTUM
SERPENS
CAUDA
υ
ρ¹
43
56
π
NGC
6716
M17
M18
M25 γ
M24
M23
-20°
M75 ECLIPTIC
ο
ξ 1,2
ν 1,2
21
μ
M21
M20
ψ
M22
Nunki
λ
M28
11
4
60
52
τ
ζ
σ
φ
M8
ω
The Teapot
X
62
59
RR
W
Sgr A*
-30°
M55
MILK
DIPPER
M54
M69
δ
γ
SAGITTARIUS
M70
Kaus
Australis
ε
η
SCORPIUS
θ¹
NGC 6723
Shaula
-40°
CORONA
AUSTRALIS
α
β¹
TELESCOPIUM
β²
INDUS

THE
ARCHER

PROFILE

📶 15

↔ 🖐🖐🖐

↕ 🖐🖐

☼ Epsilon (ε) 1.8

📅 July–August

FULLY VISIBLE 44°N–90°S

FEATURES

Sagittarii (Sgr)

Epsilon (ε) Sagittarii This star of magnitude 1.8—the constellation's brightest—has a very faint binary companion.

Beta (β) Sagittarii (Arkab) This apparent double star, consisting of two stars at about magnitude 4.0, can be split with the naked eye. But its stars are a chance alignment: in reality they are 140 and 380 light-years away.

M8 (the Lagoon Nebula) The largest and brightest deep-sky object in the constellation is visible to the naked eye as a light patch of sky. It is easy to identify with binoculars.

M22 Visible to the naked eye and a fine sight in binoculars, this is the brightest of several globular clusters on the northern edge of Sagittarius.

SAGITTARIUS

This prominent zodiacal constellation is found between Scorpius and Capricornus in the southern celestial hemisphere. It is easily identified thanks to a highly recognizable pattern of 10 stars known as the Teapot. Although old star charts depicted Sagittarius as a centaur, in Greek mythology it was

SOUTHERN HEMISPHERE

also identified as a creature called a satyr, mounted on horseback. He is usually said to be Crotus, son of the nature god Pan and inventor of archery.

Sagittarius lies in the direction of the center of our galaxy, the Milky Way, and is home to numerous star fields and clusters, including over 60 open and globular clusters and several large, bright nebulae, making the constellation a rewarding sight for binoculars or any telescope. The heart of the Milky Way itself lies hidden beyond these star clouds, some 26,000 light-years away.

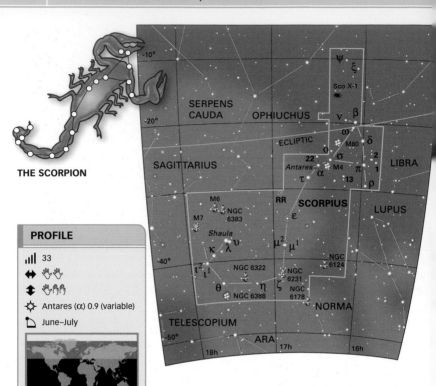

THE SCORPION

PROFILE

📶 33

↔ 🖐🖐

↕ 🖐🖐🖐

☼ Antares (α) 0.9 (variable)

🗓 June–July

FULLY VISIBLE 44°N–90°S

FEATURES

Alpha (α) Scorpii (Antares)
With a name that means "rival to Mars," Antares is a giant star 9,000 times as luminous as the Sun. It varies in brightness between magnitudes 0.9 and 1.8 in a roughly five-year cycle. A red supergiant, Antares is hundreds of times larger than the Sun and lies 600 light-years from Earth.

M4 A globular cluster, 7,000 light-years away and in orbit around the Milky Way, M4 shines at magnitude 7.4, making it a good target for binoculars or a telescope.

M6 This fine open star cluster is visible to the naked eye as a knot in the Milky Way, just above the scorpion's tail. Binoculars or a small telescope will reveal dozens of individual stars. M6 lies 2,000 light-years away.

Scorpii (Sco)

SCORPIUS

Beautiful and easily recognizable, the zodiacal constellation Scorpius is situated in the southern sky between Sagittarius and Libra. In Greek mythology, Scorpius represents a scorpion sent by the goddess Artemis to kill the hunter Orion—fittingly, Orion sets as Scorpius rises. The scorpion is usually depicted with a raised tail, marked by a curve of stars that is very prominent on a clear, dark night. Antares, the brightest star, marks the animal's heart. Libra once represented its claws.

SOUTHERN HEMISPHERE

The scorpion's tail passes through a rich area of the Milky Way. Consequently, there are a variety of targets for small telescopes, especially dense star clouds and clusters. Many of the constellation's bright stars are members of a scattered cluster, the Scorpius–Centaurus OB Association, around 430 light-years away from Earth.

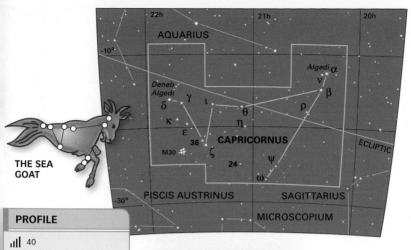

AQUARIUS

Algedi α
ν
β
Deneb
Algedi
γ ι
δ
θ
ρ
κ
η
ε
36 **CAPRICORNUS**
M30
ζ
24 ψ
ω

ECLIPTIC

**THE SEA
GOAT**

PISCIS AUSTRINUS SAGITTARIUS

MICROSCOPIUM

PROFILE

▮▮▮ 40

↔ ✋🖐

↕ 🖐🖐

☼ Deneb Algedi (δ) 2.9

🗓 August–September

FULLY VISIBLE 62°N–90°S

M30
This globular cluster, some
90 light-years across, owes its
unusual structure to a core
that has collapsed into a
dense knot, leaving extended
tendrils of stars behind.

FEATURES

Alpha (α) Capricorni (Algedi)
The brightest components
of this multiple system can be
separated with binoculars or
good eyesight. They include
a yellow supergiant (α¹) and
an orange giant (α²), 690 and
109 light-years away, with
magnitudes 4.2 and 3.6. Small
telescopes reveal the yellow
supergiant to be a double star,
while larger ones show the
orange giant as a triple star.

Beta (β) Capricorni (Dabih)
A yellow giant of magnitude
3.3, Dabih reveals a faint
companion when viewed
through binoculars. It is a
complex multiple containing
5–8 stars in orbit around one
another, 330 light-years away.

M30 This globular cluster of
magnitude 7.5 is 27,000
light-years away and visible
through a pair of binoculars.

Capricorni (Cap)

CAPRICORNUS

Lying in the southern sky

between Sagittarius and Aquarius,
Capricornus is best found by looking
to the northwest of its brighter
neighbour, the celestial archer
Sagittarius. Capricornus is
the second-faintest zodiac
constellation after Cancer. Despite
its relatively small size, it may be the

**SOUTHERN
HEMISPHERE**

oldest constellation—depictions of a strange goat with
the tail of a fish have been found on Babylonian seals
more than 4,000 years old. In Greek myth, Capricornus
represents the goat-like god Pan, who jumped into a river
and became part fish to escape from the monster Typhon.

Capricornus lies at some distance from the Milky Way
in a comparatively empty region of the sky. As a result, it
contains few deep-sky objects for amateur astronomers.
Its most interesting targets are multiple stars and a
globular cluster.

PROFILE

ıll 66

↔ ✋

↕ ✋

☼ Gamma (γ), Epsilon (ε) 4.7

🗓 August–September

FULLY VISIBLE 45°N–90°S

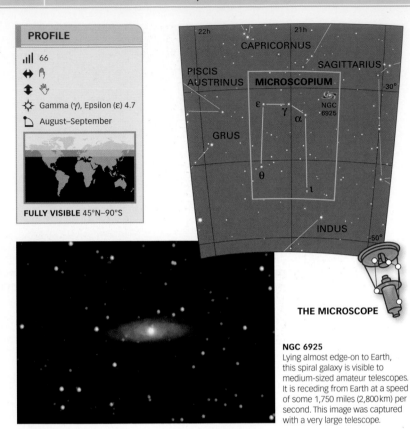

THE MICROSCOPE

NGC 6925
Lying almost edge-on to Earth, this spiral galaxy is visible to medium-sized amateur telescopes. It is receding from Earth at a speed of some 1,750 miles (2,800 km) per second. This image was captured with a very large telescope.

FEATURES

Gamma (γ) Microscopii
This yellow giant lies 245 light-years away and shines at magnitude 4.7.

Theta (θ) Microscopii
Though the brightest of several variable stars in Microscopium, Theta's (θ) variations are hard to see. It varies by only about 0.1 magnitude from its average of 4.8 in a 2-day cycle.

Alpha (α) Microscopii This double star lies 250 light-years away. The primary is a yellow giant of magnitude 5.0, while its companion is far fainter at magnitude 10 and visible only through telescopes of moderate aperture.

U Microscopii This distant red giant is a more obvious variable. Like Mira in Cetus, its magnitude varies over a cycle of 332 days.

Microscopii (Mic)

MICROSCOPIUM

A fairly insignificant constellation, Microscopium lies to the south of Capricornus in the mid-southern sky. It is best found by looking for a roughly rectangular pattern of indistinct stars between the constellations Sagittarius and Piscis Austrinus.

SOUTHERN HEMISPHERE

Microscopium is one of several small, faint groups added to the night sky by the French astronomer Nicolas Louis de Lacaille in the 1750s and mostly named after scientific instruments. It depicts an early design of the compound microscope. Its distance from the Milky Way means that it contains no deep-sky objects except for galaxies, most of which are too faint for amateur telescopes. One interesting but faint star is U Microscopii, near the border with Sagittarius. This red dwarf, 30 light-years away from Earth, is surrounded by a disk of dusty, potentially planet-forming material.

PROFILE

.ıll 60

↔ 🖐

↕ 🖐🖐

☼ Fomalhaut (α) 1.2

📖 September–October

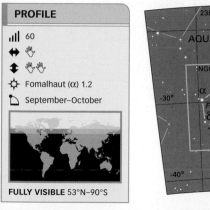

FULLY VISIBLE 53°N–90°S

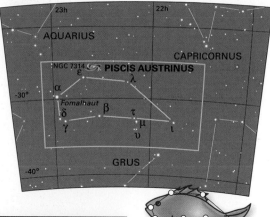

THE SOUTHERN FISH

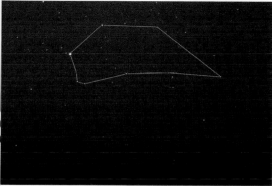

The fish's mouth
Owing to its brightness, Fomalhaut stands out from the other stars in Piscis Austrinus. Its name is derived from the Arabic for "mouth of the fish."

FEATURES

Alpha (α) Piscis Austrini (Fomalhaut) At magnitude 1.2, this blue-white star is one of the brightest in the sky. Just 25 light-years away, it was the first star found to be encircled by a ring of cold, icy material (potential building blocks for planets). This extends over twice the diameter of our own solar system.

Beta (β) Piscis Austrini This double star, 135 light-years away, consists of a primary of magnitude 4.3 with a widely spaced star of magnitude 7.7 that can easily be seen through a small telescope.

Gamma (γ) Piscis Austrini This double star, with components of magnitudes 4.5 and 8.0, is more difficult to separate from Gamma (γ) because its stars are closer. They lie 325 light-years away.

Piscis Austrini (PsA)

PISCIS AUSTRINUS

This small ring of generally faint stars is made easier to find by the presence of Fomalhaut, one of the brightest stars in the sky. Piscis Austrinus was invented in ancient times and is one of the most southerly members of a list of 48 constellations drawn up by the Greek–Egyptian astronomer

SOUTHERN HEMISPHERE

Ptolemy in the 2nd century CE. It is often depicted as a fish drinking from water poured by nearby Aquarius and has also been regarded as the parent of the more northerly fishes of Pisces. The constellation has also been called Piscis Australis.

Piscis Austrinus lacks bright clusters and nebulae, and most of its galaxies are too faint for amateur telescopes, with the exception of the spiral NGC 7314, which can be spotted with larger instruments. However, the constellation does contain some fascinating stars.

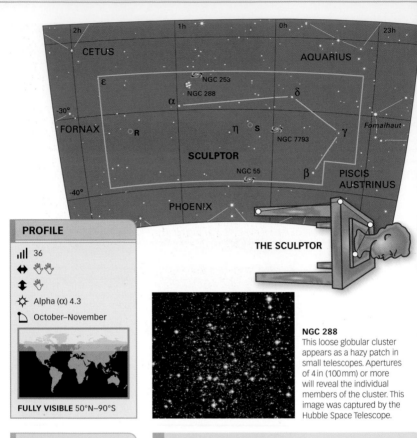

THE SCULPTOR

PROFILE

- ▂▃▅ 36
- ↔ 🖐🖐
- ↕ 🖐
- ☼ Alpha (α) 4.3
- 🗓 October–November

FULLY VISIBLE 50°N–90°S

NGC 288
This loose globular cluster appears as a hazy patch in small telescopes. Apertures of 4 in (100mm) or more will reveal the individual members of the cluster. This image was captured by the Hubble Space Telescope.

FEATURES

Alpha (α) Sculptoris The brightest star in Sculptor is this blue-white giant, 590 light-years away and shining at magnitude 4.3.

NGC 55 An 8th-magnitude spiral galaxy, NGC 55 is only 6 million light-years away, in the galaxy cluster just beyond the edge of our own Local Group (p.146). It is mottled with dust clouds and areas of star formation, and is an easy target for small telescopes.

NGC 253 The largest and brightest member of the so-called Sculptor Group, this spiral galaxy is about 9 million light-years away, at the heart of a small nearby galaxy cluster. It shines at around magnitude 7.5. Through binoculars it appears as a fuzzy oval of light with approximately the same diameter as the full Moon.

Sculptoris (Scl)

SCULPTOR

This faint and unremarkable southern constellation to the south of Cetus is easily located because it lies directly to the east of the bright star Fomalhaut in Piscis Austrinus. It is one of the more bizarre constellations added to the sky by the French astronomer Nicolas Louis de Lacaille in the 18th century, intended as it was to represent a sculptor's workshop.

SOUTHERN HEMISPHERE

The constellation has no stars brighter than magnitude 4.3 and lies far from the Milky Way, so it appears to be insignificant. However, it contains the south pole of our Galaxy—the point 90 degrees south of the plane of the Milky Way. Looking in this direction, we see straight "down," out of the plane of our galaxy, and avoid intervening stars, gas, and dust. As a result, good telescopes can reveal numerous galaxies within Sculptor, including some of our closest large galactic neighbors.

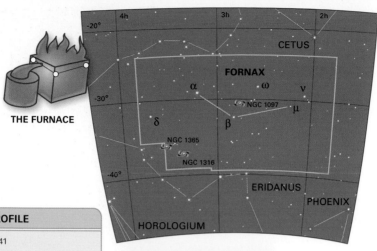

THE FURNACE

PROFILE

.ıll 41

↔ 🖐🤏

↕ 🖐

☼ Alpha (α) 3.9

🗓 November–December

FULLY VISIBLE 50°N–90°S

FEATURES

Alpha (α) Fornacis The constellation's brightest star is a double, easily split in a small telescope to reveal a magnitude-3.9 yellow star with an orange companion that has a magnitude of 6.9.

NGC 1097 Sixty million light-years away and shining at magnitude 10.3, this is one of the sky's brightest barred spiral galaxies. A small telescope will show its bright central nucleus, but a larger instrument is needed to show the barred structure and a dark column of dust through the center.

NGC 1316 This unusual galaxy is associated with a strong radio source called Fornax A. It appears to be an elliptical galaxy that recently absorbed another one. Dust and gas have awakened its central black hole, making its core active.

NGC 1365
A barred spiral and one of the brightest galaxies in the Fornax Cluster, NGC 1365 lies some 75 million light-years away from Earth. Its core can be seen with small telescopes.

Fornacis (For)

FORNAX

Situated to the south of the sea monster Cetus and enclosed within the bend of the celestial river Eridanus, this undistinguished constellation of the southern sky is made up of a handful of faint stars. Fornax was introduced by the French astronomer Nicolas Louis de Lacaille in the early 1750s. Originally named Fornax Chemica, the Chemist's Furnace, it represents a device used by chemists for distillation.

SOUTHERN HEMISPHERE

Fornax is not rich in either stars or clusters. However, it is home to the Fornax Cluster, a cluster of galaxies about 75 million light-years away from Earth. The brighter galaxies within the cluster can be observed with any amateur equipment. Another small region of the constellation was observed for more than a million seconds in 2003 to produce the Hubble Ultra Deep Field, one of our deepest views of the Universe.

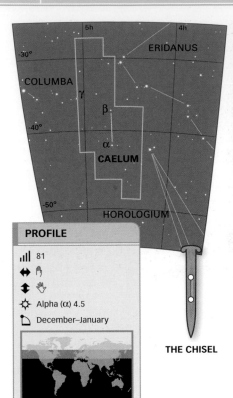

THE CHISEL

Inconspicuous shape
Caelum is relatively easy to locate in the southern sky because it lies to the west of the brilliant stars Sirius in Canis Major and Canopus in Carina.

PROFILE

📶 81

↔️ 🖐️

↕️ ✋

☼ Alpha (α) 4.5

🗓️ December–January

FULLY VISIBLE 41°N–90°S

FEATURES

Alpha (α) Caeli The brightest star in Caelum is a paltry magnitude 4.5. It is a white star 62 light-years away.

Gamma (γ) Caeli Sitting on Caelum's boundary with Columba, Gamma (γ) is an orange giant of magnitude 4.6 some 280 light-years away. A small telescope will reveal that it is a binary star, with a faint companion of magnitude 8.1.

Beta (β) Caeli Like Alpha (α) Caeli, this is a white star of average luminosity. About 65 light-years from Earth, it shines at magnitude 5.1.

R Caeli This is a variable red giant, with a long period of fluctuation of around 400 days. At a peak magnitude of 6.7, R Caeli can easily be spotted through binoculars near the northern border with Eridanus.

Caeli (Cae)

CAELUM

Sandwiched between Eridanus and Columba, this small and faint southern constellation is best found by looking to the southwest of the more distinctive Lepus (p.297). Caelum is another of the French astronomer Nicolas Louis de Lacaille's additions to the southern sky. Its pattern of just two faint stars with a line between them is supposed to represent an 18th-century stonemason's chisel, making this a companion to the equally unconvincing constellation Sculptor.

SOUTHERN HEMISPHERE

Given its distance from the Milky Way, the small patch of sky that falls within the boundary of Caelum contains little of interest to amateur astronomers aside from its stars. There are no significant nebulae or clusters, and even the brightest of the constellation's galaxies are only visible through large amateur telescopes.

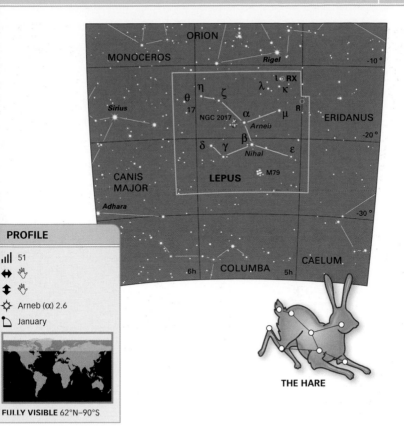

THE HARE

PROFILE

📶 51

↔ ✋

↕ ✋

☼ Arneb (α) 2.6

🗓 January

FULLY VISIBLE 62°N–90°S

FEATURES

Alpha (α) Leporis (Arneb)
At magnitude 2.6, this star seems to be of average brightness, but Arneb lies a distant 1,300 light-years away and is one of the most luminous stars visible from Earth.

Gamma (γ) Leporis This is a double star, with a yellow primary of magnitude 3.9 and an orange companion of magnitude 6.2. Both stars are at a similar distance from Earth, about 30 light-years away.

R Leporis (Hind's Crimson Star) This pulsating red giant variable is noted for its deep red color. It ranges in magnitude between 5.5 and 12.0 over a 430-day cycle.

NGC 2017 This open cluster consists of eight colorful stars in all, five of which lie between magnitudes 6 and 10.

Leporis (Lep)

LEPUS

Lying just to the south of the celestial hunter Orion, Lepus represents a hare in flight, hotly pursued by the celestial hounds Canis Major and Canis Minor. It was known to the ancient Greeks and is relatively easy to spot among its sparkling neighbors because of its distinctive bow-tie shape.

SOUTHERN HEMISPHERE

According to Greek mythology, it was placed among the stars in memory of a plague of hares that overwhelmed the island of Leros after a pregnant female hare was introduced.

Although the stars of Lepus are not bright, several of them are interesting, including attractive variables and multiples. Iota (ι) and RX Leporis, in the constellation's northwest corner, form an attractive pairing of a double star and a variable red giant. In addition, Lepus contains several other deep-sky objects.

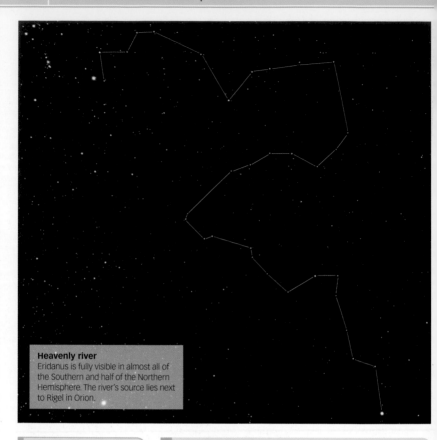

Heavenly river
Eridanus is fully visible in almost all of the Southern and half of the Northern Hemisphere. The river's source lies next to Rigel in Orion.

FEATURES

Alpha (α) Eridani (Achernar)
Eridanus's one truly bright star is a blue-white giant of magnitude 0.5. It lies some 140 light-years away from Earth.

Epsilon (ε) Eridani On the constellation's northern stretch, Epsilon (ε) Eridani is one of the closest Sun-like stars to Earth, just a little fainter and cooler than our own star. It lies 10.5 light-years away and shines at magnitude 3.7.

Omicron-2 (o²) Eridani
This triple-star system, 16 light-years from Earth, hosts the most easily seen white dwarf star in the sky. The primary is a red dwarf of magnitude 4.4, but its companion is a white dwarf of magnitude 9.5, which itself has a fainter red dwarf partner.

Eridani (Eri)

ERIDANUS

Representing a river, this large constellation meanders from the foot of Orion deep into southern skies, where it terminates at its brightest star, Achernar (meaning "river's end"). Its range in declination of 58° is the greatest of any constellation. Various mythologies have identified Eridanus with different rivers, from the Nile and Euphrates to the Po. In Greek mythology, Eridanus features in the story of Phaethon, son of the sun god Helios, who attempted to drive his father's chariot across the sky but lost control and plunged into the river below.

SOUTHERN HEMISPHERE

Despite Eridanus's size, its stars are mostly faint. It contains few bright deep-sky objects, with no star clusters visible to amateur instruments. There are a few galaxies and one planetary nebula, NGC 1535, that can be seen with binoculars under favorable conditions.

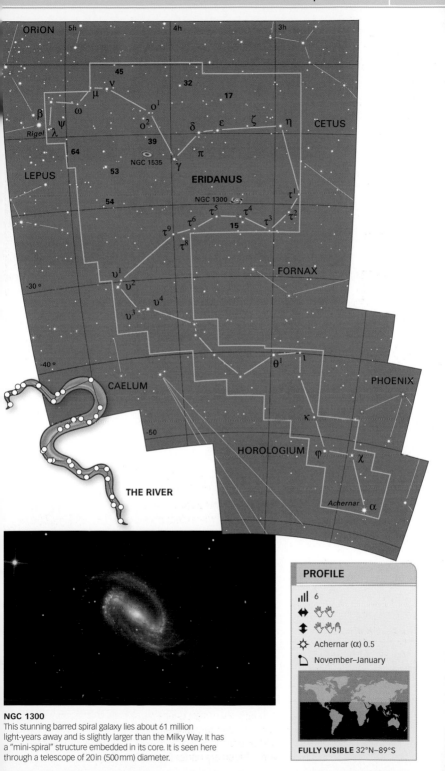

PROFILE

⏸ 6

↔ ✋✋

↕ ✋✋🖐

☼ Achernar (α) 0.5

⬛ November–January

FULLY VISIBLE 32°N–89°S

NGC 1300
This stunning barred spiral galaxy lies about 61 million
light-years away and is slightly larger than the Milky Way. It has
a "mini-spiral" structure embedded in its core. It is seen here
through a telescope of 20 in (500 mm) diameter.

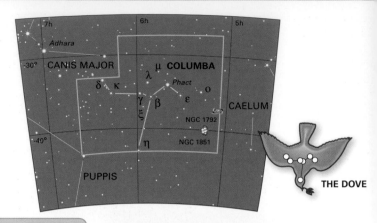

THE DOVE

PROFILE

.ıll 54

↔ 🖐

↕ 🖐

☼ Phact (α) 2.6

📖 January

FULLY VISIBLE 46°N–90°S

NGC 1792
This spiral galaxy with a bright disk appears elongated due to its tilt in relation to Earth. Moderate-sized telescopes will reveal fine details within it.

FEATURES

Alpha (α) Columbae (Phact)
Derived from the Arabic word for a collared dove, Phact is 170 light-years from Earth and shines blue-white at magnitude 2.6.

Beta (β) Columbae (Wazn)
The second-brightest star in the constellation is a yellow giant, 130 light-years away that shines at magnitude 3.1. Its Arabic name, Wazn, means "weight."

NGC 1851 This modest globular cluster is Columba's most prominent deep-sky object. It lies about 39,000 light-years away and at magnitude 7.1 is visible as a faint patch through binoculars or a small telescope.

Columbae (Col)

COLUMBA

The Dutch theologian and astronomer Petrus Plancius formed this southern constellation, representing a dove, around 1592 from stars near Lepus and Canis Major that had not previously been allocated to any constellation. Plancius was a biblical scholar, and he originally named the constellation

SOUTHERN HEMISPHERE

Columba Noachi, in reference to the bird that Noah sent from the Ark in search of dry land after the great biblical deluge. However, others have linked it to a dove from classical myth, which Jason sent ahead of his ship, the Argo, to find a safe passage into the Black Sea. Plancius may have partly had this in mind when he placed his dove close to the constellation of Puppis, a part of Argo.

Columba contains many faint galaxies, but they are too dim for most amateur telescopes. It also contains Mu (μ) Columbae, a fast-moving, 5th-magnitude star.

PROFILE

📊 65
↔ ✋
↕ ✋
☼ Alpha (α) 3.7
📄 February–March

FULLY VISIBLE 52°N–90°S

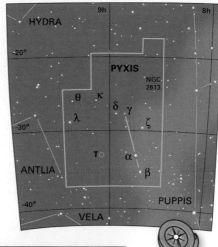

THE COMPASS

NGC 2613
An edge-on, barred spiral galaxy, NGC 2613 is a challenge for moderate instruments. However, it is a good target for long-exposure astrophotography and large telescopes.

FEATURES

Alpha (α) Pyxidis The brightest star, a blue-white supergiant, lies in the middle of a row of three linked stars. It is 18,000 times more luminous than the Sun, yet it shines at a magnitude of only 3.7 because it is more than 1,000 light-years away from Earth.

Beta (β) Pyxidis In contrast to Alpha (α), Beta (β) is a yellow giant of magnitude 4.0, 320 light-years away.

T Pyxidis This variable star lies beyond the range of most small telescopes, but occasionally it brightens to come well within binocular range and just below naked-eye visibility. The system is an unpredictable recurrent nova—a double-star system in which explosions sometimes occur on the surface of a hot, dense, white dwarf star.

Pyxidis (Pyx)

PYXIS

This faint and unremarkable southern constellation consists of three stars in a row, to the east of the brighter stars of Puppis. It was introduced in the 18th century by the French astronomer Nicolas Louis de Lacaille as part of his project to catalog the southern skies. Pyxis represents a magnetic compass, as

SOUTHERN HEMISPHERE

opposed to Circinus, which depicts the draftsman's compasses. Appropriately, its stars were originally part of Argo Navis, the great celestial ship that Lacaille broke up into the more manageable constellations Carina, Puppis, and Vela. The 19th-century British astronomer John Herschel suggested it should be renamed Malus, the mast of Argo, but his idea was not widely adopted.

Despite its position on the edge of the Milky Way, Pyxis contains no bright deep-sky objects within reach of small instruments. Its main points of interest lie in its stars.

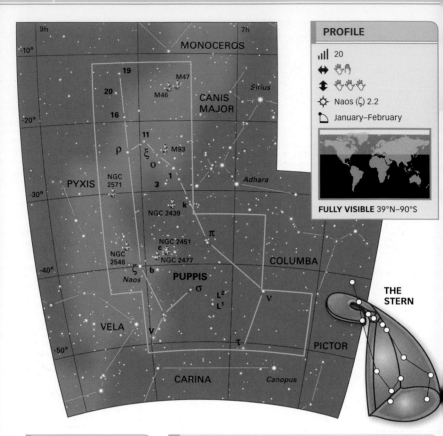

THE
STERN

FEATURES

Zeta (ζ) Puppis (Naos) The splitting of Argo has left this blue giant as the brightest star in Puppis. Shining at magnitude 2.2, it is also one of the hottest stars known, with a surface temperature six times that of the Sun. It lies 14,000 light-years from Earth.

L Puppis This line-of-sight double star consists of a blue-white star, L¹, 150 light-years from Earth, and a red giant, L², 40 light-years beyond it. L¹ is a steady magnitude 4.9, but L² is a pulsating variable, ranging in brightness between magnitude 2.6 and 6.2 over a 140-day cycle.

M47 This naked-eye object is a rich sight in binoculars. Slightly brighter than the neighboring M46, it is 1,600 light-years away and the most impressive open star cluster in Puppis.

Puppis (Pup)

PUPPIS

Straddling the Milky Way, this rich southern constellation was originally part of the former Greek constellation Argo Navis, the celestial ship. Puppis, representing the ship's stern, was the largest part. During the 18th century, the French astronomer Nicolas Louis de Lacaille divided the star pattern into

SOUTHERN HEMISPHERE

three parts, creating the separate contellations of Puppis, Carina, and Vela. The stars of each section, however, retained their original Greek letters, so that in the case of Puppis the lettering now starts with Zeta (ζ) Puppis, a star also known as Naos.

A rich part of the Milky Way passes through the constellation, making Puppis an ideal target for binocular viewing. It contains dense star fields and more than 70 open clusters, some of which can also be viewed with the naked eye.

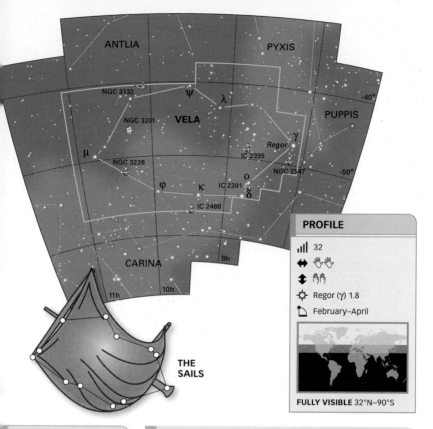

NGC 3132

ANTLIA

PYXIS

ψ

λ

-40°

NGC 3201

VELA

PUPPIS

μ

γ

Regor

NGC 3228

IC 2395

φ

κ

IC 2391

NGC 2547

-50°

O

δ

IC 2488

9h

CARINA

10h

11h

THE
SAILS

PROFILE

.ıll 32

↔ 🖐🖐

↕ 🖐🖐

☼ Regor (γ) 1.8

📖 February–April

FULLY VISIBLE 32°N–90°S

FEATURES

Gamma (γ) Velorum (Regor)
Vela's brightest star is a
multiple that contains the
brightest known Wolf–Rayet
star—a type of massive,
superhot star that is blowing
away its own outer layers
with fierce stellar winds and
exposing its ultra-hot interior.
The system shines at an overall
magnitude of 1.8.

NGC 3201 This very loose
globular cluster, roughly 15,000
light-years from Earth, can be
detected with binoculars and
small telescopes. Moderate
instruments can identify its
outlying members.

IC 2391 A beautiful open star
cluster, also known as the
Southern Pleiades, IC 2391 is
visible to the naked eye. Lying
just 400 light-years away, its
stars are a spectacular sight
in binoculars.

Velorum (Vel)

VELA

A large southern constellation,
Vela represents the sail of the great
celestial ship Argo Navis and once
formed a single huge constellation
with Puppis and Carina. Lying to
the north of Carina, Vela outlines a
dense region of the Milky Way that
includes more than 40 star clusters
and large regions of nebulosity.

**SOUTHERN
HEMISPHERE**

The labeling of Vela's stars starts with Gamma (γ)
Velorum because the stars labeled Alpha (α) and Beta
(β) in the former Argo Navis now lie in Carina. Between
the stars Gamma (γ) and Lambda (λ) Velorum lie the
gaseous strands of the Vela supernova remnant—the
product of a stellar explosion visible from Earth around
11,000 years ago. The stars Delta (δ) and Kappa (κ)
Velorum combine with two stars in Carina to form a
pattern called the False Cross, which is sometimes
mistaken for the true Southern Cross (p.305).

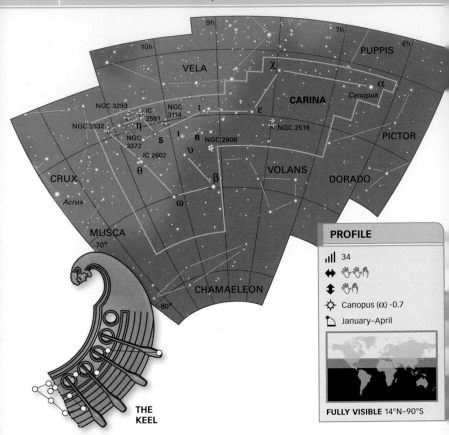

THE
KEEL

FEATURES

**Alpha (α) Carinae
(Canopus)** As the
second-brightest star in the
sky, Canopus shines at a
brilliant magnitude of -0.7.
Unlike Sirius, which outshines it
purely because of its proximity
to Earth, Canopus is a luminous
star—a yellow-white supergiant
310 light-years away.

**NGC 3372 (the Carina
Nebula)** This expansive
emission nebula is a star-
forming region 8,000 light-years
away. Four times the apparent
size of the full Moon, it is
visible to the naked eye as a
bright patch in the Milky Way.
The densest part of the nebula
surrounds Eta (η) Carinae.

**IC 2602 (the Southern
Pleiades)** This large open star
cluster, containing eight stars
brighter than magnitude 6.0,
lies about 500 light-years away.

Carinae (Car)

CARINA

This major southern constellation
lies in a rich part of the Milky Way.
It was part of the larger constellation
Argo Navis, depicting the ship of
the Argonauts, until it was split up
in the 18th century by the French
astronomer Nicolas Louis de
Lacaille. As the most southerly part
of the original constellation, Carina

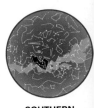

**SOUTHERN
HEMISPHERE**

represented the ship's keel. Alpha (α) Carinae (Canopus),
the second-brightest star in the night sky, marks its
rudder or steering oar. This constellation is circumpolar,
or visible above the horizon at all times, from much of
the Southern Hemisphere.

Carina contains many clusters and star fields
that provide interesting targets for binoculars. It is
also enhanced by an enormous star-forming region,
the Carina Nebula, that is larger and brighter than the
better-known Great Orion Nebula (pp.276–77).

PROFILE

📶 88

↔ 🖐

↕ 🖐

☼ Acrux (α) 0.8

📅 April–May

FULLY VISIBLE 25°N–90°S

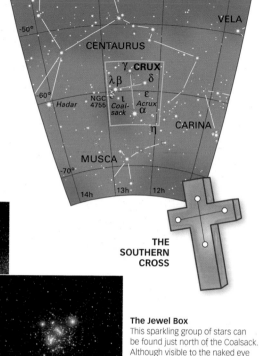

THE
SOUTHERN
CROSS

The Coalsack
This dark nebula can be seen next to Crux with the naked eye. It is a vast dust cloud that blocks the light from the stars behind it.

The Jewel Box
This sparkling group of stars can be found just north of the Coalsack. Although visible to the naked eye as a brighter patch near Beta (β) Crucis, binoculars can reveal its several blue-white stars.

FEATURES

Alpha (α) Crucis (Acrux)
Marking the southern end of the cross, Acrux is a blue-white double star of magnitude 0.8. It is divisible in a telescope into two stars of roughly equal magnitude.

Beta (β) Crucis (Becrux)
This rapidly varying blue-white star changes its brightness between magnitudes 1.25 and 1.15 every 6 hours. It lies 350 light-years away.

NGC 4755 (the Jewel Box)
This glorious star cluster is one of the gems of the southern sky. It appears to the naked eye as a single fuzzy star of magnitude 4.0. In fact, it is a cluster of stars 7,600 light-years away. Binoculars reveal that it contains several dozen blue-white stars and has a contrasting red supergiant near the center.

Crucis (Cru)

CRUX

The smallest constellation in the night sky, Crux is nevertheless one of the most famous and easily recognized star patterns of all. It was originally regarded by the ancient Greeks as a part of Centaurus, but it came to be recognized as a separate constellation, *Crux Australis* or the

SOUTHERN HEMISPHERE

Southern Cross, in the late 16th century. In more recent times, its name has been abbreviated to simply Crux.

The Milky Way is bright in this part of the sky, making the dark Coalsack nebula, a wedge-shaped cloud of dust and gas, easily visible by contrast. Other highlights of this tiny constellation include the brilliant stars themselves and a beautiful star cluster, NGC 4755 or the Jewel Box. Crux also offers a useful pointer to the south celestial pole—extending the line between Alpha (α) and Gamma (γ) reveals its location (p.35).

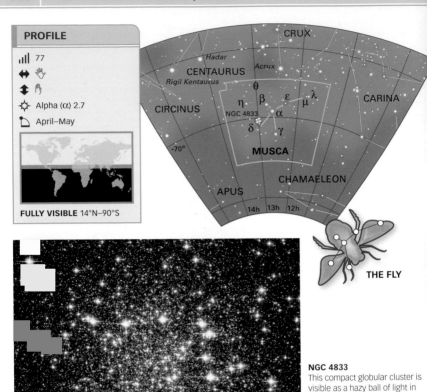

THE FLY

NGC 4833
This compact globular cluster is visible as a hazy ball of light in binoculars. Individual stars can be resolved with a telescope of 4 in (100 mm) aperture.

FEATURES

Alpha (α) Muscae A blue-white giant of magnitude 2.7, Alpha (α) lies 305 light-years away.

Beta (β) Muscae To the naked eye, Beta (β) is an almost identical twin of Alpha (α). However, a small telescope will show that it is actually a double, containing blue stars of magnitudes 3.0 and 3.7 that orbit each other in 383 years. The system lies some 310 light-years away.

Theta (θ) Muscae This double star consists of a blue supergiant of magnitude 5.7, orbited by a star of magnitude 7.3. The faint companion, visible in binoculars, is a rare Wolf–Rayet star—a fierce white star so hot that it is blasting its own outer layers away into space and aging at an accelerated rate.

Muscae (Mus)

MUSCA

Lying to the south of Crux and Centaurus, the stars of Musca are relatively bright but can be hard to identify against the background of the Milky Way. The constellation is best found by following the long axis of Crux toward the south celestial pole. Musca is the most distinctive of several constellations invented by

SOUTHERN HEMISPHERE

Dutch navigators Pieter Dirkszoon Keyser and Frederick de Houtman in the late 16th century. It was originally called Apis, the Bee, but was renamed Musca Australis, the Southern Fly, in the 1750s, partly to avoid confusion with nearby Apus, the Bird of Paradise. At the time, there was also a northern fly, Musca Borealis, but when this fell into disuse, the southern constellation became simply Musca.

It contains several interesting multiple stars, as well as a bright globular star cluster and the southern tip of the dark Coalsack Nebula, which extends into it from Crux.

PROFILE

▮▮▮ 85

↔ ✋

↕ ✋

☼ Alpha (α) 3.2

▭ May–June

FULLY VISIBLE 19°N–90°S

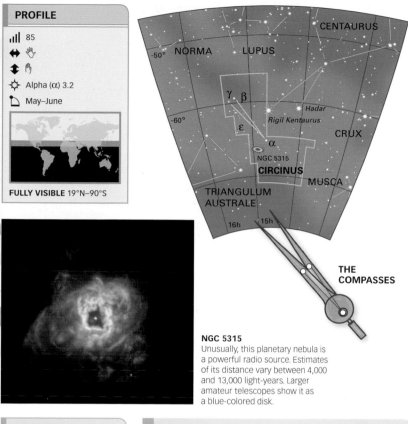

THE COMPASSES

NGC 5315
Unusually, this planetary nebula is a powerful radio source. Estimates of its distance vary between 4,000 and 13,000 light-years. Larger amateur telescopes show it as a blue-colored disk.

FEATURES

Alpha (α) Circini A white star of magnitude 3.2, Alpha (α) has a faint companion of magnitude 8.6 that can be seen through a small telescope. The system is 65 light-years away.

Gamma (γ) Circini This double can be separated only with a medium-sized telescope. It consists of blue and yellow stars of magnitudes 5.1 and 5.5, orbiting each other 500 light-years from Earth.

Theta (θ) Circini The components of this double star are too close to separate visually, but one of its stars still fluctuates unpredictably, causing its overall brightness to vary between magnitudes 5.0 and 5.4.

NGC 5315 This faint and distant planetary nebula is around 7,000 light-years away.

Circini (Cir)

CIRCINUS

A small and insignificant southern constellation, Circinus was introduced in 1756 by the French astronomer Nicolas Louis de Lacaille. It is a faint triangle of stars, supposedly resembling a set of surveyor's or navigator's compasses, as opposed to Pyxis, the magnetic compass. The constellation is squeezed in awkwardly between Centaurus and Triangulum Australe, but is easy to locate since it lies close to Alpha (α) Centauri, one of the brightest stars in the sky.

SOUTHERN HEMISPHERE

Circinus lacks bright clusters, nebulae, or galaxies but is crossed by the star clouds of the Milky Way. These clouds hide the view of one of our closest galactic neighbors, the Circinus Galaxy, discovered in the 1970s. This is a small spiral galaxy with an active, supermassive black hole, just 13 million light-years away.

PROFILE

ıll 74

↔ 🖐

↕ 🖐

☼ Gamma-2 (γ²) 4.0

🗓 June

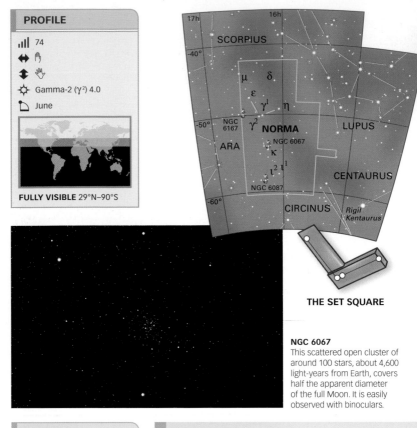

THE SET SQUARE

NGC 6067
This scattered open cluster of around 100 stars, about 4,600 light-years from Earth, covers half the apparent diameter of the full Moon. It is easily observed with binoculars.

FEATURES

Gamma (γ) Normae This line-of-sight double consists of two yellow stars at varying distances from Earth. Gamma-2 (γ²) is a giant of magnitude 4.0, some 125 light-years away, while Gamma-1 (γ¹) is a far more distant supergiant, 1,500 light-years away but still shining at magnitude 5.0.

Iota (ι) Normae This multiple star is 220 light-years away and can be separated with a small telescope into a magnitude-4.6 primary orbited by a fainter star of magnitude 8.1. Much larger telescopes can split the primary again into twin stars that orbit one another in 27 years, making the system a triple star.

NGC 6087 This open cluster of about 40 stars lies 3,000 light-years away and can be spotted with the naked eye. At its heart lies a supergiant, S Normae.

Normae (Nor)

NORMA

Sandwiched between the constellations Scorpius and Lupus, this faint triangle of stars was hived off into a separate constellation by French astronomer Nicolas Louis de Lacaille in the 1750s. Like many of his southern constellations, it is named after a scientific instrument—the surveyor's level or set square.

SOUTHERN HEMISPHERE

Originally, the constellation was called Norma et Regula, meaning the square and ruler. However, several of its stars, including its brightest members, which Lacaille used to form the ruler, were absorbed by neighboring Scorpius when the boundaries between the constellations were finally formalized.

The Milky Way runs straight through Norma, and as a result the constellation is rich in star fields worth sweeping with binoculars. However, the faint stars forming its pattern get lost among the background stars.

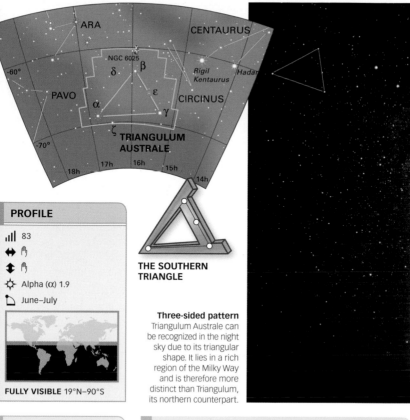

THE SOUTHERN TRIANGLE

Three-sided pattern
Triangulum Australe can be recognized in the night sky due to its triangular shape. It lies in a rich region of the Milky Way and is therefore more distinct than Triangulum, its northern counterpart.

PROFILE

📶 83

↔ 🖐

↕ 🖐

☼ Alpha (α) 1.9

📷 June–July

FULLY VISIBLE 19°N–90°S

FEATURES

Alpha (α) Trianguli Australis An orange giant of magnitude 1.9, this star marks the triangle's southeast corner. It lies 100 light-years away from Earth.

Beta (β) Trianguli Australis This white star is about 42 light-years away from Earth and shines at magnitude 2.9.

Gamma (γ) Trianguli Australis Although it has the same magnitude (2.9) as Beta (β), Gamma (γ) lies over 70 light-years away, so it is distinctly more luminous. Due to its higher luminosity its surface is hotter and blue-white.

NGC 6025 Though visible to the naked eye at magnitude 5.4, this open cluster in the Milky Way is best observed through binoculars. It lies 2,700 light-years away.

Trianguli Australis (TrA)

TRIANGULUM AUSTRALE

This distinctive pattern in the southern sky to the southeast of Centaurus is easy to spot. There are several claims to its invention—it was first recorded in *Uranometria*, a star atlas compiled by the German astronomer Johann Bayer in 1603.

SOUTHERN HEMISPHERE

However, it may have been invented by the Dutch navigators Pieter Dirkszoon Keyser and Frederick de Houtman in the 1590s. Alternatively, it could also have been discovered by the Dutch astronomer Petrus Theodorus Embdanus.

The Southern Triangle is smaller than its northern counterpart, Triangulum, but contains brighter stars and is more prominent. It is crossed by the band of the Milky Way but contains little of interest to amateur astronomers aside from a small but attractive open star cluster.

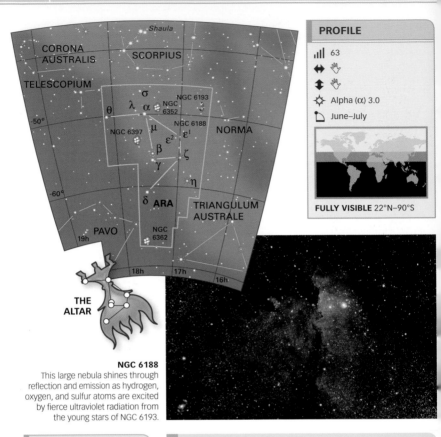

PROFILE

📶 63

↔ 🖐

↕ 🖐

☼ Alpha (α) 3.0

📅 June–July

FULLY VISIBLE 22°N–90°S

THE ALTAR

NGC 6188
This large nebula shines through reflection and emission as hydrogen, oxygen, and sulfur atoms are excited by fierce ultraviolet radiation from the young stars of NGC 6193.

FEATURES

Alpha (α) Arae A blue-white star of magnitude 3.0, Ara's brightest star lies some 460 light-years from Earth.

Gamma (γ) Arae One of the most luminous stars in our region of the galaxy, Gamma (γ) shines with the brilliance of 32,000 Suns. However, at a distance of 1,100 light-years, it only reaches magnitude 3.3.

NGC 6193 An open star cluster that is easily spotted with the naked eye, NGC 6193 is more than half the apparent width of the full Moon. It is 4,000 light-years away from Earth and still embedded in remnants of star-forming gas.

NGC 6397 A relatively nearby globular cluster, at a distance of just 7,200 light-years, NGC 6397 is easily seen with a pair of binoculars.

Arae (Ara)

ARA

Although it lies in the far south of the night sky, Ara was identified by the ancient Greeks and was listed in Ptolemy's original catalog of 48 constellations. The Greeks visualized it as the celestial altar on which the gods of Mount Olympus swore an oath of allegiance at the beginning of their ten-year war with the Titans of Mount Othrys for control of the Universe.

SOUTHERN HEMISPHERE

The constellation's pattern is obscure. Nevertheless, it is easy to locate as it lies to the south of Scorpius and is crossed by a rich band of the Milky Way's star fields. As a result, it contains several star clusters within reach of amateur instruments, as well as faint nebulae such as NGC 6188. The faint Mu (μ) Arae is a Sun-like star 50 light-years away from Earth that is orbited by a system of at least four planets.

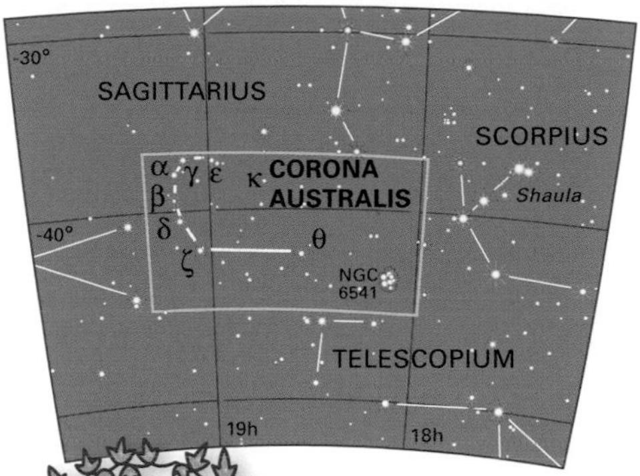

THE SOUTHERN CROWN

PROFILE

80

Alpha (α), Beta (β) 4.1

July–August

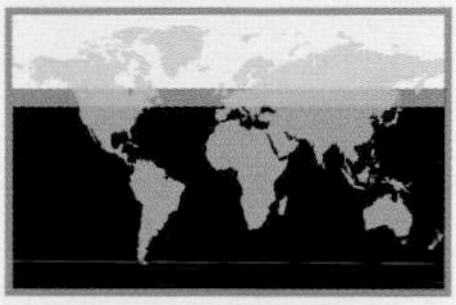

FULLY VISIBLE 44°N–90°S

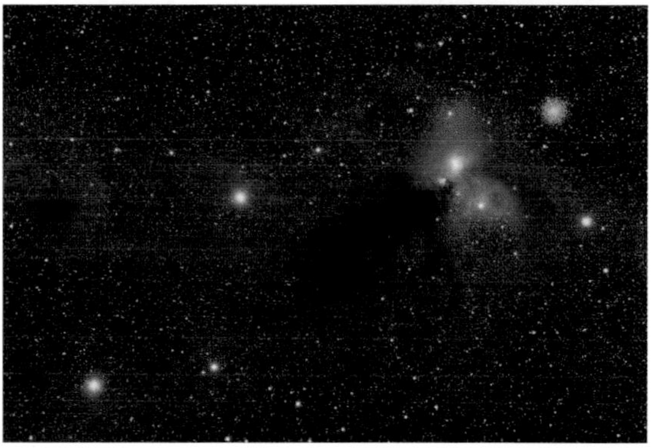

R Coronae Australis
Seen here through a 7 ft (2.2 m) telescope, the faint variable star R Coronae Australis is surrounded by a faint blue reflection nebula, with a larger, dark nebula lying nearby. R Coronae Australis lies between Gamma (γ) and Epsilon (ε).

FEATURES

Alpha (α) Coronae Australis This star of magnitude 4.1 is white in color, and lies 140 light-years from Earth.

Beta (β) Coronae Australis A yellow giant shining at magnitude 4.1, Beta (β) is as bright as Alpha (α). However, it is considerably farther away, at a distance of 510 light-years, and is around 730 times more luminous than the Sun.

Gamma (γ) Coronae Australis Both the stars in this binary system are bright enough to be seen with the naked eye. Alhough shining at magnitudes 4.8 and 5.1, a small telescope is still needed to separate them.

NGC 6541 Hovering just below naked-eye visibility, this globular cluster lies 22,000 light-years away.

Coronae Australis (CrA)

CORONA AUSTRALIS

The small southern constellation of Corona Australis lies under the feet of the constellation Sagittarius. It was one of the 48 constellations recognized by the ancient Greek–Egyptian astronomer Ptolemy and was originally depicted as a wreath. It represents a myrtle coronet placed in the sky by Dionysus, the Greek god of wine and revelry, after he had rescued his mother from the underworld. Ancient Chinese astronomers saw a turtle in the same stars.

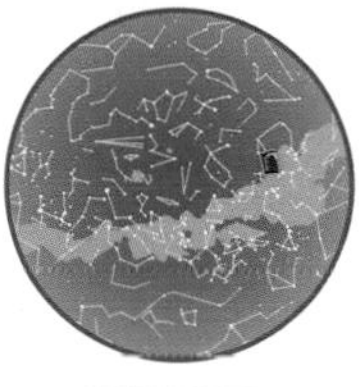

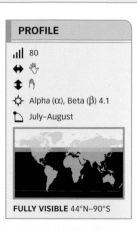

SOUTHERN HEMISPHERE

The outline of the southern crown is formed by an arc of stars, none of which is brighter than magnitude 4. Although its pattern is not as neatly spaced as its northern equivalent, it is still easy to recognize below the central Teapot star pattern in Sagittarius. Corona Australis borders on rich Milky Way star clouds and is filled with tenuous nebulosity that shows up in long-exposure photographs.

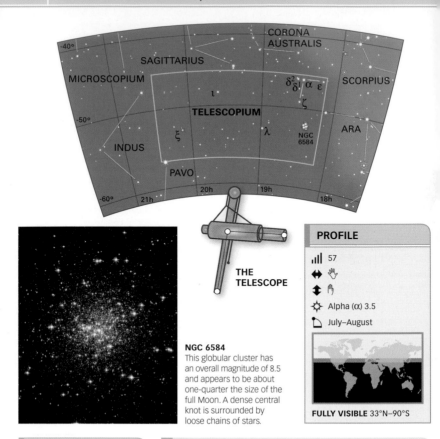

THE
TELESCOPE

PROFILE

📊 57

↔ 🖐

↕ ✋

☼ Alpha (α) 3.5

📅 July–August

FULLY VISIBLE 33°N–90°S

NGC 6584
This globular cluster has an overall magnitude of 8.5 and appears to be about one-quarter the size of the full Moon. A dense central knot is surrounded by loose chains of stars.

FEATURES

Alpha (α) Telescopii This blue-white star is around 450 light-years away from Earth and shines at magnitude 3.5.

Delta (δ) Telescopii Binoculars or sharp eyesight reveal that this star is a line-of-sight double, consisting of two unrelated blue-white stars, around 650 and 1,300 light-years away. They are of roughly equal brightness, at around magnitude 5.0.

NGC 6584 This globular cluster, 43,700 light-years away, is easy to detect with powerful binoculars and appears as a compact sphere through telescopes with small apertures. Individual stars can be resolved with larger instruments.

Telescopii (Tel)

TELESCOPIUM

One of the faintest and least recognizable constellations, Telescopium seems to have been made simply by drawing a line around an arbitrary area of sky to the south of Corona Australis. It is an 18th-century invention of the French astronomer Nicolas Louis de Lacaille, who intended its pattern of stars to represent one of the great aerial telescopes used at the Paris observatory. These were long refractors (lens-based telescopes), often more than 33ft (10m) in length, that were suspended from tall poles by ropes and pulleys.

SOUTHERN HEMISPHERE

In order to create his new constellation, Lacaille "borrowed" stars from nearby constellations, including Sagittarius, Scorpius, Ophiuchus, and Corona Australis. When the constellations were standardized in 1929, these borrowed stars were returned to their rightful owners, and Telescopium was left in its current denuded state.

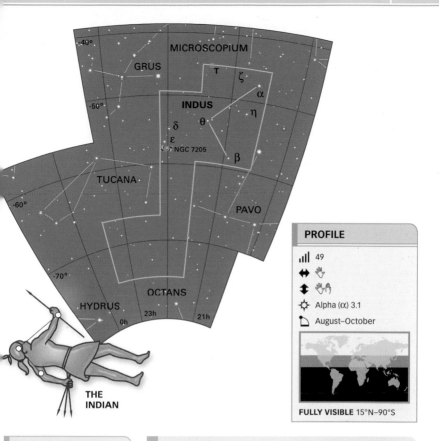

THE
INDIAN

FEATURES

Alpha (α) Indi An orange giant, this star lies 125 light-years from Earth and shines at magnitude 3.1.

Beta (β) Indi A slightly less luminous orange giant, this star is closer than Alpha (α), at just 110 light-years, but still dimmer, at magnitude 3.7.

Epsilon (ε) Indi This yellow star of magnitude 4.7 is a close stellar neighbor of the Sun, just 11.8 light-years away. It is orbited by a brown dwarf star with 45 times the mass of Jupiter.

NGC 7205 This faint spiral galaxy, of magnitude 11.4, can just be spotted with a telescope of moderate aperture.

Indi (Ind)

INDUS

The southern constellation
of Indus is generally thought to represent a North American Indian equipped with a spear and arrows. It was introduced in the 16th century by Dutch navigators Pieter Dirkszoon Keyser and Frederick de Houtman, who had been involved in the exploration of the East Indies and may have intended Indus to represent a North American Indian or Southeast Asian islander. The Dutchmen made the first record of the far-southern stars at the request of Dutch astronomer Petrus Plancius, who had already added several new constellations to the northern sky.

**SOUTHERN
HEMISPHERE**

Situated to the southeast of Sagittarius, Indus has no significant star clusters or nebulae but contains some faint galaxies within reach of amateur telescopes. It is best found by looking between Sagittarius and the more distinctive southern constellations of Grus and Tucana.

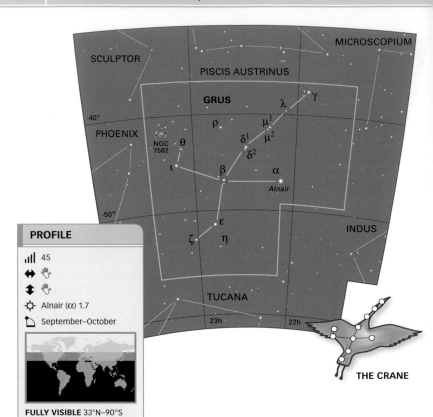

THE CRANE

PROFILE

ıll 45

↔ 🖐

↕ 🖐

☼ Alnair (α) 1.7

🗓 September–October

FULLY VISIBLE 33°N–90°S

FEATURES

Alpha (α) Gruis (Alnair) This blue-white star, whose name means "the bright one" in Arabic, is 65 light-years away and shines at magnitude 1.7.

Beta (β) Gruis A red giant, 170 light-years from Earth, Beta (β) is a variable. It oscillates unpredictably between magnitudes 2.0 and 2.3 as it swells and shrinks.

Delta (δ) Gruis Naked-eye observers can usually tell that this star is a double, but this is, in fact, just a line-of-sight effect. The two components are yellow and red giants of magnitudes 4.0 and 4.1, 150 and 420 light-years away.

NGC 7582 This barred spiral galaxy is one of three galaxies within Grus that are visible with a moderate-sized telescope.

Gruis (Gru)

GRUS

This far-southern constellation is situated between Piscis Austrinus and Tucana. It is one of several bird-shaped star patterns added to the sky by the Dutch explorers Pieter Dirkszoon Keyser and Frederick de Houtman in the 1590s. Grus was later immortalized by the German astronomer Johann Bayer in his great *Uranometria* star atlas of 1603, and represents

SOUTHERN HEMISPHERE

a long-necked wading bird. Usually it is shown as a crane, but in the past it has also been depicted as a flamingo.

Lying far from the plane of the Milky Way, this constellation lacks open clusters and nebulosity. It is nevertheless quite distinctive thanks to the row of moderately bright stars that run along the crane's body, forming a chain leading from the Small Magellanic Cloud (SMC) in Tucana (p.316) toward the bright star Fomalhaut in Piscis Austrinus.

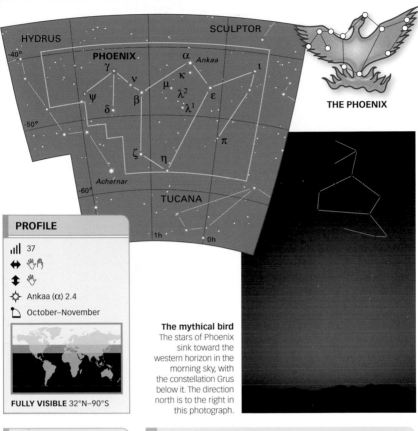

THE PHOENIX

PROFILE

ıllı 37

↔ 🖐

↕ 🖐

☼ Ankaa (α) 2.4

▱ October–November

FULLY VISIBLE 32°N–90°S

The mythical bird
The stars of Phoenix sink toward the western horizon in the morning sky, with the constellation Grus below it. The direction north is to the right in this photograph.

FEATURES

Alpha (α) Phoenicis (Ankaa) Located 88 light-years away, this yellow giant shines at magnitude 2.4.

Beta (β) Phoenicis To the naked eye, Beta (β) appears to be a yellow star of magnitude 3.3, but a medium-sized telescope will show that it is actually a double, consisting of twin yellow stars of magnitude 4.0, 130 light-years away.

Zeta (ζ) Phoenicis This interesting quadruple star lies 280 light-years away. Its brightest component mostly shines at magnitude 3.9, but is in fact an eclipsing binary that dips in brightness to magnitude 4.4 every 40 hours. A small telescope reveals a third star of magnitude 6.9, while a larger instrument will show the fourth member of the system, fainter and closer to the primary.

Phoenicis (Phe)

PHOENIX

This pattern of stars can be found at the southern end of Eridanus, next to its brightest star, Achernar. Phoenix represents a fantastic bird recorded in mythologies from classical Europe to ancient China. This mythical bird was said to live for 500 years, cremating itself on a nest of cinnamon bark and incense at the end of its life, only to be reborn from its ashes.

SOUTHERN HEMISPHERE

Phoenix is the largest of a dozen southern constellations introduced during the late 16th century by the Dutch navigator–astronomers Pieter Dirkszoon Keyser and Frederick de Houtman. However, Arabian astronomers had recognized roughly the same grouping of stars long before and considered Phoenix to represent a boat moored on the Eridanus river. This constellation is faint and insignificant, containing little of interest for amateur astronomers aside from some double stars.

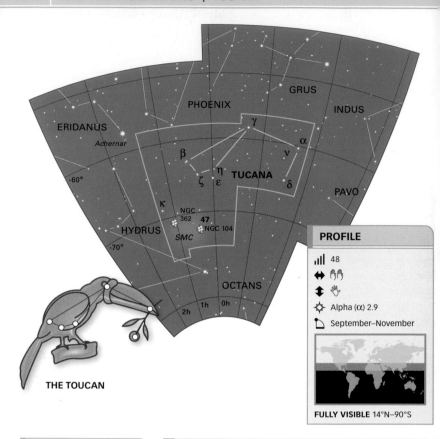

THE TOUCAN

FEATURES

NGC 104 (47 Tucanae) A compact and spectacular globular cluster, NGC 104 contains several million stars in a ball about 120 light-years across. It lies 13,500 light-years away but is easily visible as a fuzzy star with the naked eye. Small telescopes can pick up individual stars around the cluster's edge.

Small Magellanic Cloud (SMC) The smaller of the Milky Way's two major satellite galaxies, the Small Magellanic Cloud lies 210,000 light-years from Earth. It is easily visible to the naked eye, looking like a detached, wedge-shaped region of the Milky Way itself. Binoculars will reveal rich fields of stars, dust, and star-forming nebulae.

Tucanae (Tuc)

TUCANA

This far-southern constellation is found at the end of the celestial river, Eridanus, to the west of the bright star Achernar. It represents the toucan, a large-billed tropical bird native to South and Central America. Like most of the southern bird constellations, Tucana was invented in the late 16th century by the Dutch navigator–astronomers Pieter Dirkszoon Keyser and Frederick de Houtman.

SOUTHERN HEMISPHERE

Although the stars of Tucana are faint and its pattern indistinct, it contains two outstanding objects of interest to any amateur astronomer: the Small Magellanic Cloud (SMC) and NGC 104 or 47 Tucanae. Originally catalogued as a star, NGC 104 is in fact the second-brightest globular cluster. The SMC is a large satellite galaxy of the Milky Way and was first recorded by the Portuguese explorer Ferdinand Magellan around 1520.

3h
ERIDANUS
1h
Achernar
4h
HOROLOGIUM
PHOENIX
α

**THE LITTLE
WATER SNAKE**

RETICULUM
ζ π η²
ε δ
-60°
TUCANA
DORADO
HYDRUS
vw γ
ν
-70°
β
MENSA

Snake in space
The little water snake
slithers across the
southern sky between
the two Magellanic
Clouds. The brightest star
near it is Achernar in the
constellation Eridanus
(top, right).

PROFILE

📶 61

↔️ 🖐️

↕️ 🖐️

☼ Beta (β) 2.8

October–December

FULLY VISIBLE 8°N–90°S

FEATURES

Alpha (α) Hydri At magnitude
2.9, this white star lies 78
light-years away. It is Hydrus's
second-brightest member.

Beta (β) Hydri The
constellation's brightest
component is a yellow, Sunlike
star in the snake's "tail." It is
just 21 light-years away and
shines at magnitude 2.8.

Pi (π) Hydri This wide double
of unrelated red giants can be
split readily with binoculars.
Pi-1 (π¹) is 740 light-years
away, while Pi-2 (π²) lies closer,
at a distance of 470 light-years.

VW Hydri Lying near Gamma
(γ) Hydri, VW is a recurrent nova
system—a binary with a white
dwarf that has semiregular
explosions on its surface. Its
eruptions occur roughly once
a month and are easy to follow
with a small telescope.

Hydri (Hyi)

HYDRUS

Not to be confused with the
similarly named but much larger
Hydra, this constellation is a
compact group of stars nestling
in the far south of the sky. It
represents a small water snake,
and its stars are of middling
magnitude and indistinct pattern.
However, it is fairly easy to locate

**SOUTHERN
HEMISPHERE**

because Alpha (α) Hydri—the snake's head—lies close
to the brilliant Achernar in the constellation Eridanus,
the celestial river. Hydrus's body, meanwhile, sits neatly
between the Large and Small Magellanic Clouds (SMC) in
Dorado (p.321) and Tucana (p.316).

This constellation was introduced in the 16th century
by the Dutch navigator–astronomers Pieter Dirkszoon
Keyser and Frederick de Houtman. It contains no notable
deep-sky objects for amateurs, but it does have a couple
of interesting double stars, such as Pi (π) Hydri.

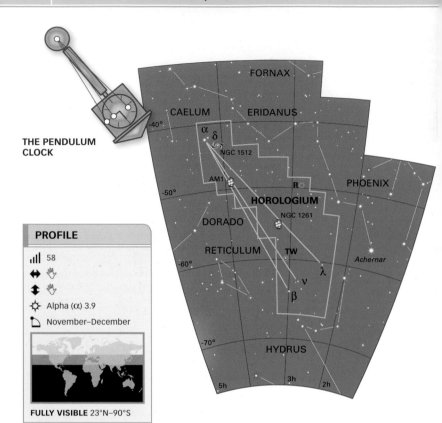

THE PENDULUM CLOCK

PROFILE

📶 58

↔ 🖐

↕ 🖐

☼ Alpha (α) 3.9

◻ November–December

FULLY VISIBLE 23°N–90°S

FEATURES

Alpha (α) Horologii The constellation's brightest star is a yellow giant of magnitude 3.9, 180 light-years from Earth.

NGC 1261 One of the more distant globular clusters orbiting the Milky Way, NGC 1261 lies 44,000 light-years away. The combined light of this huge ball of stars reaches Earth at magnitude 8.0, making it a good target for binoculars.

NGC 1512 This barred spiral galaxy is about 30 million light-years away and nearly 70,000 light-years across, two-thirds the size of our galaxy. Its bright center has a magnitude of 10 and can be seen with a small telescope. Detailed observations have shown that the center is surrounded by a huge ring of infant star clusters about 2,400 light-years across.

Horologii (Hor)

HOROLOGIUM

A faint and unremarkable constellation of the southern sky, Horologium lies close to the southern extreme of the celestial river, Eridanus. It was introduced by the French astronomer Nicolas Louis de Lacaille in the 1750s. Like many of his inventions, it is an apparently arbitrary group of faint, scattered stars named after a technological innovation. Lacaille intended it to represent a pendulum clock used for precise timekeeping in the observatories of the time. The clock can be drawn with the pendulum itself at Alpha (α) or swinging back and forth between Lambda (λ) and Beta (β).

SOUTHERN HEMISPHERE

Located at some distance from the Milky Way and occupying only a narrow strip of sky, Horologium nevertheless contains a few interesting deep-sky objects, including globular clusters and a spiral galaxy.

NGC 1313
This starburst galaxy, 20 million light-years away, is undergoing a tremendous wave of star formation. Huge star-forming nebulae stud its spiral arms and disguise its shape.

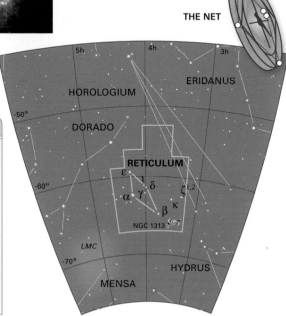

THE NET

PROFILE

.ıll 82

↔ 🖐

↕ 🖐

☼ Alpha (α) 3.4

▢ December

FULLY VISIBLE 23°N–90°S

FEATURES

Alpha (α) Reticuli The constellation's brightest star at magnitude 3.4, Alpha (α) is a yellow giant located 135 light-years away from Earth.

Beta (β) Reticuli An orange giant, some 78 light-years away, Beta (β) shines at magnitude 3.9.

Zeta (ζ) Reticuli This double star, easily resolved through binoculars, consists of near-twin yellow stars— Zeta-1 (ζ¹) and Zeta-2 (ζ²) of magnitudes 5.2 and 5.9 respectively. They lie 39 light-years from Earth and are believed to be around 8 billion years old—far more ancient than our Sun.

Reticuli (Ret)

RETICULUM

This dim but distinct diamond shape of stars lies a little way to the south of Carina's brightest star Canopus and northwest of the Large Magellanic Cloud (LMC) in Dorado. Reticulum's stars were first grouped together as a separate constellation called Rhombus (the diamond) in 1621. But the constellation received its present name from the French astronomer Nicolas Louis de Lacaille while he was in South Africa in the 1750s.

SOUTHERN HEMISPHERE

The literal meaning of *reticulum* in Latin is "net," but the constellation is supposed to represent a reticule—the gridlike set of crosshairs in the eyepiece of a telescope that Lacaille and others used for precise measurement of stellar positions. Lying at some distance from the Milky Way, Reticulum's chief attractions for amateur astronomers are a famous double star and some faint galaxies.

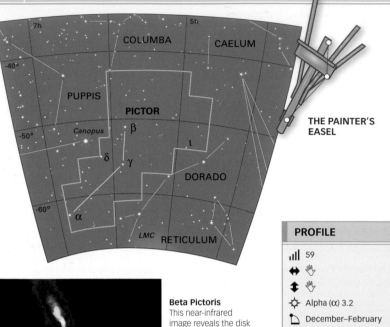

Beta Pictoris

This near-infrared image reveals the disk of planet-forming material that makes this young star shine unusually brightly at infrared wavelengths. The disk extends from the central star for more than a thousand times the distance between Earth and the Sun.

PROFILE

📶 59

↔️ ✋

↕️ ✋

☼ Alpha (α) 3.2

🗓 December–February

FULLY VISIBLE 26°N–90°S

FEATURES

Beta (β) Pictoris When viewed by infrared telescopes, this apparently unremarkable white star of magnitude 3.9, situated 63 light-years away, reveals a broad disk of planet-forming gas and dust in orbit around it. Recent studies have shown that something close to the star (most likely a newborn planet) is warping the disk out of shape.

Delta (δ) Pictoris About 2,400 light-years away, Delta (δ) is an eclipsing binary star. The two stars are too close to separate with even the most powerful telescope. They give themselves away by periodic dips in their brightness, from magnitude 4.7 to 4.9, as one passes in front of the other every 40 hours.

Pictoris (Pic)

PICTOR

Introduced in the 18th century by the French astronomer Nicolas Louis de Lacaille, Pictor lies to the south of Columba between the bright star Canopus in Carina and the Large Magellanic Cloud (LMC) (p.321) in Dorado. Lacaille imagined the stars of the constellation to represent an artist's easel, complete

SOUTHERN HEMISPHERE

with a palette, and originally called it Equuleus Pictoris.

Typically for one of Lacaille's inventions, Pictor is a group of faint stars with no obvious resemblance to the object it is supposed to represent. However, it is redeemed by the presence of some very interesting stars, including Beta (β) Pictoris, which may be host to a newly formed planetary system. In the north of the constellation, Kapteyn's Star is a red dwarf of magnitude 8.9 some 13 light-years away from Earth and one of the fastest-moving stars in the sky.

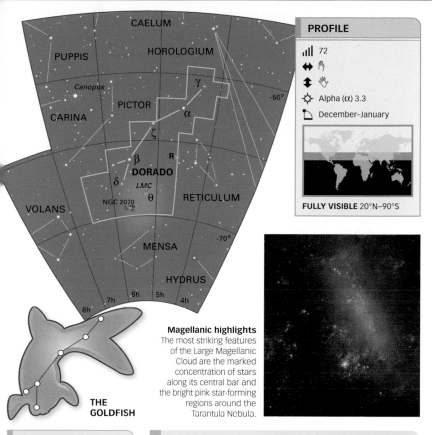

THE GOLDFISH

Magellanic highlights
The most striking features of the Large Magellanic Cloud are the marked concentration of stars along its central bar and the bright pink star-forming regions around the Tarantula Nebula.

FEATURES

Large Magellanic Cloud (LMC) The LMC is named after the Portuguese explorer Ferdinand Magellan, who recorded it in the early 1520s. It has been known, however, across the Southern Hemisphere since ancient times. Tenth-century Arab astronomers named it *Al Bakr* or the "white ox." It is an irregular satellite galaxy of the Milky Way, some 180,000 light-years away. A small telescope gives the best views of its star fields and nebulae.

NGC 2070 (the Tarantula Nebula) This nebula in the LMC is visible as a fuzzy star to the naked eye and also bears the stellar name 30 Doradus. At 800 light-years across, it is one of the largest star-forming regions known. R136, a cluster of hot blue-white supergiants, illuminates it from within.

Doradus (Dor)

DORADO

The stars of Dorado form a faint chain close to the brilliant star Canopus in Carina. The constellation was introduced in the 16th century by the Dutchmen Pieter Dirkszoon Keyser and Frederick de Houtman, and its name means goldfish. However, Dorado in fact represents the dolphinfish found in tropical waters. It is also sometimes represented as a swordfish.

SOUTHERN HEMISPHERE

While its stars are faint, Dorado is home to the spectacular Large Magellanic Cloud (LMC), a satellite galaxy of the Milky Way. The LMC is visible to the naked eye and is an excellent object for observation with any instrument. It contains NGC 2070, the Tarantula Nebula, which can also be seen without the aid of instruments. NGC 2070 is the only extra-galactic nebula that can be seen with the unaided eye.

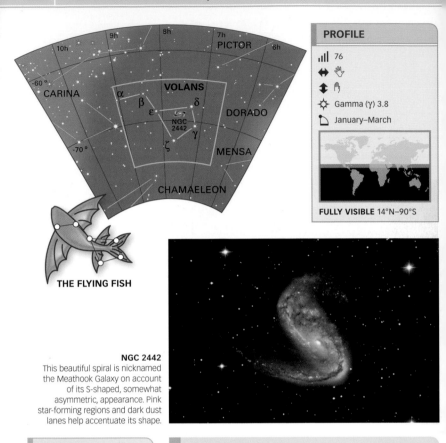

THE FLYING FISH

PROFILE

📶 76

↔ 🖐

↕ 🖐

☼ Gamma (γ) 3.8

▱ January–March

FULLY VISIBLE 14°N–90°S

NGC 2442
This beautiful spiral is nicknamed the Meathook Galaxy on account of its S-shaped, somewhat asymmetric, appearance. Pink star-forming regions and dark dust lanes help accentuate its shape.

FEATURES

Gamma (γ) Volantis The constellation's brightest star has the wrong Greek letter assigned to it. It is a double, which can be split through a small telescope to reveal a golden star of magnitude 3.8 and a yellow-white companion of magnitude 5.7. Both are 200 light-years away from Earth.

Epsilon (ε) Volantis This is another interesting double star, although it is not as colorful as Gamma (γ). The blue-white primary is 550 light-years away and shines at a magnitude of 4.4. Its companion, of magnitude 8.1, is visible only through a small telescope.

NGC 2442 This face-on, barred spiral galaxy, 50 million light-years away, is best seen with a large telescope. It is a beautiful system with spiral arms that extend out to form an "S" shape.

Volantis (Vol)

VOLANS

Although a faint and indistinct constellation, Volans can be easily found because it lies between the bright stars of Carina, the south celestial pole, and the Large Magellanic Cloud (LMC) in Dorado (p.321). The 16th-century Dutch explorers Pieter Dirkszoon Keyser and Frederick de Houtman named many of their constellations after birds, but Volans is an exception because it takes its inspiration from the bizarre flying fish of the Indian Ocean—a tropical fish that uses its outstretched fins as wings. Volans first appeared on a celestial globe published by astronomer Petrus Plancius and map-maker Jodocus Hondius around 1598.

The constellation has no bright stars, clusters, or nebulae, and even its brightest galaxies are visible only through large telescopes. Its most attractive features are probably its double stars.

SOUTHERN HEMISPHERE

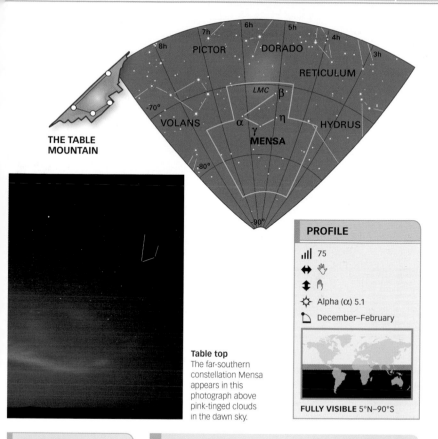

THE TABLE MOUNTAIN

PROFILE

▮▮▮ 75

↔ ✋

↕ ✋

☼ Alpha (α) 5.1

▱ December–February

FULLY VISIBLE 5°N–90°S

Table top
The far-southern constellation Mensa appears in this photograph above pink-tinged clouds in the dawn sky.

FEATURES

Alpha (α) Mensae The constellation's brightest star reaches a magnitude of only 5.1. It is an average yellow, Sun-like star, relatively close to Earth at a distance of 30 light-years.

Beta (β) Mensae The second-brightest star in Mensa shines at magnitude 5.3. Although yellow like the nearby Alpha (α), it lies much farther away. Astronomers have measured its distance at 300 light-years and believe it to be a supergiant around 100 times more luminous than Alpha (α).

Mensae (Men)

MENSA

While observing the southern skies from Cape Town, South Africa, in the 1750s, French astronomer Nicolas Louis de Lacaille named this constellation in honor of the distinctive Table Mountain that overlooks the city (*mensa* is Latin for table). It is the only one of Lacaille's constellations that is not named after a scientific or artistic tool.

SOUTHERN HEMISPHERE

Mensa has no stars brighter than magnitude 5.0, but it is easy to locate because it lies between the south celestial pole and the Large Magellanic Cloud (LMC) in Dorado. The faintest of all the 88 constellations, Mensa contains little of interest for the amateur astronomer except for a small area of the LMC that crosses into its boundary from Dorado. The Large Magellanic Cloud may have reminded Lacaille of the clouds that frequently hang over the top of the real Table Mountain.

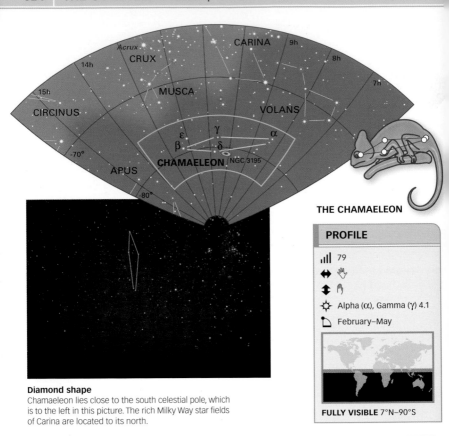

THE CHAMAELEON

Diamond shape
Chamaeleon lies close to the south celestial pole, which
is to the left in this picture. The rich Milky Way star fields
of Carina are located to its north.

PROFILE

▮▮▮	79
↔	🖐
↕	🖐
☼	Alpha (α), Gamma (γ) 4.1
⬒	February–May

FULLY VISIBLE 7°N–90°S

FEATURES

Alpha (α) Chamaeleontis
This blue-white star shines
at magnitude 4.1 and lies at a
distance of 65 light-years away
from Earth.

Delta (δ) Chamaeleontis
Binoculars will easily separate
the components of this
line-of-sight double star.
Delta-1 (δ¹), the closer of the
two stars, is an orange giant of
magnitude 5.5, 360 light-years
away from Earth. Delta-2 (δ²)
is brighter but more distant,
located 780 light-years away
and shining at magnitude 4.4.

NGC 3195 This ringlike
planetary nebula shines at
magnitude 10. Of similar
apparent size to the planet
Jupiter, it is relatively faint
and requires a medium-sized
telescope to be seen.

Chamaeleontis (Cha)

CHAMAELEON

Named after the lizard that
changes its skin color to match its
surroundings, Chamaeleon is a faint
and insignificant constellation. Lying
between the brighter stars of Crux,
Carina, Musca, and the south
celestial pole in Octans, this
constellation first appeared in
Johann Bayer's *Uranometria* star

**SOUTHERN
HEMISPHERE**

atlas of 1603. It is probably another invention of the
Dutch explorers Pieter Dirkszoon Keyser and Frederick de
Houtman from the 1590s. Its skewed diamond pattern of
stars bears little resemblance to the lizard, but it remains
above the horizon at all times for every inhabited part of
the Southern Hemisphere.

Chamaeleon contains only a handful of interesting
objects for amateur astronomers. Eta (η) Chamaeleontis
lies at the center of a recently discovered open cluster of
hot, young stars around 8 million years old.

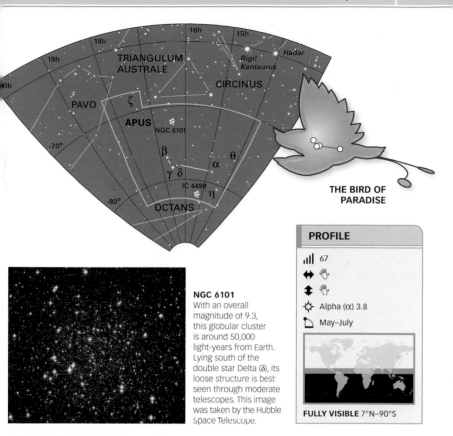

THE BIRD OF
PARADISE

NGC 6101
With an overall magnitude of 9.3, this globular cluster is around 50,000 light-years from Earth. Lying south of the double star Delta (δ), its loose structure is best seen through moderate telescopes. This image was taken by the Hubble Space Telescope.

PROFILE

ıɪll 67

↔ 🖐

↕ 🖐

☆ Alpha (α) 3.8

▭ May–July

FULLY VISIBLE 7°N–90°S

FEATURES

Alpha (α) Apodis
At magnitude 3.8, this orange giant is Apus's brightest star. It is located about 230 light years away from Earth.

Delta (δ) Apodis This wide double is the most interesting feature in Apus, consisting of orange giants with magnitudes 4.7 and 5.3. Both stars orbit each other some 310 light-years from Earth, and they can easily be separated with binoculars—or sometimes by a sharp pair of eyes.

Theta (θ) Apodis
This is a variable star with fluctuations in brightness that can easily be followed through binoculars. It ranges between magnitudes 6.4 and 8.0 in a cycle that lasts 100 days.

Apodis (Aps)

APUS

Dutch navigator–astronomers

Pieter Dirkszoon Keyser and Frederick de Houtman named this constellation after the dazzling birds of paradise they saw during their explorations of New Guinea in the 1590s. The name Apus means "no feet"– a reference to the folk belief at the time that these

SOUTHERN HEMISPHERE

beautiful tropical birds lacked feet. The constellation was at first frequently confused with the similarly named Apis, the Bee, until the latter was renamed as Musca.

Lying to the south of Triangulum Australe, Apus is not easy to identify. It is close to the south celestial pole and is permanently visible across nearly all of the Southern Hemisphere. Comparatively barren since it lies at some distance from the Milky Way, its highlights include an attractive double star and the globular clusters NGC 6101 and the much fainter IC 4499.

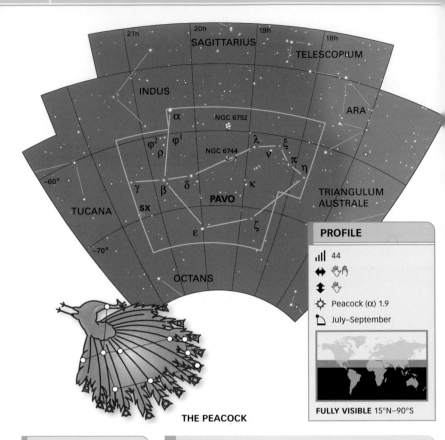

THE PEACOCK

FEATURES

Alpha (α) Pavonis (Peacock)
This is a blue-white giant star in the northeast corner of Pavo. With a magnitude of 1.9, it is brilliant enough to be seen with the naked eye although it is 360 light-years away.

Kappa (κ) Pavonis One of the brightest Cepheid variables (pp.232–33), this yellow supergiant is 550 light-years away. It expands and contracts in a 9.1-day cycle, varying in brightness between magnitudes 3.9 and 4.8.

NGC 6744 This large spiral galaxy, with its face open to Earth, is located 30 million light-years away.

NGC 6752 A globular cluster 14,000 light-years away, this is one of the largest and brightest objects of its kind. It is visible to the naked eye at magnitude 5.

Pavonis (Pav)

PAVO

One of the twelve far-southern constellations introduced at the end of the 16th century by the Dutch navigator-astronomers Pieter Dirkszoon Keyser and Frederick de Houtman, Pavo represents the exotic peacock of Southeast Asia, which the two explorers encountered on their travels.

SOUTHERN HEMISPHERE

In more recent times, its brightest star, the 2nd-magnitude Alpha (α) Pavonis, has been named Peacock. In Greek mythology, the peacock was the sacred bird of Hera, wife of Zeus, who traveled through the air in a chariot drawn by these birds. It was Hera who placed the famous eyelike markings on the tail of the peacock.

This constellation is best found by looking on the edge of the Milky Way, south of Sagittarius. It lies in a fairly featureless area but does contain an impressively bright globular cluster.

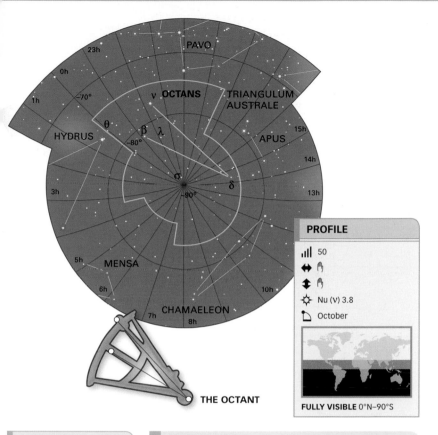

THE OCTANT

PROFILE

📶 50

↔ 🖐

↕ 🖐

☼ Nu (ν) 3.8

🗓 October

FULLY VISIBLE 0°N–90°S

FEATURES

Beta (β) Octantis A white star, Beta (β) lies 110 light-years away. It shines at magnitude 4.1, only outshone by Nu (ν) at magnitude 3.8.

Gamma (γ) Octantis This chain of three stars, whose members lie at different distances from Earth, is usually separable with the naked eye. Gamma-1 (γ^1) and Gamma-3 (γ^3) are yellow giants, 270 and 240 light-years away, and shine at magnitudes 5.1 and 5.3. Between them lies Gamma-2 (γ^2), an orange giant of magnitude 5.7 at a distance of 310 light-years.

Sigma (σ) Octantis Sitting 300 light-years away, this southern "pole star" is dim and yellow-white. Its only noteworthy feature is its location about one degree off the south celestial pole.

Octantis (Oct)

OCTANS

Originally known as Octans Nautica or Octans Hadleianus, this constellation was introduced in the 18th century by the French astronomer Nicolas Louis de Lacaille. It represents the octant, a navigational device invented by the English instrument maker John Hadley in around 1730. Other than its location at the south celestial pole, this faint constellation is of little interest to amateur astronomers.

SOUTHERN HEMISPHERE

Within naked-eye range, Sigma (σ) Octantis is the nearest star to the south celestial pole. Lying 300 light-years away, with a magnitude of 5.4, it is far from prominent. Due to the slow effect of precession, the south celestial pole is moving farther away from Sigma toward the constellation Chamaeleon. In around 1,500 years, the pole will be just over a degree away from Delta (δ) Chamaeleontis, a star of magnitude 4.

DIRECTORY

The following tables provide detailed information about the celestial bodies of the Solar System—such as their sizes, orbits, and physical properties. They also serve as a guide to the stars, nebulae, and galaxies that are most easily seen from Earth. A complete Messier catalog is also provided on pp.336–37.

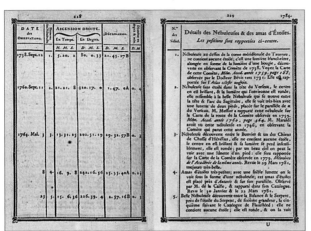

Messier catalog
Throughout history there have been many reasons for cataloging celestial objects. French astronomer Charles Messier's catalog, published in 1781, was a list of 103 objects that could be mistaken for comets when viewed through a small telescope.

THE PLANETS

There are eight planets in our solar system. The four closest to the Sun (Mercury, Venus, Earth, and Mars) are relatively small and rocky, while the four outer planets (Jupiter, Saturn, Uranus, and Neptune) are larger and consist mostly of gas. This table gives information about the orbital properties of each planet, followed by details of its physical characteristics.

ORBITAL PROPERTIES

Planet	Average distance from the Sun	Orbital period	Rotational period	Inclination of axis of rotation	Apparent magnitude
Mercury	36 million miles (57.9 million km)	88 days	59 days	00.1°	-1.9 to 5.7
Venus	67.2 million miles (108.2 million km)	224.7 days	243 days	177.4°	-4.6 to -3.8
Earth	93 million miles (149.6 million km)	365.2 days	23.9 hours	23.4°	n/a
Mars	141.6 million miles (227.9 million km)	687 days	24.6 hours	25.2°	-2.9 to -4.5
Jupiter	483.8 million miles (778.6 million km)	11.9 years	9.9 hours	3.1°	-2.9
Saturn	890.8 million miles (1,434 million km)	29.5 years	10.7 hours	26.7°	-0.2 to 1.2
Uranus	1,785 million miles (2,873 million km)	84 years	17.2 hours	97.8°	5.5
Neptune	2,793 million miles (4,496 million km)	164.8 years	16.1 hours	28.3°	7.8

PHYSICAL PROPERTIES

Planet	Equatorial diameter	Mass in relation to Earth	Volume in relation to Earth	Average surface or cloud-top temperature	Gravity in relation to Earth	Number of moons
Mercury	3,032 miles (4,879 km)	0.1	0.1	333°F (167°C) (surface)	0.4	0
Venus	7,521 miles (12,104 km)	0.8	0.9	867°F (464°C) (surface)	0.9	0
Earth	7,926 miles (12,756 km)	1	1	59°F (15°C) (surface)	1	1
Mars	4,221 miles (6,792 km)	0.1	0.2	-81°F (-63°C) (surface)	0.4	2
Jupiter	88,846 miles (142,984 km)	318	1,321	-162°F (-108°C) (cloud-top)	2.5	64
Saturn	74,898 miles (120,536 km)	95.0	763.7	-218°F (-139°C) (cloud-top)	1.1	62
Uranus	31,763 miles (51,118 km)	14.5	63.1	-319°F (-195°C)	0.9	27
Neptune	30,775 miles (49,528 km)	17.1	57.7	-364°F (-220°C)	1.1	13

NOTABLE ASTEROIDS

Millions of asteroids lie in a belt between the orbits of Mars and Jupiter. Apart from Vesta, none can be seen with the naked eye, although some are visible through binoculars. The largest asteroid, Ceres, is also classified as a dwarf planet.

Asteroid	Average distance from the Sun	Orbital speed	Orbital period	Rotation period	Length
Eros	135 million miles (218 million km)	54,492 mph (87,696 kph)	1.8 years	5.3 hours	21.1 miles (34 km)
Gaspra	206 million miles (331 million km)	44,470 mph (71,568 kph)	3.3 years	7.0 hours	11.2 miles (18 km)
Annefrank	206 million miles (331 million km)	Unknown	3.3 years	Unknown	3.7 miles (6 km)
Vesta	219 million miles (353 million km)	43,262 mph (69,624 kph)	3.6 years	5.3 hours	348 miles (560 km)
Toutatis	234 million miles (376 million km)	37,334 mph (60,084 kph)	4.0 years	5.4 and 7.3 days (two axes)	2.8 miles (4.5 km)
Mathilde	246 million miles (396 million km)	40,220 mph (64,728 kph)	4.3 years	418 hours	41 miles (66 km)
Ceres	257 million miles (414 million km)	40,000 mph (64,375 kph)	4.6 years	9.1 hours	Diameter: 596.5 miles (960 km)
Ida	266 million miles (428 million km)	Unknown	4.8 years	4.6 hours	36 miles (58 km)

ANNUAL METEOR SHOWERS

Meteor showers occur when Earth crosses a stream of debris that has accumulated in the orbit of a comet. The part of the night sky from which they appear is called the radiant.

Shower	Peak date(s)	Date range	Maximum number per hour	Notes
Quadrantids	Jan 3–4	Jan 1–6	100	Yellow and blue meteors traveling at medium speed.
Alpha Centaurids	Feb 8	Jan 28–Feb 21	20	A southern shower with some very bright, swift meteors.
Virginids	Apr 7–15	Mar 10–Apr 21	5	Slow, long trails with multiple radiants.
Lyrids	Apr 21–22	Apr 16–28	About 12	Fairly fast meteors from Comet Thatcher.
Eta Aquarids	May 6	Apr 21–May 24	35	A good southern shower with very fast, bright meteors from Comet Halley.
Arietids	Jun 7	May 22–Jun 30	55	A daytime shower, with some visible just before dawn.
June Boötids	Jun 28	Jun 27–30	Variable	A slow-speed variable shower with occasional strong outbursts. From Comet Pons–Winnecke.
Capricornids	Jul 5–20	Jun 10–Jul 30	5	Slow, yellow and blue bright meteors. Several peak dates and radiants.
Delta Aquarids	Jul 29 Aug 13–14	Jul 14–Aug 18 Jul 16–Sep 10	20	A southern shower with a double peak.
Piscis Austrinids	Jul 28	Jul 16–Aug 8	5	Southern-Hemisphere shower of quite slow meteors.
Alpha Capricornids	Aug 1	Jul 15–Aug 25	5	Produces slow fireballs visible for many seconds.
Perseids	Aug 12	Jul 23–Aug 22	80	Many bright meteors with trails, from Comet Swift–Tuttle.
Alpha Aurigids	Sep 1	Aug 25–Sep 7	7	Occasional bursts of over 100 per hour have been seen.
Giacobinids or Draconids	Oct 8	Oct 6–10	Variable	Slow meteors from Comet Giacobini–Zinner.
Orionids	Oct 20	Oct 5–30	25	Fast, with many trails. From Halley's Comet.
Taurids	Oct 30–Nov 7 Nov 4–7	Sep 17–Nov 27 Oct 12–Dec 2	10	Bright and slow meteors with two peaks from Comet Encke.
Leonids	Nov 17	Nov 14–21	Variable	Very fast meteors with trails, from Comet Temple–Tuttle.
Geminids	Dec 14	Dec 6–18	100	Medium speed, bright meteors, from asteroid Phaethon (3,200).
Ursids	Dec 22	Dec 17–25	10	Slow meteors from Comet Tuttle.

BRIGHTEST STARS VIEWED FROM EARTH

The brightness of a star seen from Earth is known as its apparent magnitude. This depends on two factors: the star's actual brightness (its absolute magnitude) and its distance from Earth. This table lists the 20 brightest stars in our night sky, with their apparent magnitudes and their distances from Earth.

Rank	Name	Apparent magnitude	Bayer designation	Distance from Earth (light-years)
1	Sirius	-1.4	Alpha (α) Canis Majoris	8.6
2	Canopus	-0.7	Alpha (α) Carinae	310
3	Rigil Kentaurus	-0.3	Alpha (α) Centauri	4.3
4	Arcturus	-0.04	Alpha (α) Boötis	36
5	Vega	0.0	Alpha (α) Lyrae	25
6	Capella	0.1	Alphae (α) Aurigae	42
7	Rigel	0.1	Beta (β) Orionis	770
8	Procyon	0.4	Alpha (α) Canis Minoris	11.4
9	Achernar	0.5	Alpha (α) Eridani	140
10	Betelgeuse	0.0–1.3	Alpha (α) Orionis	430
11	Hadar	0.6	Beta (β) Centauri	525
12	Altair	0.8	Alpha (α) Aquilae	17
13	Aldebaran	0.8–1.0	Alpha (α) Tauri	65
14	Acrux	0.8	Alpha (α) Crucis	320
15	Spica	0.9–1.2	Alpha (α) Virginis	260
16	Antares	0.9–1.8	Alpha (α) Scorpii	600
17	Pollux	1.1	Beta (β) Geminorum	34
18	Fomalhaut	1.2	Alpha (α) Piscis Austrini	25
19	Mimosa/Becrux	1.15–1.25	Beta (β) Crucis	350
20	Deneb	1.3	Alpha (α) Cygni	1,500

CLOSEST STARS TO EARTH

The brightest stars are not necessarily those closest to Earth. Canopus, for example, is the second-brightest star, but is 310 light-years away. This table lists the 20 stars closest to Earth, showing their distances and various apparent magnitudes.

Rank	Star	Distance from Earth (light-years)	Constellation	Apparent magnitude
1	Alpha (α) Centauri system: Proxima Centauri Alpha (α) Centauri	4.2 4.3	Centaurus	11.1 -0.3
2	Barnard's Star	6.0	Ophiuchus	9.6
3	Wolf 359	7.8	Leo	13.4
4	Lalande 21185	8.3	Ursa Major	7.5
5	Sirius	8.6	Canis Major	-1.4
6	Luyten 726–8	8.7	Cetus	12.5
7	Ross 154	9.7	Sagittarius	10.4
8	Ross 248	10.3	Andromeda	12.3
9	Epsilon (ε) Eridani	10.5	Eridanus	3.7
10	Lacaille 9352	10.7	Piscis Austrinus	7.3
11	Ross 128	10.9	Virgo	11.1
12	EZ Aquarii	11.3	Aquarius	13.3
13	Procyon	11.4	Canis Minor	0.4
14	Bessel's Star	11.4	Cygnus	5.2
15	Struve 2398	11.5	Draco	8.9
16	Groombridge 34	11.6	Andromeda	8.1
17	Epsilon (ε) Indi	11.8	Indus	4.7
18	DX Cancri	11.8	Cancer	14.8
19	Tau (τ) Ceti	11.9	Cetus	3.5
20	GJ 1061	12.1	Horologium	13.1

STAR CLUSTERS TO OBSERVE

There are of two types of star cluster: globular clusters, which are tight groups of hundreds or thousands of very old stars, and open clusters, which are looser groups of only a few hundred young stars. Open clusters can be particularly stunning.

Cluster name	Cluster type	Constellation	Apparent magnitude	Distance from Earth (light-years)
47 Tucanae	Globular	Tucana	4.0	13,500
M44 (Beehive Cluster)	Open	Cancer	3.7	577
M6 (Butterfly Cluster)	Open	Scorpius	4.2	2,000
Hyades	Open	Taurus	4.2	160
Jewel Box	Open	Crux	4.0	7,600
M4	Globular	Scorpius	7.4	7,000

Cluster name	Cluster type	Constellation	Apparent magnitude	Distance from Earth (light-years)
M12	Globular	Ophiuchus	6.6	16,000–18,000
M14	Globular	Ophiuchus	6.4	23,000–30,000
M15	Globular	Pegasus	6.2	33,600
M52	Open	Cassiopeia	7.3	5,000
M68	Globular	Hydra	7.5	33,000–44,000
M93	Open	Puppis	6.2	3,600
M107	Globular	Ophiuchus	8.9	20,900
NGC 3201	Globular	Vela	8.2	15,000
NGC 4833	Globular	Musca	5.3	17,000
Omega (ω) Centauri	Globular	Centaurus	3.7	17,000
Pleiades	Open	Taurus	1.5	400

VARIABLE STARS TO OBSERVE

This table highlights a wide range of variable stars—stars whose magnitudes vary over time.

Name	Variable type	Constellation	Minimum apparent magnitude	Maximum apparent magnitude	Period	Distance from Earth (light-years)
Rasalgethi	Eclipsing binary	Hercules	4.0	2.8	128 days	380
Delta (δ) Cephei	Pulsating variable	Cephus	4.4	3.5	5.3 days	982
Eta (η) Aquilae	Pulsating variable	Aquila	3.9	3.5	7.2 days	1,173
Eta (η) Geminorum	Eclipsing binary	Gemini	4.2	3.3	233 days	349
Gamma (γ) Cassiopeiae	Irregular variable	Cassiopeia	3.0	1.6	variable	613
Lambda (λ) Tauri	Eclipsing binary	Taurus	3.9	3.4	4 days	370
Mira	Pulsating variable	Cetus	10.0	2.0	332 days	418
Mu (μ) Cephei	Pulsating variable	Cepheus	5.1	3.4	about 730 days	5,258
W Virginis	Cepheid variable	Virgo	10.8	9.6	17 days	10,000
R Coronae Borealis	Irregular variable	Corona Borealis	14.8	5.8	variable	6,000
RR Lyrae	Pulsating variable	Lyra	8.1	7.1	0.6 days	744
Algol	Eclipsing binary	Perseus	3.4	2.1	2.87 days	93
Zeta (ξ) Geminorum	Pulsating variable	Gemini	4.2	3.6	10.2 days	1,168

MULTIPLE STARS TO OBSERVE

A multiple star is a group of stars that, from Earth, appear close to each other. This closeness can be a physical closeness, or an optical (or apparent) closeness.

Name	Constellation	Number of stars	Apparent magnitude	Distance from Earth (light-years)
Castor	Gemini	6	1.6	50
Sigma (σ) Orionis	Orion	5	3.8	1,150
Alcyone	Taurus	4	2.9	368
Algol	Perseus	3	2.1–3.4	93
Almach	Andromeda	4	2.3	355
Epsilon (ε) Lyrae	Lyra	4	4.7	160
Mizar & Alcor	Ursa Major	4	2.3	81
Theta (θ) Orionis	Orion	4	4.7	1,800
Albireo	Cygnus	3	3.1	385
Beta (β) Monocerotis	Monoceros	3	5.0	700
Omicron (ο) Eridani	Eridanus	3	4.4	16
Rigel	Orion	3	0.1	770
15 Monocerotis	Monoceros	2	4.7	1,020
Epsilon (ε) Aurigae	Auriga	2	3.1	2,040
Epsilon (ε) Boötis	Boötes	2	2.7	203
M40	Ursa Major	2	8.4	385
Polaris	Ursa Minor	2	2.0	430
Zeta (ξ) Boötis	Boötes	2	3.8	180

NEBULAE TO OBSERVE

Nebulae are mainly of two types: emission nebulae (including planetary nebulae), which are clouds of ionized gas that emit light; and dark nebulae, which emit no light and obscure the light of the stars behind them. Another type of nebula is a supernova remnant.

Name	Nebula type	Constellation	Apparent magnitude	Distance from Earth (light-years)
Cat's Eye Nebula	Planetary	Draco	8.1	3,600
Cone Nebula	Dark	Monoceros	3.9	2,500
Crescent Nebula	Planetary	Cygnus	7.4	4,700
Dumbbell Nebula	Planetary	Vulpecula	7.6	1,000
Eagle Nebula	Emission	Serpens Cauda	6.0	7,000
Eight-Burst Nebula	Planetary	Vela	8.1	2,000
Eskimo Nebula	Planetary	Gemini	8.6	3,800
Carina Nebula	Emission	Carina	1.0	8,000

Name	Nebula type	Constellation	Apparent magnitude	Distance from Earth (light-years)
Helix Nebula	Planetary	Aquarius	6.5	700
Hourglass Nebula	Planetary	Musca	11.8	8,000
IC 2944	Emission	Centaurus	4.5	5,900
Lagoon Nebula	Emission	Sagittarius	5.8	5,200
Orion Nebula	Emission	Orion	4.0	1,500
Owl Nebula	Planetary	Ursa Major	9.8	2,600
Ring Nebula	Planetary	Lyra	8.8	2,000
Saturn Nebula	Planetary	Aquarius	8.0	5,200
Stingray Nebula	Planetary	Ara	10.8	18,000
Tarantula Nebula	Emission	Dorado	5.0	160,000
Trifid Nebula	Emission	Sagittarius	9.0	7,000
Veil Nebula	Supernova remnant	Cygnus	7.0	1,500

GALAXIES TO OBSERVE

Of the galaxies listed here, Andromeda Galaxy, Bode's Galaxy, the Triangulum Galaxy, and the Small and Large Magellanic Clouds are just visible to the naked eye. The Whirlpool Galaxy can be seen through binoculars; the remaining galaxies can only be seen through a telescope.

Name and catalogue number	Galaxy type	Constellation	Apparent magnitude	Distance from Earth (light-years)
Andromeda Galaxy (M31)	Spiral	Andromeda	4.5	2.5 million
Black Eye Galaxy (M64)	Spiral	Coma Berenices	8.5	17 million
Bode's Galaxy (M81)	Spiral	Ursa Major	6.9	10 million
Cigar Galaxy (M82)	Spiral	Ursa Major	8.4	12 million
Large Magellanic Cloud	Irregular	Dorado	0.4	180,000
Small Magellanic Cloud (NGC 292)	Irregular	Tucana	2.3	210,000
Sombrero Galaxy (M104)	Spiral	Virgo	9.0	28 million
Triangulum Galaxy (M33)	Spiral	Triangulum	5.7	2.7 million
Whirlpool Galaxy (M51)	Spiral	Canes Venatici	8.4	23 million

MESSIER OBJECTS

In 1781 French astronomer Charles Messier published a list of objects that could be mistaken for comets. His list of 103 objects has since been expanded to include 110 objects, all of which are visible through binoculars or a small telescope.

Messier number	Type (popular name)	Constellation
M1	Supernova remnant (Crab Nebula)	Taurus
M2	Globular cluster	Aquarius
M3	Globular cluster	Canes Venatici
M4	Globular cluster	Scorpius
M5	Globular cluster	Serpens Caput
M6	Open cluster (Butterfly Cluster)	Scorpius
M7	Open cluster (Ptolemy Cluster)	Scorpius
M8	Emission nebula (Lagoon Nebula)	Sagittarius
M9	Globular cluster	Ophiuchus
M10	Globular cluster	Ophiuchus
M11	Open cluster (Wild Duck Cluster)	Scutum
M12	Globular cluster	Ophiuchus
M13	Globular cluster (The Great Globular Cluster)	Hercules
M14	Globular cluster	Ophiuchus
M15	Globular cluster	Pegasus
M16	Open cluster/ emission nebula (The Eagle Nebula)	Serpens Cauda
M17	Emission nebula (Omega/Swan/ Lobster/Horseshoe Nebula)	Sagittarius
M18	Open cluster	Sagittarius
M19	Globular cluster	Ophiuchus
M20	Emission/reflection/ dark nebula (Trifid Nebula)	Sagittarius
M21	Open cluster	Sagittarius
M22	Globular cluster	Sagittarius
M23	Open cluster	Sagittarius
M24	Starfield (Sagittarius Star Cloud)	Sagittarius

Messier number	Type (popular name)	Constellation
M25	Open cluster	Sagittarius
M26	Open cluster	Scutum
M27	Planetary nebula (Dumbbell Nebula)	Vulpecula
M28	Globular cluster	Sagittarius
M29	Open cluster	Cygnus
M30	Globular cluster	Capricornus
M31	Spiral galaxy (Andromeda Galaxy)	Andromeda
M32	Dwarf elliptical galaxy	Andromeda
M33	Spiral galaxy (Triangulum Galaxy)	Triangulum
M34	Open cluster	Perseus
M35	Open cluster	Gemini
M36	Open cluster	Auriga
M37	Open cluster	Auriga
M38	Open cluster	Auriga
M39	Open cluster	Cygnus
M40	Double star (Winnecke 4)	Ursa Major
M41	Open cluster	Canis Major
M42	Emission/reflection (Orion Nebula)	Orion
M43	Emission/reflection nebula (De Mairan's Nebula)	Orion
M44	Open cluster (Beehive Cluster)	Cancer
M45	Open cluster (Pleiades/Seven Sisters)	Taurus
M46	Open cluster	Puppis
M47	Open cluster	Puppis
M48	Open cluster	Hydra
M49	Elliptical galaxy	Virgo
M50	Open cluster	Monoceros
M51	Spiral galaxy (Whirlpool Galaxy)	Canes Venatici
M52	Open cluster	Cassiopeia
M53	Globular cluster	Coma Berenices

Messier number	Type (popular name)	Constellation
M54	Globular cluster	Sagittarius
M55	Globular cluster	Sagittarius
M56	Globular cluster	Lyra
M57	Planetary nebula (Ring Nebula)	Lyra
M58	Barred spiral galaxy	Virgo
M59	Elliptical galaxy	Virgo
M60	Elliptical galaxy	Virgo
M61	Spiral galaxy	Virgo
M62	Globular cluster	Ophiuchus
M63	Spiral galaxy (Sunflower Galaxy)	Canes Venatici
M64	Spiral galaxy (Black Eye Galaxy)	Coma Berenices
M65	Spiral galaxy	Leo
M66	Spiral galaxy	Leo
M67	Open cluster	Cancer
M68	Globular cluster	Hydra
M69	Globular cluster	Sagittarius
M70	Globular cluster	Sagittarius
M71	Globular cluster	Sagitta
M72	Globular cluster	Aquarius
M73	Asterism	Aquarius
M74	Spiral galaxy	Pisces
M75	Globular cluster	Sagittarius
M76	Planetary nebula (Little Dumbbell Nebula)	Perseus
M77	Barred spiral galaxy	Cetus
M78	Reflection nebula	Orion
M79	Globular cluster	Lepus
M80	Globular cluster	Scorpius
M81	Spiral galaxy (Bode's Galaxy)	Ursa Major
M82	Spiral galaxy (Cigar Galaxy)	Ursa Major
M83	Barred spiral galaxy (Southern Pinwheel Galaxy)	Hydra
M84	Lenticular galaxy	Virgo

Messier number	Type (popular name)	Constellation
M85	Lenticular galaxy	Coma Berenices
M86	Lenticular galaxy	Virgo
M87	Elliptical galaxy (Virgo A)	Virgo
M88	Spiral galaxy	Coma Berenices
M89	Elliptical galaxy	Virgo
M90	Spiral galaxy	Virgo
M91	Barred spiral galaxy	Coma Berenices
M92	Globular cluster	Hercules
M93	Open cluster	Puppis
M94	Spiral galaxy	Canes Venatici
M95	Barred spiral galaxy	Leo
M96	Spiral galaxy	Leo
M97	Planetary nebula (Owl Nebula)	Ursa Major
M98	Spiral galaxy	Coma Berenices
M99	Spiral galaxy	Coma Berenices
M100	Spiral galaxy	Coma Berenices
M101	Spiral galaxy (Pinwheel Galaxy)	Ursa Major
M102		Not unambiguously identified
M103	Open cluster	Cassiopeia
M104	Spiral galaxy (Sombrero Galaxy)	Virgo
M105	Elliptical galaxy	Leo
M106	Spiral galaxy	Canes Venatici
M107	Globular cluster	Ophiuchus
M108	Spiral galaxy	Ursa Major
M109	Barred spiral galaxy	Ursa Major
M110	Dwarf elliptical galaxy	Andromeda

GLOSSARY

Terms in *italics* have their own entries in this glossary.

ACCRETION The colliding and sticking together of small, solid objects and particles.

ACCRETION DISK A disk of gas that revolves around a star or *black hole*.

ACTIVE GALAXY A galaxy that emits an exceptional amount of radiant energy over a wide range of wavelengths.

ALTITUDE The angle between the horizon and a celestial object. An object on the horizon has an altitude of 0°; one directly overhead is at 90°. See also *azimuth*.

APHELION The point on its elliptical orbit at which a planet or other Solar System body is at its furthest from the Sun. See also *perihelion*.

APOGEE The point on its orbit at which a body orbiting Earth is at its furthest from Earth. See also *perigee*.

ASTERISM A conspicuous pattern of stars that is not itself a *constellation*. An example is the Big Dipper, or Plough, part of the constellation Ursa Major (the Great Bear).

ASTEROID A small, irregular Solar System object made of rock and/or metal with a diameter of less than 600 miles (1,000 km). See also *Main Belt*, *near-Earth asteroid*.

AURORA A glowing display of light in Earth's upper atmosphere (and the atmospheres of some other planets) caused by particles in the *solar wind* colliding with gas atoms in the atmosphere, stimulating them to emit light.

AZIMUTH The angle between the north point on an observer's horizon and a celestial object, measured clockwise around the horizon. The azimuth of due north is 0°, due east is 90°, due south is 180°, and due west is 270°. See also *altitude*.

BARRED SPIRAL GALAXY A galaxy that has spiral arms emanating from the ends of an elongated, bar-shaped nucleus. See also *spiral galaxy*.

BIG BANG The event in which the Universe we see around us today had its origin.

BINARY STAR A pair of stars, bound together by gravity, that orbits a single point (the common center of mass).

BLACK HOLE A compact region of space, surrounding a collapsed mass, within which gravity is so powerful that no object or radiation can escape.

BROWN DWARF STAR A body that forms out of a contracting cloud of gas in the same way as a star, but which, because it contains too little mass, never becomes hot enough to initiate the nuclear *fusion* reactions that power a normal star.

CATADIOPTRIC TELESCOPE A telescope that collects and focuses light using both lenses and mirrors. See also *reflecting telescope*, *refracting telescope*.

CELESTIAL EQUATOR A great circle on the *celestial sphere* that is a projection of Earth's own equator onto the celestial sphere.

CELESTIAL MERIDIAN The imaginary line on the *celestial sphere* that passes through both *celestial poles* and the vernal *equinoctial point*. The celestial meridian is the line of 0° *right ascension*.

CELESTIAL POLES The celestial equivalent of Earth's poles. The night sky appears to rotate on an axis through the celestial poles.

CELESTIAL SPHERE An imaginary sphere, surrounding Earth, on which all celestial objects appear to lie.

CEPHEID VARIABLE A type of *variable star* with a regular pattern of brightness changes linked to the star's luminosity. The more luminous the Cepheid, the longer its period of variation.

COMET A small body composed mainly of dust-laden ice that orbits the Sun.

CONJUNCTION A close alignment in the sky of two or more celestial bodies that occurs when they lie in the same direction as viewed from Earth. When a planet lies directly on the opposite side of the Sun from Earth, it is said to be at superior conjunction. When either Mercury or Venus

passes between Earth and the Sun, the planet is said to be at inferior conjunction. See also *opposition*.

CONSTELLATION 1) A named pattern of stars. **2)** An area of the night sky with boundaries that are determined by the International Astronomical Union. See also *asterism*.

CORONA The outermost region of the atmosphere of a star.

COSMIC RAYS Highly energetic subatomic particles, such as electrons, protons, and atomic nuclei, that hurtle through space at close to the speed of light.

DARK NEBULA A dust-laden cloud that blocks out the light from background stars.

DECLINATION (DEC) The angular distance of a celestial object north or south of the *celestial equator*; the equivalent of latitude on Earth. Declination is positive (+) if the object is north of the celestial equator and negative (-) if it is south of the celestial equator. An object on the celestial equator has a declination of 0°; an object at one of the celestial poles has a declination of 90°. See also *right ascension*.

DEEP-SKY OBJECT Any celestial object external to the Solar System but excluding stars.

DIFFUSE NEBULA A luminous cloud of gas and dust. The term "diffuse" refers to the cloud's fuzzy appearance.

DOUBLE STAR Two stars that are close together in the sky. If they orbit each other, the system is called a *binary star*. An optical double star consists of two stars that appear close together only because they lie in the same direction when viewed from Earth.

DWARF PLANET A body orbiting the Sun that is massive enough to be nearly spherical but which has not cleared its neighborhood of other objects and is not a moon.

DWARF STAR A star with a mass similar to, or less than, the Sun's.

ECLIPSE An alignment of a planet or moon with the Sun that casts a shadow on another body.

ECLIPSING BINARY A binary star system in which each star alternately passes in front of the other, causing a periodic variation in the combined light of the two stars as seen from Earth.

ECLIPTIC 1) The plane on which Earth's orbit around the Sun is situated. **2)** The track along which the Sun travels around the *celestial sphere*, relative to the background stars, in the course of a year.

ELECTROMAGNETIC RADIATION Oscillating electric and magnetic disturbances that propagate energy in the form of waves (electromagnetic waves). Examples include visible light and radio waves.

ELECTROMAGNETIC SPECTRUM The entire range of energy emitted by different objects in the Universe, from the shortest wavelengths (gamma rays) to the longest wavelengths (radio waves).

ELLIPTICAL GALAXY A *galaxy* that is round or elliptical in shape.

ELONGATION The angular separation between the Sun and a planet or other Solar System body, as viewed from Earth. Greatest elongation is the maximum possible elongation of a body that lies inside Earth's orbit, such as Mercury or Venus.

EMISSION NEBULA A cloud of gas and dust that contains one or more extremely hot, young, high-luminosity stars; ultraviolet radiation emitted by these stars causes the surrounding gas to glow.

EQUINOCTIAL POINT Either of two points on the *celestial sphere* where the *celestial equator* and *ecliptic* intersect. The Northern Hemisphere's vernal equinoctial point—also known as the first point of Aries—is the origin for *right ascension* measurements.

EQUINOX The occasion when the Sun is vertically overhead at a planet's equator and day and night have equal duration for the whole planet. See also *solstice*.

FINDER An optical device that can be fitted to a telescope to help locate celestial objects in the telescope's field of view.

FUSION (NUCLEAR FUSION) A process whereby atomic nuclei join together to form heavier atomic nuclei, releasing large amounts of energy. Stars are powered by fusion reactions in their cores.

GALAXY A large aggregation of stars and clouds of gas and dust, held together by gravity. Galaxies may be elliptical, spiral, or irregular in shape. See also *Milky Way*.

GALAXY CLUSTER An aggregation of approximately 50 to 1,000 galaxies held together by gravity. See also *galaxy supercluster*.

GALAXY SUPERCLUSTER A cluster of galaxy clusters. A supercluster may contain up to about 10,000 galaxies.

GALILEAN MOON Any of the four largest of Jupiter's moons that were discovered in 1610 by the Italian scientist Galileo Galilei (1564–1642). From largest to smallest, they are Ganymede, Callisto, Io, and Europa

GAMMA RADIATION *Electromagnetic radiation* with extremely short wavelengths (shorter than X-rays). See also *electromagnetic spectrum*.

GIANT PLANET A large planet composed mainly of hydrogen and helium. The giant planets in the Solar System are Jupiter, Saturn, Uranus, and Neptune. See also *rocky planet*.

GIANT STAR A star that is larger and much more luminous than a *main-sequence star* of the same surface temperature. See also *supergiant star*.

GLOBULAR CLUSTER A near-spherical cluster of between 10,000 and more than 1 million stars, held together by gravity.

HALO A spherical region that surrounds a *galaxy* and contains *globular clusters*, thinly scattered stars, and some gas. A dark-matter halo is an accumulation of dark matter within which a galaxy is embedded.

INDEX CATALOG (IC) See *New General Catalog*.

INFERIOR CONJUNCTION See *conjunction*.

INFERIOR PLANET A Solar System planet whose orbit is inside Earth's. The inferior planets are Mercury and Venus. See also *superior planet*.

INFRARED RADIATION *Electromagnetic radiation* with wavelengths longer than visible light but shorter than microwaves or radio waves. See also *electromagnetic spectrum*.

IRREGULAR GALAXY A *galaxy* that has no well-defined structure or symmetry.

KUIPER BELT A region of the Solar System outside the orbit of the planet Neptune that contains icy *planetesimals*. See also *Oort Cloud*.

LENTICULAR GALAXY A *galaxy* shaped like a convex lens. It has a central bulge that merges into a disk but no spiral arms.

LIGHT-YEAR The distance that light travels through a vacuum in one year (365.25 days): 5,878.6 billion miles (9,460.7 billion km).

LIMB The outer edge of the observed disk of the Sun, a moon, or a planet.

LOCAL GROUP A small cluster of over 40 galaxies that includes our own galaxy, the *Milky Way*. See also *galaxy cluster*.

MAGNETOSPHERE The region of space around a planet within which the motion of charged particles is affected by the planetary magnetic field.

MAGNITUDE A measure of the brightness of a celestial object. Absolute magnitude is a measure of the intrinsic brightness of an object. Apparent magnitude is a measure of the brightness of an object as seen from Earth. The brighter the object, the smaller the numerical value of its apparent magnitude. Very bright objects have negative apparent magnitudes. A star said to be of 1st magnitude has a magnitude of 1.49 or less, a star of 2nd magnitude has a value of 1.50 to 2.49, and so on.

MAIN BELT A region of the Solar System lying between the orbits of Mars and Jupiter that contains a high concentration of *asteroids*.

MAIN-SEQUENCE STAR A star that is in the stage of its evolution when it is converting hydrogen to helium by nuclear *fusion* in its core. The Sun is currently a main-sequence star.

MARE (PLURAL: MARIA) A dark, low-lying area of the Moon, filled with lava.

MESSIER CATALOG A catalog of nebulous objects (mostly nebulae, star clusters, and galaxies) that was first published in 1781. Objects in this catalog are designated by the letter "M" followed by a number. See also *New General Catalog*.

METEOR The short-lived streak of light, also known as a shooting star, seen when a *meteoroid* enters Earth's atmosphere and is heated by friction. See also *meteorite*.

METEORITE A *meteoroid* that reaches the ground and survives impact. See also *meteor*.

METEOROID A lump or small particle of rock, metal, or ice orbiting the Sun in interplanetary space. See also *asteroid, comet, meteor, meteorite*.

MICROWAVE RADIATION *Electromagnetic radiation* with wavelengths longer than infrared and visible light but shorter than radio waves. See also *electromagnetic spectrum*.

MILKY WAY 1) The *spiral galaxy* that contains the Sun. **2)** A band of light across the night sky that consists of the combined light of vast numbers of stars and nebulae in our galaxy.

MIRA VARIABLE A class of *variable stars* named after the star Mira. A Mira variable is a cool, giant, pulsating star that varies in brightness over a period ranging from 100 to more than 500 days.

MOON A natural satellite orbiting a planet. The Moon is Earth's natural satellite.

MULTIPLE STAR A system consisting of two or more stars bound together by gravity and orbiting around each other. See also *binary star*.

NEAR-EARTH ASTEROID (NEA) An asteroid whose orbit comes close to or intersects Earth's orbit.

NEBULA A cloud of gas and dust in interstellar space, visible either because it is illuminated by nearby or embedded stars, or because it is obscuring more distant stars.

NEUTRON STAR An exceedingly dense, compact star that is composed almost entirely of neutrons (a type of subatomic particle with zero electric charge).

NEW GENERAL CATALOG (NGC) A catalog of nebulae, clusters, and galaxies that was first published in 1888. Objects in this catalog are denoted by "NGC" followed by a number. The Index Catalog (IC) is a supplement to the New General Catalog that lists additional objects, denoted by "IC" followed by a number. See also *Messier catalog*.

NOVA A star that suddenly brightens, then fades back to its original brightness over a period of weeks or months. See also *supernova*.

NUCLEAR FUSION See *fusion*.

NUCLEUS 1) The compact central core of an atom. **2)** The solid, ice-rich body of a *comet*. **3)** The central core of a *galaxy*.

OCCULTATION The passage of one body in front of another, causing the more distant one to be wholly or partially hidden.

OORT CLOUD A vast spherical region in the outer reaches of the Solar System, thought to contain a huge number of icy *planetesimals* and *comets*.

OPEN CLUSTER A loose group of stars that formed at the same time. Open clusters are found in the arms of *spiral galaxies*.

OPPOSITION The time a *superior planet* lies on the exact opposite side of Earth from the Sun. The planet is closest to Earth at this time. See also *conjunction*.

ORBIT The path a celestial body takes in space under the influence of the gravity of other, relatively nearby objects.

PARALLAX The apparent shift in the position of an object when it is observed from different locations. Annual parallax is the maximum angular shift of a star from its average position due to parallax.

PENUMBRA 1) The lighter, outer part of the shadow cast by an opaque body. **2)** The less-dark and less-cool outer region of a *sunspot*. See also *umbra*.

PERIGEE The point on its orbit at which a body orbiting Earth is at its closest to Earth. See also *apogee*.

PERIHELION The point on its elliptical orbit at which a planet or other Solar System body is at its closest to the Sun. See also *aphelion*.

PHASE The proportion of the visible hemisphere of the Moon or a planet that is illuminated by the Sun at any particular instant.

PLANET A celestial body that orbits around a star, is massive enough to have cleared away any debris from its orbital path, and is approximately spherical. See also *dwarf planet*.

PLANETARY NEBULA A glowing shell of gas ejected by a star of similar mass to the Sun toward the end of its evolutionary development.

PLANETESIMAL One of the large number of small bodies, composed of rock or ice, that formed within the *solar nebula* and from which the planets were eventually assembled by *accretion*.

PLASMA A mixture of ions and electrons that behaves like a gas but which conducts electricity and is affected by magnetic fields.

PRECESSION A slow change in the orientation of a body's rotational axis.

PULSAR A rapidly rotating *neutron star* with a strong magnetic field. If the poles of the pulsar's magnetic field are not aligned with the rotation axis of the star, jets of radiation sweep around space at high speed.

QUASAR A compact but extremely powerful source of radiation that is almost star-like in appearance, but which is believed to be the most luminous kind of active galactic nucleus.

RADIANT The point in the sky from which the *meteors* in a shower appear to originate.

RADIO GALAXY A *galaxy* that is exceptionally luminous at radio wavelengths.

RADIO TELESCOPE An instrument designed to detect radio waves from astronomical sources.

RADIO WAVES *Electromagnetic radiation* with wavelengths longer than microwaves. See also *electromagnetic spectrum*.

RED DWARF STAR A cool, red, low-luminosity star.

RED GIANT STAR A large, highly luminous star with a low surface temperature and a reddish color.

RED SHIFT The displacement of *spectral lines* to longer wavelengths that is observed when a light source is moving away from an observer. The shift in wavelength is proportional to the speed at which the source is receding.

RED SUPERGIANT STAR An extremely large star of very high luminosity and low surface temperature.

REFLECTING TELESCOPE (REFLECTOR) A telescope that uses a concave mirror to collect and focus light. See also *catadioptric telescope, refracting telescope*.

REFLECTION NEBULA A *nebula* containing tiny dust particles that are lit up by light from a neighbouring bright star.

REFRACTING TELESCOPE (REFRACTOR) A telescope that uses a lens to collect and focus light. See also *catadioptric telescope, reflecting telescope*.

RETROGRADE MOTION 1) An apparent temporary reversal in the direction of motion of a planet, such as Mars, when it is being overtaken in its orbital motion by Earth. **2)** Orbital motion in the opposite direction to that of Earth and all the other planets of the Solar System. **3)** The motion of a satellite along its orbit in the opposite direction to the rotation of its parent planet.

RIGHT ASCENSION (RA) The angular distance, measured eastward, between the *celestial meridian* and a celestial object; the equivalent of longitude on Earth. Together with *declination*, right ascension specifies the position of an object on the *celestial sphere*.

ROCKY PLANET A planet composed mainly of rock, with similar basic characteristics to Earth. The rocky planets in the Solar System are Mercury, Venus, Earth, and Mars. See also *giant planet*.

SATELLITE A body that orbits a planet; a natural satellite is also known as a moon.

SEYFERT GALAXY A *spiral galaxy* with an unusually bright, compact nucleus that in many cases fluctuates in brightness. Seyfert galaxies are one of several categories of *active galaxy*.

SOLAR CYCLE A cyclic variation in solar activity. The level of activity reaches a maximum at intervals of about 11 years.

SOLAR NEBULA The cloud of gas and dust from which the Sun and planets formed.

SOLAR SYSTEM The Sun together with the family of four *rocky planets*, four *giant planets*, and smaller bodies (*dwarf planets*, *moons, asteroids, meteoroids, comets,* dust, and gas) that orbit the Sun.

SOLAR WIND A continuous stream of fast-moving charged particles that escapes from the Sun and flows out through the Solar System.

SOLSTICE One of the two points on the *ecliptic* at which the Sun is at its maximum *declination* north or south of the *celestial equator*. See also *equinox*.

SPECTRAL CLASS A class into which a star is placed according to the lines that appear in its spectrum. See also *spectral line*.

SPECTRAL LINE A bright or dark line that appears at a particular wavelength in an object's *spectrum*, due to emission or absorption of radiation by that object at a distinct wavelength.

SPECTRUM The range of wavelengths of light emitted by a celestial object. See also *electromagnetic spectrum*.

SPIRAL GALAXY A *galaxy* that consists of a central concentration of stars surrounded by a flattened disk of stars, gas, and dust, within which the major visible features are clumped together into spiral arms. The *Milky Way* is an example of a spiral galaxy.

STAR A huge sphere of glowing *plasma* that generates energy by nuclear *fusion* reactions at its center.

STARBURST GALAXY A *galaxy* within which star formation is taking place at an exceptionally rapid rate.

STAR CLUSTER A gravitationally bound group of between a few tens and approximately 1 million stars, all of which are thought to have formed from the same original massive cloud of gas and dust.

STELLAR WIND An outflow of charged particles from the atmosphere of a star. See also *solar wind*.

SUNSPOT A region of the Sun's photosphere (visible outer layer) that appears dark because it is cooler than its surroundings. Sunspots result from localized disturbances in the Sun's magnetic field.

SUPERGIANT STAR An exceptionally large, luminous star. Supergiant stars can be many hundreds of times larger than the Sun and thousands of times brighter. See also *giant star*.

SUPERIOR CONJUNCTION See *conjunction*.

SUPERIOR PLANET A Solar System planet whose orbit around the Sun is outside Earth's. The superior planets are Mars, Jupiter, Saturn, Uranus, and Neptune. See also *inferior planet*.

SUPERNOVA An exceptionally violent explosion of a star during which it expels the great bulk of its material and its brightness increases enormously.

SUPERNOVA REMNANT The expanding cloud of debris created by a *supernova* explosion.

SYNCHRONOUS ROTATION The rotation of a body around its axis in the same period of time that it takes to orbit another body.

TERMINATOR The edge of the sunlit area of the surface of a moon or planet.

TRANSIT The passage of a smaller body in front of a larger one; for example, the passage of Venus across the face of the Sun.

ULTRAVIOLET RADIATION *Electromagnetic radiation* with wavelengths shorter than visible light but longer than X-rays. See also *electromagnetic spectrum*.

UMBRA 1) The dark central part of the shadow cast by an opaque body. **2)** The darker, cooler, central region of a *sunspot*. See also *penumbra*.

VARIABLE STAR A star that varies in brightness. A pulsating variable star contracts and expands in a periodic way, varying in brightness as it does so. An eruptive variable star brightens and fades abruptly. See also *Cepheid variable, nova*.

WHITE DWARF STAR The dense, intensely hot, glowing star left when a star of similar mass to our Sun dies. See also *planetary nebula*.

X-RAY *Electromagnetic radiation* with wavelengths shorter than ultraviolet radiation but longer than gamma rays. See also *electromagnetic spectrum*.

ZODIAC The area of the *celestial sphere* around the *ecliptic* through which the Sun, the Moon, and the planets move. Over the course of each year, the Sun passes through 13 constellations in this region; of these, 12 correspond to the signs of the zodiac.

INDEX

Page numbers in **bold** are for main page entries.

ACKNOWLEDGMENTS

Dorling Kindersley would like to thank the following people from Smithsonian Enterprises at the **Smithsonian Institution** in Washington D.C: Carol LeBlanc, Vice President Brigid Ferraro, Director of Licensing Ellen Nanney, Licensing Manager Kealy Wilson, Product Development Coordinator

The publisher would like to thank the following people: Janet Mohun and Miezan van Zyl, for editorial work; Steve Setford for proof-reading; Jane Parker for the index; Simon Murrell, Jacqui Swan, and Anna Reinbold, for design work; and Sophia Tampakopoulos, Jacket Design Development Manager.

The publisher would like to thank the following for their kind permission to reproduce their photographs:

(Key: a-above; b-below/bottom; c-center; f-far; l-left; r-right; t-top)

1 Galaxy Picture Library: Robin Scagell. **2 ESO:** J. Emerson Vista Cambridge Astronomical Survey Unit. **6–7 NASA:** 2MASS/G.Kopan R. Hurt Atlas Image courtesy of 2MASS/UMass/IPAC-Caltech/NASA/NSF. **8 Corbis:** Stocktrek Images/Jay GaBany (tr). **Science Photo Library:** David A Hardy, Futures: 50 Years in Space (ca); Detlev van Ravensroaay (clb). **Anatoly Klypin (NMSU) and Joel Primack (UCSC):** Stefan Gottloeber (AIP) (br). **10 Dorling Kindersley:** Imagewerks/Imagewerks Japan (c) Getty (br). **NASA:** Andrea Dupree (Harvard-Smithsonian CfA), Ronald Gilliland (STScI), NASA and ESA (tr); Hubble Heritage Team (AURA/STScI/NASA/ESA) (c). **11 Getty Images:** Flickr/Matteo Colombo (tr). **12 Alamy Images:** AlamyCelebrity (t). **ESO:** (b). **NASA:** JPL-Caltech/Cornell (c). **13 Corbis:** (tr, br). **ESO:** H. H. Heyer (bl). **NASA:** ESA, H. Weaver (JHU/APL), A. Stern (SwRI), and the HST Pluto Companion Search Team (tl).

14 Corbis: Roth Ritter/Stocktrek Images (b). **ESO:** (t). **NASA:** ESA, Hubble & NASA (c). **15 Corbis:** (t). **NASA:** ESA, R. O'Connell (University of Virginia), B. Whitmore (Space Telescope Science Institute), M. Dopita (Australian National University), and the Wide Field Camera 3 Science Oversight Committee (bl); N. Benitez (JHU), T. Broadhurst (Racah Institute of Physics/The Hebrew University), H. Ford (JHU), M. Clampin (STScI), G. Hartig (STScI), G. Illingworth (UCO/Lick Observatory), the ACS Science Team and ESA (tr). **18 Corbis:** Stapleton Collection (t). **19 Corbis:** Radius Images (t). **22 Corbis:** Kerrick James (cra). **Dreamstime.com:** Windsteel (cla, clb, bl). **ESO:** (crb). **Getty Images:** Takanori Yamakawa/Sebun Photo (br). **23 Getty Images:** Chris Walsh (t). **Science Photo Library:** Frank Zullo (b). **24–25 NASA:** ESA, J. Hester and A. Loll (Arizona State University) (b). **25 Getty Images:** William James Warren/Science Faction (tl). **Science Photo Library:** NASA (tr). **26 Corbis:** Bettmann (b/Venus). **Dreamstime.com:** Yiannos1 (b/earth). **NASA:** JPL/USGS (b/moon); JPL/Space Science Institute (b/Saturn). **27 Dorling Kindersley:** Luciano Corbella (b/Andromeda galaxy). **NASA:** A Fujii/ESA/HST (t). **Science Photo Library:** Celestial Image Co. (bc); Mark Garlick (br). **28–29 Science Photo Library:** BabakTafreshi. **31 Corbis:** Reuters (t). **Science Photo Library:** Pekka Parviainen (b). **33 Fotolia:** Anton Balazh (t). **36 Galaxy Picture Library:** Robin Scagell (t). **37 Galaxy Picture Library:** Robin Scagell (cl, tc, br). **38 Galaxy Picture Library:** Robin Scagell (cb, br, c, bc). **39 Galaxy Picture Library:** Robin Scagell (cr, br). **43 Getty Images:** Travel Ink (bl). **45 Corbis:** Dennis di Cicco (tc, tr). **48 Corbis:** Keren Su (c, fcr, br). **50 Corbis:** HO/Reuters (cl). **52 Alamy Images:** Adam van Bunners (cl). **ESO:** (cr). **(c) Robin Jackson:** (tr). **Stuart Macintosh:** (bl). **53 Corbis:** Richard Crisp (bl); Roger Ressmeyer (tr, tl/detail). **54 Galaxy Picture Library:** Robin Scagell (cl). **55 Science Photo Library:** J-P Metsavainio (tr). **57 Galaxy Picture Library:** Dave Tyler (br). **Science Photo Library:** John Chumack (cr). **58 www.nightskyhunter.com:** (bl). **59 Krzysztof Jastrzebski, Poland - Skawina:** (cl). **Serge Veillard:** (crb). **Roel Weijenberg, www.roelblog.nl:** (br). **60–61 Galaxy Picture Library:** Dave Taylor. **63 Corbis:** Bettmann (cla/Venus); Walter Myers/Stocktrek Images (ca/Jupiter). **Dreamstime.com:** Yiannos1 (clb/Earth). **Getty Images:** Yoshinori Watabe (bc). **NASA:** Johns Hopkins University Applied Physics Laboratory/Arizona State University/Carnegie Institution of Washington (cla/Mercury); NEAR-JHUAPL (br); JPL/Space Science Institute (cra/Saturn); JPL (cb/Uranus); Erich Karkoschka, University of Arizona (crb/Neptune). **U.S. Geological Survey:** Astrogeology Research Program (cb/Mars). **64 Science Photo Library:** Babak Tafreshi (t). **65 Alamy Images:** Galaxy Picture Library (tc). **Will Gater:** (tl, tr). **66 Corbis:** Bettmann (c/Venus); Walter Myers/Stocktrek Images (c/Jupiter). **Dreamstime.com:** Yiannos1 (clb, c/Earth). **NASA:** Johns Hopkins University Applied Physics Laboratory/Arizona State University/Carnegie Institution of Washington (c/Mercury); SDO (t, bl, c/Sun); Erich Karkoschka, University of Arizona (c/Neptune); JPL (c/Uranus). **U.S. Geological Survey:** Astrogeology Research Program (c/Mars). **67 Corbis:** Jay Pasachoff/Science Faction (br). **NASA:** SDO/AIA (tr/corona); SOHO (ESA & NASA) (bl). **68 Corbis:** Michael Benson/Kinetikon Pictures (cr). **NASA:** SOHO (ESA & NASA) (cl). **Science Photo Library:** Greg Piepol (b). **69 NASA:** SOHO (ESA & NASA) (c, br). **Science Photo Library:** NOAO (tl). **70 Galaxy Picture Library:** Robin Scagell (bl). **71 Corbis:**

Francesc Muntantada (t). **NASA:** SDO and the AIA, EVE, and HMI science teams (b). **72 Corbis:** Bettmann (c/Venus); Walter Myers/Stocktrek Images (c/Jupiter). **Dreamstime.com:** Yiannos1 (c/Earth, bc). **NASA:** Johns Hopkins University Applied Physics Laboratory/Arizona State University/Carnegie Institution of Washington (c/Mercury); JPL/USGS (t, bl); JPL (c/Uranus); Erich Karkoschka, University of Arizona (c/Neptune); JPL/Space Science Institute (c/Saturn); SDO/GSFC (c/Sun). **U.S. Geological Survey:** Astrogeology Research Program (c/Mars). **74 Corbis:** Mikael Svensson/Johnér Images (bl). **Galaxy Picture Library:** (bc/binocular view). **NASA:** JPL/USGS (bc/telescope). **Science Photo Library:** John Sanford (br); Eckhard Slawik (t). **75 Alamy Images:** Ria Novosti (t). **Getty Images:** Time & Life Pictures (b). **NASA:** (c). **76 NASA:** GSFC/Arizona State University (b, t). **77 Alamy Images:** Galaxy Picture Library (tr). **NASA:** Johnson Space Center (br); nssdc/gsfc (cl, bl). **78 Galaxy Picture Library:** Jamie Cooper (tl). **NASA:** JPL/USGS (b). **SuperStock:** science & society (c). **79 Galaxy Picture Library:** Damian Peach (b). **Getty Images:** (t); Space frontiers/Stringer (cl). **80 Galaxy Picture Library:** NASA (clb). **NASA:** (b); NSSDC/GSFC (tr); (cr). **81 Getty Images:** Jamie Cooper/SSPL (tr). **Science Photo Library:** NASA (tl). **SuperStock:** Science and Society (br, cl). **82 Science Photo Library:** Thierry Legault/Eurelios (b). **83 Science Photo Library:** John Bova (cl, cr, bl, br). **84 Corbis:** Jean-Christophe Bott/epa (cb). **E. Israel:** (tr). **85 Corbis:** Jay Pasachoff/Science Faction (br). **Getty Images:** AFP (bl); SPL/Rev. Ronald Royer (tr); SSPL (cr). **Science Photo Library:** H R Bramaz, ISM (cl). **86 Corbis:** Bettmann (c/Venus); Walter Myers/Stocktrek Images (c/Jupiter). **Dreamstime.com:** Yiannos1 (bc, c/Earth). **NASA:** Johns Hopkins University Applied Physics Laboratory/Arizona State University/Carnegie Institution of Washington (tl, c/Mercury, bl); JPL/Space Science Institute (c/Saturn); JPL (c/Uranus); Erich Karkoschka, University of Arizona (c/Neptune); SDO/GSFC (c/Sun). **U.S. Geological Survey:** Astrogeology Research Program (c/Mars). **87 Galaxy Picture Library:** Maurice Gavin (b). **88 Corbis:** Bettmann (t, hl, c/Venus); Walter Myers/Stocktrek Images (c/Jupiter). **Dreamstime.com:** Yiannos1 (bc, c/Earth). **NASA:** Johns Hopkins University Applied Physics Laboratory/Arizona State University/Carnegie Institution of Washington (c/Mercury); Erich Karkoschka, University of Arizona (c/Neptune); JPL/Space Science Institute (c/Saturn); JPL (c/Uranus); SDO (c/Sun). **U.S. Geological Survey:** Astrogeology Research Program (c/Mars). **89 Will Gater:** (b). **90 Corbis:** Bettmann (c/Venus); Walter Myers/Stocktrek Images (c/Jupiter). **Dreamstime.com:** Yiannos1 (br, c/earth). **NASA:** Johns Hopkins University

Applied Physics Laboratory/Arizona State University/Carnegie Institution of Washington (c/Mercury); Erich Karkoschka, University of Arizona (c/Neptune); JPL/Space Science Institute (c/Saturn); JPL (c/Uranus); SDO/GSFC (c/Sun). **U.S. Geological Survey:** Astrogeology Research Program (t, bl, c/Mars). **92 NASA:** JPL (tr); JPL/Cornell (c); JPL-Caltech/University of Arizona (br); JPL/University of Arizona (bc). **93 Galaxy Picture Library:** Robin Scagell (cl, clb); Dave Tyler (bl). **94 Corbis:** Bettmann (c/Venus); Walter Myers/Stocktrek Images (t, bl, c/Jupiter). **Dreamstime.com:** Yiannos1 (c/earth, bc). **NASA:** Johns Hopkins University Applied Physics Laboratory/Arizona State University/Carnegie Institution of Washington (c/Mercury); SDO/GSFC (c/Sun)); JPL/Space Science Institute (c/Saturn); JPL (c/Uranus); Erich Karkoschka, University of Arizona (c/Neptune). **U.S. Geological Survey:** Astrogeology Research Program (c/Mars). **96 NASA:** Johns Hopkins University Applied Physics Laboratory Southwest Research Institute (clb); JPL/DLR (t, cra, cl); JPL/University of Arizona (c); JPL (b). **97 Alamy Images:** Galaxy Picture Library (bl, bc). **Corbis:** Walter Myers/Stocktrek Images (tl). **Will Gater:** (br). **98 Corbis:** Bettmann (c/Venus); Walter Myers/Stocktrek Images (c/Jupiter). **Dreamstime.com:** Yiannos1 (bc, c/Earth). **NASA:** Johns Hopkins University Applied Physics Laboratory/Arizona State University/Carnegie Institution of Washington (c/Mercury); Erich Karkoschka, University of Arizona (c/Neptune); JPL/Space Science Institute (c/Saturn); JPL (c/Uranus). **U.S. Geological Survey:** Astrogeology Research Program (c/Mars). **100 NASA:** JPL/Space Science Institute (tc, tr, cl, cr, bl, bc, br, tl); JPL (c). **101 Galaxy Picture Library:** Robin Scagell (bl, bc); Dave Tyler (br). **NASA:** JPL/Space Science Institute (c); The Hubble Heritage Team (STScI/AURA) Acknowledgment: R.G. French (Wellesley College), J. Cuzzi (NASA/Ames), L. Dones (SwRI), and J. Lissauer (NASA/Ames) (tr). **102 Corbis:** Bettmann (c/Venus); Walter Myers/Stocktrek Images (c/Jupiter). **Dreamstime.com:** Yiannos1 (bc, c/Earth). **NASA:** Johns Hopkins University Applied Physics Laboratory/Arizona State University/Carnegie Institution of Washington (c/Mercury); Erich Karkoschka, University of Arizona (c/Neptune); JPL/Space Science Institute (c/Saturn); JPL (t, bl, c/Uranus); SDO/GSFC (c/Sun). **U.S. Geological Survey:** Astrogeology Research Program (c/Mars). **103 Corbis:** (br). **W.M. Keck Observatory:** (tc). **NASA:** JPL (tr). **104 Corbis:** Bettmann (c/Venus); Walter Myers/Stocktrek Images (c/Jupiter). **Dreamstime.com:** Yiannos1 (bc, c/Earth). **NASA:** Johns Hopkins University Applied Physics Laboratory/Arizona State University/Carnegie Institution of Washington (c/Mercury); Erich Karkoschka, University of Arizona (t, bl, c/Neptune);

JPL/Space Science Institute (c/Saturn); JPL (c/Uranus); SDO/GSFC (c/Sun). **U.S. Geological Survey:** Astrogeology Research Program (c/Mars). **105 Corbis:** NASA/Roger Ressmeyer (tr). **Science Photo Library:** John Chumack (br). **106 NASA:** ESA, and M. Showatter (SETI Institute) (tl). **107 Lowell Observatory Archives:** (b). **NASA:** ESA, J. Parker (southwest Research Institute), P. Thomas (Cornell University), L. Mc Fadden (University of Maryland, College Park) and M. Mutchler and Z. Levay (STScI) (tl); ESA, and M. Buie/Southwest Research Institute (tr). **108 Corbis:** Dennis di Cicco (t). **110 Corbis:** Tony Moow/ amanaimages (tr). **Getty Images:** Yoshinori Watabe (tl). **NASA:** (b). **111 ESO:** S. Deiries (cr). **Getty Images:** Malcolm Park (tr). **NASA:** JPL (tl); JPL-Caltech/UMD (bl). **Science Photo Library:** Rev. Ronald Royer (br). **112 Science Photo Library:** Tony and Daphne Hallas (t). **113 Dorling Kindersley:** Colin Keates/Natural History Museum (cl, clb, bl). **Getty Images:** Barcroft Media (br). **114 NASA:** JPL/USGS (t). **115 NASA:** R Evans and K Stapelfeldt (JPL) (br). **116 NASA:** JPL-Caltech/UCLA/MPS/ DLR/IDA (t); JPL/JHUAPL (bl); JPL/USGS (br). **117 ESA:** MPS/UPD/.LAM/IAA/ RSSD/INTA/UPM/DASP/IDA (bl). **Japan Aerospace Exploration Agency (JAXA):** ISAS (cr). **NASA:** JHUAPL (tr); JPL/USGS (cr); JPL/Steve Ostro (br). **118 Alamy Images:** Design Pics Inc (t). **119 Science Photo Library:** NASA/ESA/CXS/STScI (br). **120 Alamy Images:** Sunpix Travel (t). **121 NAOJ:** H. Fukushima, D. Kinoshita, & J. Watanabe (br). **Science Photo Library:** Stephen J Krasemann (tl); Pekka Parviainen (bl); Babak Tafreshi (bc). **122-123 NASA and The Hubble Heritage Team (AURA/STScI):** NASA, ESA, N. Smith (University of California, Berkeley),. **124 ESO:** (c). **126 Galaxy Picture Library:** Damian Peach (t). **127 The Art Agency:** Terry Pastor (b). **128 NASA:** ESA M. Roberto Space Telescope Science Institute/ESA and the Hubble Space Telescope Orion Treasury Project Team (t); ESA AURA/ Caltech (c). **Jeremy Perez:** (b). **129 Corbis:** Stocktrek Images (t). **ESO:** (cl). **Getty Images:** Stocktrek Images (b). **130 Corbis:** Robert Gendler/Stocktrek Images (b); Stocktrek Images (c). **Will Gater:** (t). **131 Corbis:** Stocktrek Images (t); Visuals Unlimited (b). **NASA:** JPL-Caltech/UCLA (c). **132 Pedro Ré. . Laboratorio Maritimo da Guia, Portugal. . :** (bc, br). **133 ESO:** L. Calçada (cl). **134 ESO:** (t). **Science Photo Library:** Celestial Image Co. (b). **135 Will Gater:** (t, ca, c, cl). **136 Corbis:** Visuals Unlimited (cl). **ESO:** (bl). **Will Gater:** (t). **NASA:** ESA, and G. Meylan (Ecole Polytechnique Federale de Lausanne) (bc). **137 Corbis:** Stocktrek Images (bl). **Will Gater:** (tl, br). **NASA:** JPL-Caltech, J Stauffer (SSC/Caltech) (tr). **138-139 The Art Agency:** Stuart Jackson-Carter. **140 Corbis:** Momatiuk - Eastcott (t). **Will Gater:** (b).

141 NASA: Ames Research Center, W. Stenzel (br); ESA, P. Kalas, J. Graham, E. Chiang, E. Kite (University of California, Berkeley), M. Clampin (NASA Goddard Space Flight Center), M. Fitzgerald (Lawrence Livermore National Laboratory), and K. Stapelfeldt and J. Krist (NASA Jet Propulsion Laboratory) (bc, bl). **142 NASA:** NASA, ESA, and N. Pirzkal (European Space Agency/STScI) and the HUDF Team (STScI) (cl, cr); NASA, ESA, and The Hubble Heritage Team (STScI/AURA) (br, bl). **143 ESA:** NASA (br). **ESO:** (t). **NASA:** NASA, Holland Ford (JHU), the ACS Science Team and ESA (tl). **144 Corbis:** Tony Hallas/Science Faction (b); Stocktrek Images (t). **NASA:** JPL-Caltech/ Harvard-Smithsonian CfA/NOAO (c). **145 Corbis:** Stocktrek Images (b, tr); Visuals Unlimited (t). **146 Getty Images:** Stocktrek Images (b). **The Art Agency:** Terry Pastor (t). **147 NASA:** ESA and R. Massey (California Institute of Technology) (tl, tr). **Sloan Digital Sky Survey (SDSS):** M. Blanton & SDSS Collaboration, www.sdss.org (b). **148 NASA:** WMAP Science Team (t). **150–151 Galaxy Picture Library:** Nigel Evans. **152 Alamy Images:** Galaxy Picture Library (bl, cb). **155 NASA:** Atlas Image courtesy of 2MASS/UMass/IPAC-Caltech/NASA/ NSF (br). **160 Alamy Images:** Galaxy Picture Library (br). **167 Corbis:** Roger Ressmeyer (br). **173 ESO:** IDA/Danish 1.5 m/R.Gendler, J-E. Ovaldsen, and C. Feron. (br). **179 Alamy Images:** Galaxy Picture Library (bl). **185 Corbis:** Takashi Katahira/ amanaimages (br). **190 ESA:** A. Fuji (bl). **197 Corbis:** Stocktrek Images (br). **203 Corbis:** Stocktrek Images (bl). **208 Alamy Images:** Galaxy Picture Library (br). **214 Corbis:** Tony Hallas (br). **220 Corbis:** Stocktrek Images (bl). **228 Till Credner/AlltheSky.com:** (cr). **231 Till Credner/AlltheSky. com:** (br). **NASA:** ESA, HEIC, and The Hubble Heritage Team (STScI/AURA) (tr). **233 Till Credner/AlltheSky.com:** (b). **236 NOAO/AURA/NSF:** T.A.Rector and B.A.Wolpa, Copyright WIYN Consortium, Inc., all rights reserved (cl). **238 Till Credner/AlltheSky.com:** (t). **239 NASA:** NASA, ESA, and The Hubble Heritage Team (STScI/AURA) (bl). **243 Till Credner/AlltheSky.com:** (t). **244 NASA:** ESA, the Hubble Heritage (STScI/AURA)-ESA/Hubble Collaboration, and the Digitized Sky Survey 2. J. Hester (Arizona State University) and Davide De Martin (ESA/ Hubble) (cl). **245 Till Credner/ AlltheSky.com:** (cl). **246 NASA:** JPL-Caltech (cr). **249 NASA:** 2002 R. Gendler (b). **Hunter Wilson:** (t). **253 Science Photo Library:** Eckhard Slawik (b). **Hunter Wilson:** (tr). **255 NOAO/AURA/NSF:** Peter Kukol/ Adam Block (cl). **256 NASA:** The Hubble Heritage Team (STScI/AURA) Acknowledgment: R.G. French (Wellesley College), J. Cuzzi (NASA/ Ames), L. Dones (SwRI), and J. Lissauer (NASA/Ames) (cr). **259 Corbis:** Stocktrek Images (b). **ESA:** Image by C. Carreau (t). **260 Till Credner/**

AlltheSky.com: (cr). **261 Till Credner/AlltheSky.com:** (t). **262 Hunter Wilson:** (tl). **263 ESO:** (cr). **NASA:** ESA, STScI, J. Hester and P. Scowen (Arizona State University) (b). **265 Corbis:** Rolf Geissinger/Stocktrek Images (cl). **266 Till Credner/ AlltheSky.com:** (tl). **266–277 Corbis:** © Stocktrek Images/Miguel Claro/Stocktrek Images. **268 ESO:** (cr). **269 Till Credner/AlltheSky.com:** (cr). **270 Till Credner/AlltheSky. com:** (cr). **275 NASA:** JPL-Caltech/ Harvard-Smithsonian CfA (cr). **277 NASA:** ESA, M. Robberto (Space Telescope Science Institute/ESA) and the Hubble Space Telescope Orion Treasury Project Team (b). **278 Till Credner/AlltheSky.com:** (cl). **280 Hunter Wilson:** (tl). **281 ESO:** IDA/Danish 1.5 m/R. Gendler, S. Guisard (www.eso.org/~sguisard) and C. Thöne (b). **282 Daniel Verschatse - Observatorio Antilhue - Chile:** © Copyright 2001–2008 by Daniel Verschatse (cr). **283 Galaxy Picture Library:** Gordon Garradd (cr). **284 NOAO/AURA/NSF:** Bill and Sean Kelly/Adam Block (cl). **285 NASA:** ESA, and the Hubble Heritage Team (STScI/ AURA)-ESA/Hubble Collaboration (cr). **287 ESO:** (t). **NASA:** A Fujii (b). **291 NASA:** ESA (cr). **292 Joseph Brimacombe:** (cl). **293 Till Credner/ AlltheSky.com:** (cl). **294 NASA:** ESA, Hubble & NASA (cr). **295 ESO:** (cr). **296 Galaxy Picture Library:** Robin Scagell (tr). **298 Till Credner/ AlltheSky.com:** (t). **299 Alamy Images:** Stocktrek Images, Inc. (bl). **300 ESO:** P. Barthel (cr). **301 ESO:** IDA/ Danish 1.5 m/R. Gendler, J-E. Ovaldsen, C. Thöne and C. Féron (cl). **305 ESO:** S. Brunier (cl); (c). **306 NASA:** NASA, ESA, and The Hubble Heritage Team (STScI/AURA) (cl). **307 NASA:** (cl). **308 Guillermo Yanez, Lo Barnechea, Chile:** (cl). **309 Till Credner/AlltheSky.com:** (tr). **310 Science Photo Library:** Robert Gendler, Martin Pugh (cr). **311 ESO:** Loke Kun Tan (StarryScapes) (cr). **312 Science Photo Library:** NASA/ ESA/STScI (cl). **315 Till Credner/ AlltheSky.com:** (cl). **317 Till Credner/AlltheSky.com:** (cl). **319 ESO:** (tl). **320 ESO:** (cl). **321 Corbis:** Stocktrek Images (cr). **322 ESO:** (cr). **323 Till Credner/ AlltheSky.com:** (cl). **324 Till Credner/AlltheSky.com:** (cl). **325 Science Photo Library:** NASA/ESA/STSCL (cl)

Jacket images:
Front: **ESO:** tr; **NASA:** ftl, JPL/Space Science Institute b; *Back:* **Corbis:** Tony Hallas/Science Faction tr; **U.S. Geological Survey:** Astrogeology Research Program tl; *Spine:* **Corbis:** Robert Gendler/Stocktrek Images cb; **NASA:** JPL/Space Science Institute t, JPL/University of Arizona ca.

All other images © Dorling Kindersley
For further information see:
www.dkimages.com